D1432084

Fundamentals of Dimensional Analysis

Alberto N. Conejo

Fundamentals of Dimensional Analysis

Theory and Applications in Metallurgy

 Springer

Alberto N. Conejo
School of Metallurgical and Ecological Engineering
University of Science and Technology Beijing
(USTB)
Beijing, China

ISBN 978-981-16-1601-3   ISBN 978-981-16-1602-0   (eBook)
https://doi.org/10.1007/978-981-16-1602-0

This Springer imprint is published by the registered company Springer Nature Singapore Pte Ltd.
The registered company address is: 152 Beach Road, #21-01/04 Gateway East, Singapore 189721,
Singapore

This small book is dedicated to the memory of my mother Ma. Cruz Nava† and to my family: Carmen, Alberto and Roberto.

Preface

As engineers and researchers, we are confronted with the application of mathematical tools to understand natural phenomena. Galileo put it in these words: "The universe cannot be read until we have learned the language and become familiar with the characters in which it is written. It is written in mathematical language, without which it is humanly impossible to comprehend a single word. Without these, one is wandering about in a dark labyrinth". Therefore, one way to measure the technical evolution of human beings can be defined in terms of the mathematical tools created. The list of milestones in mathematics is extensive: creation of the number systems, geometry, algebra, probability theory, calculus, etc. Using these tools makes it possible to understand and predict the behavior of natural phenomena of daily life. Finally, knowledge of the natural world is expressed by mathematical relationships that involve variables. Looking into the historical development of those relationships, it is easy to discover a large amount of hard work and time consumed to reach our current knowledge.

Dimensional analysis is the study of physical quantities that describe a natural phenomenon using dimensionless variables, replacing in some cases complex numerical modeling, involving the main fundamental forces with the least amount of experimental work.

The first time that dimensional analysis is employed to understand a natural phenomenon and a mathematical relationship is developed with simple mathematical operations that usually take a few minutes; it is difficult to believe that many of those same equations were derived after many years of hard experimental work. Once we find the benefits of dimensional analysis, there is a mandatory question: why do we keep using dimensional variables in our experimental work if using dimensionless groups requires not only less experimental work but also the results have a broader application? I would assume that one reason is that engineering students lack this knowledge but in addition to this, I consider that the way dimensional analysis is presented in most textbooks is limited to the definition of general relationships. If dimensional analysis is presented only as a tool to reduce the number of variables, using dimensionless groups, its full potential becomes limited. It is like planting a fruit tree, taking care of the tree until it grows and then, never eating the fruit. In this book, the main objective is to provide general knowledge about the different methods of dimensional analysis to define a metallurgical problem in terms of dimensionless groups and in the second part, provide a description of several metallurgical problems and how the main variables can be experimentally measured. The experimental part and the definition of a particular equation is the ultimate step to evaluate the whole process. This is the part where the user of dimensional analysis can recognize if the variables employed fulfilled the condition of being the relevant variables, otherwise, we have reached a point that requires to re-evaluate the problem and search for missing variables and the need to get a better understanding of the problem.

The metallurgical problems presented in this book have been taken from the literature, giving all the credits to the original authors. My contribution comprises the following elements: a complete introduction to the metallurgical problem, step-by-step development of the general relationship with dimensionless groups, a systematic and homogeneous representation of these results, an explanation of the experimental work required to define the values of the variables of interest and plots that describe the final experimental results. All the metallurgical problems in Chap. 11 include this treatment.

The following is a short description of the chapters in this book: Chap. 1 is the introduction which describes the importance of dimensional analysis. Chapter 2 is a short review of the history of dimensional analysis. Chapter 3 describes units and dimensions, two important concepts in the study of dimensional analysis. Chapter 4 gives some details about dimensionless numbers, what they represent and short biographies of the important scientists who bear the name of some of those numbers. Chapters 5–9 describe in detail five methods of dimensional analysis. Two concepts that are the pillars of dimensional analysis, the principle of dimensional homogeneity and the π-theorem, are proved in Chaps. 5 and 6, respectively. Simple problems involving two dimensionless groups can be solved quickly using the inspection method and for a larger number of variables, the matrix method is more convenient. For the common problems found in engineering, the π-method of repeating variables is highly recommended. Chapter 10 provides some general engineering applications. Chapter 11 is the central part of this book. In this chapter, a series of metallurgical problems are described in detail, with an introduction to the subject which is important to understand the definition of the relevant variables and the use of experimental data to define the particular solution with dimensionless groups. Some of the databases were kindly provided by Prof. Pilon (UCLA—USA) and Dr. Kamaraj (NML—India). Chapter 12 describes scaling and similarity. One of the applications of the results of dimensional analysis is to provide the tools to define the value of the variables in a scaled model (scaling-down or scaling-up). Finally, in Chap. 13 is presented another way to define dimensionless groups using the governing laws in the form of differential equations; additionally, there is a discussion on strategies to further decrease the number of dimensionless groups.

The conventional approach to study engineering problems in our time is still marked by the use of dimensional variables. The reader will find that using methods of dimensional analysis leads to solutions that are not only less consuming in experimental work and provide more insight into the forces behind a natural phenomenon but also its scope of application is more universal. I am confident that our current knowledge of the natural world would accelerate using dimensional analysis.

Beijing, China   Alberto N. Conejo

Contents

1 Introduction 1
   1.1 Nature Is Dimensionless 1
   1.2 Advantages and Disadvantages of Dimensional Analysis 2
   References 5

2 Origin and Historic Evolution of Dimensional Analysis 7
   2.1 Origins of Dimensional Analysis 7
   2.2 Historic Evolution of Dimensional Analysis 9
   References 11

3 Units, Dimensions and Dimensional Homogeneity 13
   3.1 Units 13
      3.1.1 Origin and Evolution of Units 13
      3.1.2 Metric System, Base Units and SI 15
      3.1.3 New Developments on the Definition of Units 19
      3.1.4 Previous Systems of Units 20
      3.1.5 Derived Units 20
   3.2 Dimensions 21
   3.3 Principle of Dimensional Homogeneity 25
      3.3.1 Derivation of the Principle of Dimensional
           Homogeneity [16, 17] 26
   References 30

4 Dimensionless Numbers 31
   4.1 Concept of a Dimensionless Number 31
   4.2 Common Dimensionless Numbers 33
   References 52

5 Rayleigh's Method 55
   5.1 John William Strutt [1–4] 55
   5.2 Rayleigh's Method 56
   5.3 Examples of Application of Rayleigh's Method 58
   References 66

6   Buckingham's π Theorem 69
   6.1   Edgar Buckingham 69
   6.2   General π-Theorem 70
   6.3   Examples of Application of π-Theorem 74
   References 76

7   Step by Step Method (Ipsen's Method) 77
   7.1   Step by Step Method 77
   7.2   Examples of Application of the Step by Step Method 78
   References 82

8   Matrix Method 83
   8.1   Matrix Algebra 83
   8.2   Matrix Method 87
   8.3   Examples of Application of Matrix Method 90
   References 93

9   Inspection Method 95
   9.1   Inspection Method 95
   9.2   Examples of Application of the Inspection Method 95
   Reference 97

10 Engineering Problems with and without Experimental Data 99
   10.1 General Engineering Problems
       with and without Experimental Data 99
       10.1.1 Bubble Shape Dynamics Rising in a Liquid 99
       10.1.2 Motion of Non-Metallic Inclusions in Liquid
            Metals 106
       10.1.3 Thermal Stresses in Continuous Casting 112
       10.1.4 Mass Transfer Coefficient Under Natural
            Convection 118
       10.1.5 Blast Furnace Raceway Dimensions 121
   10.2 Engineering Problems with Experimental Data 128
       10.2.1 Deformation of an Elastic Ball Striking a Wall 128
       10.2.2 Pressure Drop of a Fluid in a Long Circular Pipe 131
   References 137

11 Applications of Dimensional Analysis in Metallurgy 139
   11.1 Slag Foaming 139
       11.1.1 Introduction to the Problem 139
       11.1.2 Experimental Data on Foaming Index
            from Fruehan et al.: [1–9] 141
       11.1.3 Experimental Data Including the Effect of Bubble
            Diameter from Fruehan et al. [4, 5] 147
       11.1.4 Experimental Data Including the Effect of Bubble
            Diameter from Ghag et al. [16, 17] 149
       11.1.5 Experimental Database from Pilon et al. [18–22] 156

11.2   Slag Open Eye Area 164
       11.2.1   Introduction to the Problem 164
       11.2.2   Experimental Data from Yonezawa
                and Schwerdtfeger [24] 165
       11.2.3   Experimental Data from Peranandhanthan
                and Mazumdar [25] 174
       11.2.4   Ladle Eye as a Function of Nozzle Radial Position 184
11.3   Mixing Time in the Ladle Furnace 199
       11.3.1   Experimental Data from Pan et al. [32] 199
       11.3.2   Experimental Data from Mandal and Mazumdar
                [33, 34] 212
11.4   Entrainment of Iron Droplets at Liquid/Liquid Interfaces
       Due to Rising Gas Bubbles 218
11.5   Tundish Open Slag Eye (TOSE) 226
11.6   Liquid Steel Discharge from a Ladle to the Tundish Using
       a Protective Shroud [57–62] 242
11.7   Teeming of Liquid Steel 262
       11.7.1   Experimental Work from Kuwana [65] 263
       11.7.2   Experimental Work from Kamaraj et al. [66] 268
11.8   Rate of Consumption of Mold Fluxes 278
11.9   Penetration of a High-Velocity Gas Jet Through a Liquid
       Surface 290
References 316

12 Scaling and Similarity 323
    12.1   Introduction 323
    12.2   Similarity Criteria 323
           12.2.1   Geometric Similarity 324
           12.2.2   Materials Similarity 326
           12.2.3   Mechanical Similarity 326
           12.2.4   Thermal Similarity 341
           12.2.5   Chemical Similarity 343
    References 344

13 Non-dimensionalization of Differential Equations 347
    13.1   Introduction 347
    13.2   Scaling Differential Equations 351
    13.3   How to Decrease the Number of Dimensionless Groups 370
    13.4   Discriminated Dimensional Analysis (DDA) 372
    References 377

Epilogue 379

About the Author

Alberto N. Conejo (Cuitzeo, Mexico, September 25, 1959) is a Mexican Professor who has been working at the University of Science and Technology Beijing (USTB) at the School of Metallurgical and Ecological Engineering since September 2018. Before this, he was a professor of ironmaking and steelmaking in Mexico for 30 years. He also worked for several metallurgical plants from 1982 until 1988 in several positions: Welding Supervisor at TEISA, a producer of turbines; Head of the Heat Treatment Department at ENCO, a producer of gears for the auto industry; and manager of Research and Development at NKS, a producer of large castings and forged products. He did his Ph.D. in metallurgy at Colorado School of Mines in Golden, CO, USA. His Ph.D. thesis was related to gas–solid reactions and the production of iron carbide from iron oxides.

For more than 15 years, he had a strong collaboration with the steel industry in Mexico, in particular with ArcelorMittal Lazaro Cardenas. He provided more than 1200 men-hours of technical training on ironmaking and steelmaking to union workers, supervisors and process engineers.

He has been visiting professor at Tohoku University, Japan, with Prof. Shin-Ya Kitamura, for 3 months from July to September 2009; Arcelor-Mittal laboratories in Aviles, Spain, for 6 months from February to July 2011; School of Metallurgical and Ecological Engineering at the University of Science and Technology Beijing (USTB), P. R. China, with Prof. Lifeng Zhang, for one year, August 2014–July 2015; Indian Institute of Technology Kanpur, India, with Prof. Dipak Mazumdar, for 6 months, December 2017–May 2018.

He has received several national and international awards: Charles W. Briggs award 2002, Iron and steel society, USA; Best paper award on EAF steelmaking; Michoacán State award: Technology award, 2005; Work on EAF slag foaming. National award (2nd place): Technology and Science award granted by the Mexican Steel Producers Association (Canacero), 2010/2011. Work on EAF modeling. Member of the National System of Researchers (SNI). Level II (top 18% nationwide). Current period: 2018–2021. Level II from 2014 to 2017. Previously level I from 1989 to 1993, 2002–2013.

He is Editorial Member of the journal *Metallurgical Research and Technology* (previously known as *Revue de Metallurgie*). Since January 2017, he is Editorial Member of the International Journal of Minerals, Metallurgy and Materials, Beijing, China. Since July 2019, he is a reviewer of more than 10 international journals.

He has published more than 60 technical papers in journals, with an index from Scopus, h = 15. He has supervised or co-supervised 18 master and two Ph.D. students.

In 2018, he founded Ferrous Metallurgy Research Institute (FeMRI), a provider of consulting, training and research services for the steel industry worldwide.

Chapter 1
Introduction

1.1 Nature Is Dimensionless

Any natural phenomena can be described in the form of equations which involve variables, usually one dependent variable and a number of independent variables. These variables or physical quantities are measured using dimensions and units. Nature doesn't express based on any system invented by mankind but in a dimensionless way. Its frame of reference is absolute not relative. The concepts of units and dimensions are a human invention to understand natural phenomena, relative concepts that permanently change as a function of our understanding on the units of measurement. One example is the meter as a unit of length. Its definition has changed four times since 1795.

Ever since the first civilizations, many natural phenomena have been studied and the knowledge resulting from their observations and experimental work has been summarized in terms of relationships among a dependent variable and one or multiple independent variables. In many of those cases it took hundreds of years to define a given relationship. Today, with the aid of the method of dimensional analysis many of those laws and relationships can be derived in its general structure with a simple manipulation of dimensions and units regarding the variables involved in a given natural phenomenon. The method of dimensional analysis not only reduces significantly the amount of experimental work but also defines the forces in which nature expresses.

The method is usually too simple "if" the variables that affect a natural phenomena are properly identified. All we need is a solid understanding of the natural phenomena, in particular the definition of the variables involved, then, the principle of dimensional homogeneity which involves dimensions is applied and finally, the result can be expressed as a relationship in terms of dimensionless groups.

Butterfield [1] define the objective of dimensional analysis as follows: "To minimize the dimensional space of the n primary variables, $V = \{v_1, v_2, \ldots v_n\}$, in which the behaviour of a specific system might be studied, by combining them into N dimensionless groups (DGs), where N is less than n". By reducing the number of

© The Author(s), under exclusive license to Springer Nature Singapore Pte Ltd. 2021
A. N. Conejo, *Fundamentals of Dimensional Analysis*,
https://doi.org/10.1007/978-981-16-1602-0_1

variables from n to N the amount of experimental work is decreased. This is possible because dimensional analysis uses the minimum number of independent variables that characterize a system.

The result when a method of dimensional analysis is applied is a general functional relationship that defines a dimensionless group containing the dependent variable with other dimensionless groups containing the independent variables. In a second and final step, the final form of the functional relationship is defined carrying out experimental work.

1.2 Advantages and Disadvantages of Dimensional Analysis

Which are the benefits of dimensional analysis and the final equation in dimensionless form?; The following is a list of the many benefits of dimensional analysis:

1. A decrease in the number of dimensionless variables: A decrease in the number of variables reduces the complexity of the problem reducing the amount of experimental work. In some cases, depending on the number of variables, the decrease in the number of experiments is reduced by orders of magnitude.
2. Fast response to define a relationship between the physical quantities involved.
3. Dimensional variables with a small exponent in a dimensionless group are not relevant and can be neglected.
4. The dimensionless equation can be applied in wider range of physical conditions.
5. Derivation of physical laws: In many cases, the final result is a familiar equation that describes a natural phenomenon, derived in a few simple steps.
6. Derivation of scaling laws: Equations in dimensionless form can be applied to scale up or scale down a given process. Usually, its main application is to scale down a real process to obtain experimental data under laboratory conditions which are valid at the industrial scale.
7. Many complex metallurgical problems that cannot be solved analytically or using numerical methods can be studied with dimensional analysis.
8. There is no need to define the governing equations of the phenomena investigated.

The main disadvantages of dimensional analysis are:

1. Requires solid experience in the problem to be investigated in order to define the minimum number of independent variables. Without this experience dimensional analysis can provide useless or erroneous results. Selection of the important variables requires, in some problems, considerable judgment and experience.
2. It doesn't provide details into the mechanisms involved in the physical phenomena investigated.
3. Provides a general expression that minimizes the number of variables but in order to define a final expression some experimental work is still needed.

4. The conventional methods of dimensional analysis can result in multiple sets of results. In many cases it is not possible to define the forces involved. In order to get familiar dimensionless numbers, the results should be algebraically modified.

Dimensional analysis can be summarized in three steps;

1. Definition of relevant variables that affect the phenomena of interest.
2. Define variables in terms of its main dimensions.
3. Apply the principle of dimensional homogeneity.

The dimensionless numbers can be obtained in two ways [2]:

1. Methods of dimensional analysis
2. Scaling differential equations.

This book is primarily focused with methods of dimensional analysis. The alternative approach to define the dimensionless groups is transforming a known governing equation in dimensionless form. Heat, mass and momentum transport have their own governing equations, in a general form and specific for different particular cases. The great advantage to use this method is that the problem of choosing the right variables and the right number which are truly independent, is avoided. Scaling of differential equations involves reference values for the variables involved and the transformation from dimensional into dimensionless variables.

A complex problem involving numerical solution of governing equations and their boundary conditions is sometimes extremely time demanding and requires strong basis on mathematical modeling. In other cases, the problem is so complex that even current computer capacity is not enough but dimensional analysis can deal with complex problems. Using dimensional analysis, the solution will not require that knowledge.

What is the scope of application and what type of problems can be solved with dimensional analysis? In engineering, dimensional analysis can be applied to study any natural phenomena and to explore the relationship between variables, defining the effect of a number of dimensionless independent variables on a given dimensionless dependent variable. In this sense their applications are unlimited, covering any field, for example mechanics, economics, electric engineering, etc. The difference with conventional analysis of engineering problems using dimensional variables, which was the conventional method employed for thousands of years, is that the method of dimensional analysis provides a more universal solution with the least amount of experimental work. This book will focus on the application of DA to study physicochemical phenomena occurring in metallurgy however the scope of dimensional analysis is beyond process engineering, including other fields such as Operations Management (OM) and social sciences. The units and dimensions in process engineering and OM have important differences. Miragliotta [3] has described the application of DA to design a flexible manufacturing system with 13 dimensional quantities and then reduced to 9 dimensionless groups. In order to study social sciences, such as economics, the variables should be properly measured. Barnett [4] describes the need to reach a consistent and correct use of dimensions in economic studies.

Perhaps, in spite of all the advantages described using dimensional analysis, if you are learning for the first time this subject, a preliminary example can show you the usefulness of this tool. One classical example is to derive the Hagen–Poiseulle law that gives the pressure drop in an incompressible and Newtonian fluid flowing through a long cylindrical pipe of constant cross section. This law was experimentally derived independently by Jean Poiseuille and Gotthilf Hagen and published by Poiseuille in 1840–1841 [5]. The first two steps in dimensional analysis consist in defining both the relevant variables and their dimensions. In this example, if the motion of the liquid through the tube occurs due to a pressure difference gradient between the two ends of the tube, the geometric variable is the tube length and tube radius. The material property is the fluid viscosity. The motion of the liquid is represented by the flow rate.

Variables and dimensions:

	Variable	Dimensions (FLT)
Pressure drop	ΔP	F/L^3
Tube radius	R	L
Tube length	l	L
Fluid viscosity	μ	FT/L^2
Flow rate	Q	L^3/T

The pressure drop depends on four independent dimensional variables:

$$\Delta P = f(R, l, \mu, Q)$$

The amount of experimental work can be estimated on the basis that a minimum of 5 experiments are needed with each variable, keeping each one of the other variables constant. This amount can be much higher depending on the range of each variable. Since we have four independent variables, for each change in one variable we need 5 experiments with each of the other variables. The total number of minimum experiments is $5^4 = 625$. It grows exponentially if we have to increase the number of experiments. By using dimensional analysis, the general form of the Hagen-Poiseulle law can be derived in a matter of a few minutes. The result shows the following general equation:

$$\Delta P = f\left(\frac{Ql\mu}{R^4}\right)$$

By dimensional analysis the problem has been reduced to one independent variable, therefore, with a minimum of 5 experiments it would be possible to define the exact value of this relationship. Perhaps one of the most valuable aspects of dimensional analysis is that without a good knowledge of a physical phenomenon we can derive its governing laws with a minimum amount of experimental work.

In most of the textbooks on dimensional analysis, the independent or relevant number of variables is always provided. The problem is then reduced to apply any of the methods of dimensional analysis to define the problem in terms of dimensionless groups. In spite that all the literature on dimensional analysis indicate that the definition of the relevant variables is the most important step, there is a lack of textbooks which study a natural phenomenon and start with the condition that the relevant variables are unknown. Its definition requires a full understanding of the driving forces for the problem under investigation. The main purpose of this book is to apply dimensional analysis in several problems of metallurgical engineering, including some cases where the number of relevant variables is defined after a long and exhaustive amount of experimental work.

The choice of variables is the most important step in dimensional analysis and the only way to choose those variables that are truly independent is by a rigorous knowledge of the physical phenomenon. As a guideline, the following properties can be used to define the minimum number of independent variables: the system geometry (dimensions), material properties (thermochemical and thermophysical properties), external conditions, knowledge of governing equations, etc.

The fact that the purpose of dimensional analysis is to represent a physical phenomenon with dimensionless numbers does not invalidate the use of dimensional variables because those dimensionless numbers depend on the numerical value of the dimensional variables. In this sense, dimensional analysis represents a method employed to reduce the amount of experimental work and defines relationships between the variables investigated using dimensionless groups which have a broader range of application in comparison with the conventional method using dimensional variables, furthermore, the resulting dimensionless groups can be used to scale the forces acting on the phenomena investigated. In simple words, dimensional analysis is a method to simplify the experimental work and make their results more general.

The variables treated in this book are treated as scalars. It has been pointed out, as early as 1892 the vectorial nature of some variables [6]. In spite that considering the vectorial nature of variables decreases the number of dimensionless groups [7], this subject is still subject of some controversy.

References

1. R. Butterfield, Dimensional analysis revisited. Proc. Inst. Mech. Eng. Part C **215**, 1365–1375 (2001)
2. M.C. Ruzicka, On dimensionless numbers. Chem. Eng. Res. Des. **86**, 835–868 (2008)
3. G. Miragliotta, The power of dimensional analysis in production systems design. Int. J. Prod. Econ. **131**, 175–182 (2011)
4. W. Barnett II., Dimensions and economics: some problems. Q. J. Austrian Econ. **7**, 95–104 (2004)
5. S. Sutera, R. Skalak, The history of Poiseuille's Law. Annu. Rev. Fluid Mech. **25**, 1–19 (1993)

6. W. Williams, On the relation of the dimensions of physical quantities to directions in space. Phylosophical Mag. Ser. **5**(34), 234–271 (1892)
7. C.N. Madrid, F. Alhama, Discriminated dimensional analysis of the energy equation: application to laminar forced convection along a flat plate. Int. J. Therm. Sci. **44**, 333–341 (2005)

Chapter 2
Origin and Historic Evolution of Dimensional Analysis

2.1 Origins of Dimensional Analysis

In antiquity dimensional analysis was ignored. Dimensions and units were not part of the discussion when a technical problem was investigated. Dimensional analysis has evolved with our understanding of mathematics. Geometry in antiquity had few principles involving only the dimension of length and angles. The earliest records on arithmetic and geometry came from Egypt (Rhind papyrus, 2000–1800 BC) and Mesopotamia (Babylonian clay tablets, 1900 BC). Euclid of Alexandria around 300 BC wrote *The Elements*, the most influential book on mathematics from antiquity. The oldest decimal multiplication table was invented by the Chinese, found in the Tsinghua Bamboo Slips dated 305 BC. In India the Shulba sutras correspond to texts of mathematics written from 800 BC to 200 AC, including geometry and algebra. Algebra is particularly important in the historic evolution of dimensional analysis because once a natural phenomenon is understood it is possible to define the variables involved in the form of equations. Equations were defined by words in full sentences, including the numbers. This is called rhetorical algebra. It was first developed in Mesopotamia and remained dominant up to the fifteenth century. An example of rhetorical algebra is the solution of a quadratic equation [1]: "A square and ten of its roots are equal to nine-and-thirty units, that is, if you add ten roots to one square, the sum is equal to nine-and-thirty. The solution is as follows: halve the number of roots, that is, in this case, five; then multiply this by itself, and the result is five-and-twenty. Add this to the nine-and-thirty, which gives four-and-sixty; take the square root, or eight, and subtract from it half the number of roots, namely, five, and there remains three: this is the root of the square which was required and the square itself is nine". In modern terms, it translates as:

$$x^2 + 10x = 39$$
$$x^2 + 10x + 25 = 64$$
$$\therefore x + 5 = 8$$

A. N. Conejo, *Fundamentals of Dimensional Analysis*,
https://doi.org/10.1007/978-981-16-1602-0_2

$$\therefore x = 3$$

Symbolic algebra, was introduced by Arab mathematicians (14th–15th century). The French mathematicians François Viète (1540–1603) and René Descartes (seventeenth century) introduced the modern notation of algebra. Descartes introduced the notation using the first letters of the alphabet for known and the last for unknown quantities (x, y).

Galileo Galilei (1564–1642) said in his book Il Saggiatore in 1623, "The universe cannot be read until we have learned the language and become familiar with the characters in which it is written. It is written in mathematical language, and the letters are triangles, circles and other geometrical figures, without which it is humanly impossible to comprehend a single word. Without these, one is wandering about in a dark labyrinth." A major step to explain natural phenomenon was the unification by Rene Descartes (1596–1650) of algebra and geometry into a single subject called analytic geometry published in *La Geometrie* in 1637; an appendix to his *Discours de la methode*. In spite of the progress to describe a natural phenomenon using the modern concepts of algebra introduced by Descartes, a solid definition of basic dimensions was still absent. One example is the second law of motion. Sharma [2] indicates that what we know today as Newton's second law of motion was originally defined in *The Mathematical Principles of Natural Philosophy* by words and not with a mathematical equation, as follows: "The alteration of motion is ever proportional to the motive force impressed and is made in the direction of the right line in which that force is impressed". The original definition describes force in terms of velocity and not in terms of acceleration:

$$F = k(u - v)$$

At the time of Newton (1642–1727), the terms acceleration and the second derivative were still undefined. Leonhard Euler (1703–1787), in 1775, 48 years after the death of Newton, defined force in terms of acceleration, the familiar equation for Newton´s second law of motion

$$F = ma$$

Another contribution from Euler was the use of the expression f(x) to define a mathematical function. Euler was the first mathematician to describe the importance of dimensional homogeneity in 1765 in his *Theoria motus corporum solidorum seu rigidorum*, however for almost six decades his ideas were neglected [3].

Macagno [4] in his historical review of dimensional analysis indicates that Newton was the first to recognize the concept of derived units and also attributes to Joseph Fourier (1768–1830) the origin of dimensional analysis. Fourier established the principle of dimensional homogeneity which states that units in either side of the equation should be the same. The principle of dimensional homogeneity was reported in his book *Théorie analytique de la chaleur* from 1822. Martins [5] disagrees with

Macagno stating that the first application of dimensional analysis was made by Francois Daviet de Foncenex in 1761. Today, it is generally accepted that Fourier was the founder of dimensional analysis. In spite of the definition of the principle of dimensional homogeneity by Fourier, its systematic application was ignored for the next 50 years.

2.2 Historic Evolution of Dimensional Analysis

The next important contribution in the field of dimensional analysis was due to Maxwell (1831–1879) in 1871 who introduced the modern symbolism to denote dimensions [6]. Maxwell employed brackets to define a dimension, for example [M] for mass, [L] for length and [T] for time. He was part of a committee who reviewed various systems of units for electricity and magnetism. This committee was funded by the British Association for the Advancement of Science (BAAS) in 1861. The committee was chaired by William Thompson.

William Thompson (1824–1907) developed a thermodynamic scale of temperatures and also made important contributions in the field of electricity and on the system of electrical units. He was also a strong advocate to the development of precision measuring instruments. He was given the title of Lord Kelvin by Queen Victoria in 1792. This title refers to the River Kelvin, which flows close to his laboratory at the University of Glasgow. William Thompson and Lord Kelvin are the same person.

The first method of dimensional analysis was proposed by the British Physicist Lord Rayleigh (1842–1919) in his book "The Theory of sound" published in 1877. In this method the dependent variable is proportional to a product of independent variables elevated to unknown exponents. The value of these exponents is obtained by solving a system of linear algebraic equations which results from the application of the principle of dimensional homogeneity on both sides of the original equation. In 1899 Lord Rayleigh reported the method of indices to assess the effect of temperature on the viscosity of argon [7]. The real name of Lord Rayleigh was John William Strutt, Baron Rayleigh. He was a Professor of experimental physics who won the Nobel prize in physics in 1904 for the discovery of argon. Rayleigh was appointed president of the Advisory committee for aeronautics, part of of the National Physical Laboratory, in 1909. In this position he was able to explore the use of dimensional analysis into the field of aeronautics. All of these developments were reviewed in detail by Rott [8]. Lord Rayleigh predicted the true value of dimensional analysis when he wrote [7]: "It happens not infrequently that results in the form of 'laws' are put forward as novelties on the basis of elaborate experiments, which might have been predicted a priori after a few minutes' consideration." It is indeed true that many laws who took centuries to get developed can be easily derived using the method of dimensional analysis, in spite of ignoring the mechanisms controlling a given natural phenomenon. He also anticipated that the real challenge is a true knowledge of the variables involved: "It is necessary as a preliminary step to specify clearly all the quantities on which the desired result may reasonably supposed to depend".

The fundamental theorem of dimensional analysis is that an equation that relates different variables can also be expressed in an equivalent form which is comprised entirely of dimensionless quantities. The great importance of this theorem is that a functional relationship between dependent and independent variables, collected in the form of dimensionless groups, can be defined only on the basis of dimensional considerations with very simple mathematical operations. This theorem was employed by Rayleigh but its demonstration is attributed to different researchers. It is commonly accepted that the French engineer Aimé Vaschy in 1892 and the Russian mathematician Dimitri Riabouchinsky in 1911, in an independent way, did the demonstration. The Russian mathematician, Federman has also been credited with its first derivation in 1911 [9]. However, the final credit for the diffusion and application of the fundamental theorem of dimensional analysis was for the American physicist Edgar Buckingham in 1914 [10–12]. Buckhingam promoted the application of dimensional analysis extensively. Two of his publications in particular [10, 12] describe in more detail some of the challenges of this method because he said: "*the results of dimensional reasoning are subject to the same limitations as those of any other theory*". Any real phenomena can be affected by a large number of variables or quantities but in order to make a practical representation, the system can be idealized or simplified with the minimum number of variables that have the largest effect. In the end, dimensional analysis will be an approximation between the ideal picture and its real prototype.

Buckingham (1867–1940) made great contributions applying physics to soil science [13]. His work on the movement of soil moisture, published in 1907 helped to improve agriculture through quantitative studies of soil moisture. He earned a bachelor's degree from Harvard and a doctoral degree from Leipzig. He joined the Bureau of Soils in 1902 for four years and then moved to the National Bureau of Standards until his retirement in 1937. In 1914 published and coined the π-theorem of dimensional analysis.

The method reported by Buckingham is known as the method of repeated variables. The π-theorem of dimensional analysis is associated with the method of Buckingham of repeated variables, however, there are a number of variations to the original method, for example Langhaar [14] in 1951 proposed a matrix method and Ipsen [15] in 1961 proposed a method that is based on making dimensionless the dependent variable.

The adoption of dimensional analysis was a slow process. It took more than a hundred years after Fourier to recognize its value. Buckingham in 1924 had to defend the method of dimensional analysis from physicists who claimed it was cumbersome and superfluous and therefore it had to be buried and forgotten [16]. At this time quotes a letter of October 30, 1915 he received from Rayleigh in regard with his publication from 1914 where he warns him about the difficulties to implement the method; "It is the best exposition of the subject I have come across; the only doubt is whether it is not too good for a majority of physicists".

The first textbook on dimensional analysis was made by Bridgman [17] in 1922. This book was the result of a collection of five lectures in physics at Harvard University in the spring of 1920.

References

1. American Assn. Adv. Science, The evolution of algebra. Science **18**(452), 183–187 (1891)
2. A. Sharma, Isaac Newton, Leonhard Euler and F=ma. Phys. Essays **27**(3), 503–509 (2014)
3. T. Zirtes, *Applied Dimensional Analysis and Modeling*, 2nd edn. (UK, Elsevier-BH, 2007), p. 102
4. E.O. Macagno, Historico-critical review of dimensional analysis. J. Franklin Inst. **292**(6), 391–402 (1971)
5. R. Martins, The origin of dimensional analysis. J. Franklin Inst. **311**(5), 331–337 (1981)
6. J. Clerk-Maxwell, Remarks on the mathematical classification of physical quantities. Proc. Lond. Math. Soc. **3**(34), 224–233 (1871)
7. L. Rayleigh, On the viscosity of argon as affected by temperature. Proc. R. Soc. Lond. **66**(1), 68–74 (1899)
8. N. Rott, Lord Rayleigh and hydrodynamic similarity. Phys. Fluids A **4**, 2595–2600 (1992)
9. B.E. Pobedrya, D.V. Georgievskii, On the proof of the Π-theorem in dimension theory. Russ. J. Math. Phys. **13**, 431–437 (2006)
10. E. Buckingham, On physically similar systems: illustrations of the use of dimensional equations. Phys. Rev. **4**, 345–376 (1914)
11. E. Buckingham, The principle of similitude. Nature **96**, 396–397 (1915)
12. E. Buckingham, Notes on the method of dimensions. Phil. Mag. Ser. **6**(42), 696–719 (1921)
13. T.N. Narasimhan, Buckingham, 1907: an appreciation. Vadose Zone J. **4**, 434–441 (2005)
14. H.L. Langhaar, *Dimensional Analysis and Theory of Models* (Wiley, USA, 1951)
15. D.C. Ipsen, *Units, Dimensions, and Dimensionless Numbers* (McGraw-Hill, USA, 1960)
16. E. Buckingham, Dimensional analysis. Phil. Mag. Ser. **6**(48), 141–145 (1924)
17. P.W. Bridgman, *Dimensional Analysis*, 1st edn. (Yale University Press, New Haven and London, USA, 1922)

Chapter 3
Units, Dimensions and Dimensional Homogeneity

Any method of dimensionless analysis starts by the definition of an engineering problem indicating the relevant variables. Once these variables have been defined, the solution is based on the analysis of dimensions applying the principle of dimensional homogeneity. In order to understand the method of dimensional analysis it is important to have a good understanding on the concepts of units and dimensions.

3.1 Units

Units and dimensions are associated with the concept of measurement. Ipsen [1] reviewed in detail these concepts in his book published in 1960 "Units, dimensions, and dimensionless numbers". The value of a quantity, a physical variable, is expressed as a number times a unit but it can also be mathematically described as the product of two numbers. A unit is not only an accepted standard name for measuring or quantifying a certain amount or quantity but it is also a number. For example, the quantity 10 m indicates a distance, it includes a number and a unit. The number is 10 and the unit is 1 m. The expression 10 m indicates 10 times multiplied by 1 m. This interpretation is important in order to justify a conversion of units.

3.1.1 Origin and Evolution of Units

Metrology is the science of measurement. The Proto-Indo-European word for moon is *me* which means "to measure". Metrology derives from the Greek word *metron*. The sun and the moon were the first references in antiquity for measuring time. In antiquity the units of length were based on the human body, perfectly summarized by Protagoras (490–420 BC) who said, "Man is the measure of all things". In antiquity the human body was the reference for short lengths; the digit (finger), palm, foot,

© The Author(s), under exclusive license to Springer Nature Singapore Pte Ltd. 2021   13
A. N. Conejo, *Fundamentals of Dimensional Analysis*,
https://doi.org/10.1007/978-981-16-1602-0_3

Fig. 3.1 Maya Cubit. Louvre Museum. Egyptian antiquities (Wikipedia)

cubit (fore-arm) and fathom (arms-span). Larger lengths were based on the pace. One phatom is equivalent to six feet or two yards. This word derives from old German which means pair of outstretched arms.

Williams [2] provides an important review on the origin of units; The oldest unit of length was the cubit. Egyptians, Babylonians, Indians and Chinese people defined the cubit using as a reference the length of the forearm, from the elbow to the tip of the middle finger. Egyptians had two variations of the cubit, the royal and the ordinary cubit. The royal cubit was approximately 523.5 mm and the ordinary cubit approximately 500 mm. This difference was used by the Pharaoh in his benefit when buying and selling goods. The royal standard cubit was carved out from a block of wood or granite. The Cubit was employed from antiquity to the middle ages. Figure 3.1 shows a cubit displayed in the Louvre which belonged to Maya, a royal treasurer under Tutankhamun and Horemheb. In the bible the cubit is mentioned several times, for example in Genesis 6:13–15 it says about Noah´s ark: "I am going to put an end to all people, for the earth is filled with violence because of them. I am surely going to destroy both them and the earth. So make yourself an ark of cypress wood; make rooms in it and coat it with pitch inside and out. This is how you are to build it: The ark is to be three hundred cubits long, fifty cubits wide and thirty cubits high".

Stone [3] summarized the value of the cubit in different countries, through time as well as the change of this length based on the anatomy of different people, giving an average value of 18.5 inches. Figure 3.2 shows different length scales based on the anatomy of the human body.

In China the hand was the reference for several units of length. A Span was the distance between the tip of the thumb and the tip of the little finger. A Chi was the distance between the tip of the thumb and the tip of the index finger. This unit is more than 3000 years old.

Greeks and Romans inherited the foot from the Egyptians. The Roman foot was divided into 12 unciae. The inch is derived from the Latin word *uncia*, which means one-twelfth. An uncia was 2.46 cm. Ounce is also derived from the same word. Julius Cesar (100–44 BC) introduced the mille, defined as 1000 paces, each pace consisting of two steps, equivalent to 5000 Roman feet or 1480 m. Rounding off a mille or mile as 1500 m, each Roman legionary had to make single paces of about 75 cm.

The cubit provides a convenient middle unit between the foot and the yard. The English yard is approximately a double cubit, about 0.1 m or 36 inches, measured from the center of a man's body to the tip of the fingers of an outstretched arm.

Furlong means furrow length and it was defined as the length of the furrow in one acre of a ploughed open field, about 660 feet. An acre was the amount of land arable by one man behind one ox in one day, measuring 1 furrow times 1 chain (66 feet).

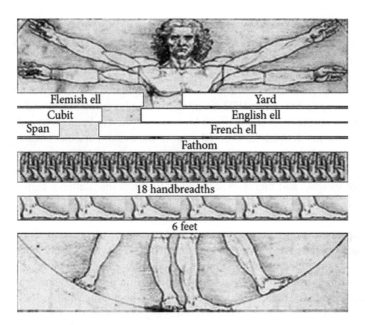

Fig. 3.2 Length scales based on the human body (Wikipedia)

The unit of mass was the *libra pondo*. In Latin, libra is a scale and pondo the weight placed on one side of the scale. The word *pound* is derived from the word pond and the symbol *lb*, from the word libra.

3.1.2 Metric System, Base Units and SI

Simon Stevin first suggested in 1585 to apply a decimal system for weights and measures. The decimal system was developed in Egypt. Gabriel Mouton in 1670 with his book, *Observationes diametrorum Solis et Lunae apparentium*, had the largest influence to adopt the metric system in France. Estrada et al. [4] explains that it was during the French revolution (1789–1799) that the metric system was finally adopted. The need for an international system of weight and measures for scientific purposes was pointed out by James Watt from England since 1783. In 1790, Charles Maurice de Talleyrand Bishop of Autun and Deputy at the National Assembly led a project to unify weight s and measures [5]. The project was adopted on May 8th, 1790. The committee was integrated by French scientists like Count Luois de Lagrange and Marquis Pierre Simon de Laplace. Members of the Royal Society in London were invited but due to the French Revolutionary wars they declined and consequently the final report excluded England. The unit of length was defined as the *metre*. On April 7th, 1795 the metric system (*Systeme Metrique Decimal*) was officially adopted in France. This system included the *litre* for volume and *gram* for weight. The final adoption of the metric system took longer time due to opposition to this system. Napoleon opposed to the metric system and in 1812 he re-established the former units. After the defeat of Napoleon, the metric system was reinstated on July 4th 1837, declaring obligatory from January 1840.

The internationalization of the metric system originated from a convention held in Paris from August 8th–13th, 1872, forming the Comision International du Metre. A treaty, known as the Convention of the Metre was signed in Paris on May 20th, 1875 by 18 nations that defined the establishment of the *Conference Generale des Pois et Mesures* (CGPM) or General Conference on Weights and Measures and the *Bureau International des Poids et Mesures* (BIPM) devoted to international agreement on matters of weights and measures. Subsequently, each international conference has resulted in new contributions to the system of weight and measures, as is shown in Table 3.1.

Thompson was a strong supporter of an International System of Units (SI). He stated the followong [6]: "I often say that when you can measure what you are speaking about, and express it in numbers, you know something about it; but when you cannot measure it, when you cannot express it in numbers, your knowledge is of a meagre and unsatisfactory kind: it may be the beginning of knowledge but you

Table 3.1 Relevant contribution from the conferences on weights and measures

	Year	CGPM	Relevant contribution
1	1889	1st	International standard prototype metre
3	1901	3rd	Definition of the litre. Declaration of the kilogram as the unit of mass. Weight was defined as a unit of force. Adoption of the of the gravitational constant, $g = 980.665$ cm s^{-2}
6	1921	6th	Units extended to include electrical and photometrical units
7	1927	7th	Definition of electric units: ampere, volt, ohm, coulomb, faraday, henry and weber
8	1933	8th	Idea to replace international units by absolute units
9	1948	9th	Replacement of the melting point of ice by the triple point of water. Adoption of Joule as a quantity of heat and Ampere for electric current
10	1954	10th	Set of six base units: length, mass, time, electric current, thermodynamic temperature and luminous intensity. Definition of standard atmosphere and the unit of time
11	1960	11th	The CGPM adopted the name Systeme International d´unites (SI). New definition of meter and second
12	1964	12th	Standardization of the curie as 3.7×10^{10} s^{-1}
13	1967	13th	New definitions for the second, kelvin and candela
14	1972	14th	Definition of a new unit; the mole, for the amount of substance. Adoption of two SI derived units; Pascal and Siemens
17	1995	17th	New definition of the meter based on the velocity of light in vacuum
24	2011	24th	Resolution for a possible future revision of the international system of units

Table 3.2 International system of units (SI). Dimensions and base units

Physical quantity	Dimension	Unit name	Symbol
Mass	M	Kilogram	kg
Length	L	Meter	m
Time	T	Second	s
Temperature	θ	Kelvin	K
Amount of substance	N	Mole	mol
Electric current intensity	I	Ampere	A
Luminous intensity	J	Candela	cd

have scarcely, in your thoughts, advances to the stage of science, whatever the matter may be".

The international System of Units (SI) has seven basic units, as shown in Table 3.2.

The base units of the SI are defined as follows:

Meter: One meter is the unit of length. French *metre* derives from latin "*metrum*", then from greek "*metron*" which means "to measure". The definition of one meter has been changed several times:

1. It was originally defined as 1/40,000,000 part of the earth´s meridian passing through Paris, on December 10 h, 1799. A sintered platinum bar was made with this length and deposited in the Archives de France.
2. At the 11th CGPM, in 1960, a new standard was defined based on the wavelength of the visible radiation of krypton, equal to 1,650,763.73 wavelengths of the radiation in vacuum due to the transition between the levels $2p_{10}$ and $5d_5$. This definition found difficulties because the coherence length of the radiation was less than 50 cm and a double scan was needed.
3. In 1983, at the 17th CGPM one meter was defined taking as a reference the speed of light previously defined in 1975 as 299,792,458 ms^{-1} as follows: One meter is he length of the path travelled by light in vacuum during a time interval of 1/299,792,458 s.

Kilogram: One kilogram is the unit of mass. Initially Lavoisier (1743–1794) proposed water as a reference of mass. The kilogram derives from the French kilogramme, from the Greek kilo (one thousand) and gramme from Latin that means "a small weight". One kilogram was defined as the mass of water of one cubic decimeter at the temperature of its maximum density (freezing point). A cylinder of pure sintered platinum was made by the French Academy on June 22nd, 1799. In 1878 new standards were prepared using an alloy containing 90% platinum-10% iridium. In 1880 these new standards were compared with the original reference at the archives and the one whose mass was closest was chosen as the new international prototype. In 1901 at the 3rd CGPM the kilogram was defined as a unit of mass rather than weight. The standard is kept on a glass plate covered by three bell jars, however due to accumulation of dust on the surface, its weight increases over time, equivalent to one microgram per year. The BIPM has a procedure for its cleaning and keep the original mass.

Second: second is the unit of time. The earliest clocks to display seconds appeared during the last half of the sixteenth century. The first clock (the pendulum clock) with an accurate measurement of seconds was invented by Huygens in 1656. Gauss in 1832 proposed the second as a unit of time. The word second is derived from the Latin word secundus which means "next in order".

1. At one time it was defined as the fraction of 1/86,400 of the mean solar day. This definition was based on the assumption that a day has exactly 24 h.
2. In 1960 at the 11th CGPM the second was defined as 1/315,569,259,747 of the tropical year for 1900.
3. Later on, it was found that an atomic clock was more precise. In 1967 during the 13th CGPM one second was defined as the duration of 9 192 631 770 periods of the radiation corresponding to the transition between the two hyperfine levels of the ground state of the Cesium atom.

Kelvin: kelvin is the unit of thermodynamic temperature. The word temperature derives from the Latin temperare, "to mix in due due proportion". Temperature is the average kinetic energy of a molecule. The kelvin temperature scale is an absolute temperature scale.

1. In 1954 at the 10th CGPM the temperature at the triple point of water was defined as 273.16 K.
2. In 1967 at the 13th CGPM the Kelvin was defined as the fraction 1/273.16 of the thermodynamic temperature of the triple point of water.

In contrast to the relative scales of temperature, expressed for example as °C, the absolute scale is defined only as K.

William Thompson in June 1847 made a report to the Cambridge Philosophical Society with the title; "On an absolute thermometric scale founded on Carnot´s theory of the motive power of heat and calculated from Regnault´s observations". This scale of temperature defined the absolute zero of temperature at a point marked as -273 on the air-thermometer scale. This scale is independent of the properties of any particular kind of substance.

Mole: mole is the unit for the amount of substance, defined in the 14th CGPM in 1972 as the amount of substance of a system which contains as many elementary entities as there are atoms in 0.012 kg of carbon in its ground state. The amount of substance is a general term and should be replaced in any particular case.

Ampere: One ampere is the unit of electric current intensity. It was defined in 1948 at the 9th CGPM as a constant current which, if maintained in two straight parallel conductors of infinite length, of negligible circular cross-section, and placed 1 m apart in vacuum, would produce between these conductors a force equal to 2×10^{-7} N per meter of length.

Charles-Augustin de Coulomb (1736–1806) was the first to measure the electromotive force in 1785, indicating that "the attractive force between two oppositely charged spheres is proportional to the product of the quantities of charge on the spheres and is inversely proportional to the square of the distance between the spheres". André-Marie Ampère (1775–1836) published a book in 1827 defining the basic theory that provides the relationship between electricity and magnetism. In 1892 Lord Kelvin rejected a proposal to replace the name kilowatt-hour for kelvin-hour [7] arguing that the unit kwh had more connection with a supply-meter.

Candela: candela is the unit of luminous intensity. Initially the luminous intensity was based on incandescent filament lamps. It was formally defined in 1948 and 1979.

1. In 1948 at the 9th CGPM adopted the name "new candle" and defined the brightness of the full radiator at the temperature of solidification of platinum equivalent to 60 new candles per square centimeter.
2. In 1968 at the 13th CGPM the name changed to candela and defined as the luminous intensity, in the perpendicular direction, of a surface of 1/600,000 square meter of a black body at the temperature of freezing platinum under normal standard pressure of 101,325 newtons per square meter.
3. In 1979 at the 16th CGPM the candela was redefined as the luminous intensity in a given direction of a source that emits monochromatic radiation of frequency 540×10^{12} Hz with a radiant intensity in that direction of 1/683 W per steradian.

Before the definition of the international system of units, every country had its local version of a system of units. Cardarelli [5] makes a full review of older systems of units in many countries. It was common that empires forced their colonies to follow a common system of units. England enforced their British system in the commonwealth and similarly Spain in America. In México, in some parts of the countryside the cuarteron is still used as unit of volume to trade corn and beans.

3.1.3 New Developments on the Definition of Units

Among the seven base units, the kilogram is still defined in terms of a material arte-fact and other base units; ampere, mole and candela also depend on the value of the mass of the international prototype of the kilogram. The most important limitation of this prototype is that its mass is not explicitly linked to an invariant of nature (universal constant) and in consequence its long-term stability is not fixed. To over-come this limitation, several natural constants have been proposed [8–10]. Kose et al. [8] proposed in 2003 to use hydrogen as the only source of those constants. Mills et al. [9, 10] have suggested the use of the Planck´s constant, the Boltzmann´s constant and Avogadro´s number to define kilogram, second, ampere, kelvin and mole. Andres et al. [11] disputed the use of Avogadro´s number to redefine the mole on the basis that Avogadro´s number is not a true natural constant but a number defined from a mass balance. The debate on a formal new definition of units is still in progress. The meeting in 2014 concluded that: "despite this progress the data do not yet appear to be sufficiently robust for the CGPM to adopt the revised SI at its 25th meeting". On November 16th, 2018 it was finally announced that "all SI units will now be defined in terms of constants that describe the natural world" [12]. The changes, which will

come into force on May 20th, 2019, will bring an end to the use of physical objects
to define measurement units. The definition of the kilogram will be replaced by the
Planck constant. The ampere will be defined by the elementary electrical charge (e).
The kelvin will be defined by the Boltzmann constant (k). The mole will be defined
by Avogadro's constant (N).

3.1.4   Previous Systems of Units

MKS: It is based on three fundamental units, meter, kilogram-force and second. It
lacks of the unit of mass, electro-magnetic and thermodynamic units.

MTS: It is based on three fundamental units, meter, metric tonne and second. It
was a legal system in France between 1919 and 1961. It lacks of electro-magnetic
and thermodynamic units.

CGS: It is based on three fundamental units, centimeter, gram and second. It lacks
of the unit of mass, electro-magnetic and thermodynamic units. It was proposed in
1873 by Lord Kelvin, James C. Maxwell and Ernst Werner von Siemens. It has a
clear difference between mass and force. Some of its base and derived units are still
used because the order of magnitude of some physical phenomena. It was divided
into three sub-systems; electromagnetic units (emu), electrostatic units (esu) and the
system of practical units. The Gauss system is a combination of the emu and esu
subsystems.

MKSA: Giovanni Giorgi in 1904 proposed a system on five fundamental units;
meter, kilogram, second, second, ampere and vacuum magnetic permeability. It is a
precursor of the SI system.

IEUS: The International Electrical Units System was a separate system employed
by US electric engineers until 1947.

BSU: The British System of Units has been employed in the UK and the USA.
It was established by the Weight and Measures Act of June 17th, 1824. It has three
base units; pound, yard and second. The yard was defined in 1878 as the distance
between two lines set in a bonze bar. The imperial unit of capacity was the gallon,
defined as the volume of 10 lb of distilled water, weighed in air against brass weights
at a temperature of 62°F and atmospheric pressure of 30 inches of mercury.

3.1.5   Derived Units

The derived units are those that can be expressed in terms of base units, my multiplication or division, excluding subtraction or addition. Table 3.3 shows some derived
units.

Base units define the number of dimensions. Dimensions are the basis of dimensional analysis. It will be shown later that force, a derived unit, can also be employed
as a fundamental dimension. In this sense, Bridgman argues that a derived unit can

Table 3.3 Derived units

Derived unit	Symbol	Units
Area	A	m^2
Volume	V	m^3
Velocity	U	$m\,s^{-1}$
Density	ρ	$kg\,m^{-3}$
Force	N	$kg\,m\,s^{-2}$
Pressure	Pa	$N\,m^{-2}$
Energy	J	$N\,m$
Dynamic viscosity	μ	$Pa\,s$
Surface tension	σ	$N\,m^{-1}$

also be used as a base unit and consequently they also can define dimensions: "physically, force is perfectly well adapted to be used as a primary quantity, since we know what we mean by saying that one force is twice another, and the physical processes are known by which force may be measured in terms of units of its own kind. It is the same way with velocities; it is possible to set up a physical procedure by which velocities may be added together directly, and which makes it possible to measure velocity in terms of units of its own kind, and so to regard velocity as a primary quantity" [13].

Figure 3.3 illustrates relationships of the SI derived units with special names and symbols and the SI base units Taken from NIST [14].

3.2 Dimensions

Ipsen [1] defines a dimension as "*a generalized unit.* Anything that could be measured in length units, for example, is said to have the dimensions of length; anything that could be measured in mass units is said to have the dimensions of mass: in general, the specific unit merely has to be replaced by the general concept to get the dimension". A dimension is a general description of one property of a physical variable without a numerical value. In the international System of Units with seven base units there are also seven base or fundamental dimensions. Maxwell suggested the use of one capital letter enclosed in brackets to designate a dimension. The fundamental dimensions are:

- Length: [L]
- Mass: [M]
- Time: [T]
- Temperature:[θ]
- Amount of substance[N]
- Electric current intensity[I]
- Luminous intensity[J]

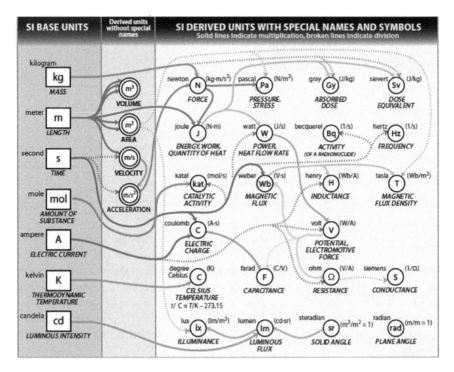

Fig. 3.3 Relationships of the SI derived units with special names and symbols and the SI base units [14]

Derived dimensions are obtained from base or fundamental dimensions. One example is Force. Force is defined from Newton´s second law of motion:

$$F = ma = m\frac{du}{dt}$$
$$[F] = M\frac{L}{T}\frac{1}{T} = MLT^{-2}$$

F in brackets can be read as "dimensions of force". The dimensions of force are MLT^{-2}. It can also be observed that a physical variable is represented with their dimensions in the form of power-law monomials, as follows:

$$[MLT^{-2}] = [M]\left[\frac{LT^{-1}}{T}\right]$$
$$[MLT^{-2}] = [MLT^{-2}]$$

A physical variable with dimensions of unity is called dimensionless.

Not all the dimensions are involved in one physical variable, therefore a given engineering field can be described by one particular set of dimensions, for example in mechanics, all physical variables can be defined by three dimensions: Length, Mass and Time [LMT].

Since mass and force are related by Newton´s second law of motion, an alternate set of dimensions in mechanics can be [LFT]. This is obtained from the following relationships:

$$[F] = MLT^{-2}$$

$$[M] = \frac{FT^2}{L}$$

For example, the dimensions of dynamic viscosity in the [LMT] system are:

$$[\mu] \equiv \frac{Mass}{Length \cdot Time} = \frac{M}{LT} = ML^{-1}T^{-1}$$

Using the conversion between mass and force from Newton´s second law of motion:

$$[\mu] \equiv \frac{M}{LT} = \frac{1}{LT}\frac{FT^2}{L} = FTL^{-2}$$

Therefore, dimensions for dynamic viscosity can be $ML^{-1}T^{-1}$ or FTL^{-2}.

In dimensional analysis every variable is defined in terms of their dimensions and not on their units. Table 3.4 summarizes the dimensions of some of the most common variables for systems of dimensions in fluid mechanics and thermal engineering.

Dimensions and units involving heat and temperature: Heat (Q) and temperature (θ) are two additional dimensions in heat transfer problems. Heat is thermal energy in motion due to a gradient of temperature. On the basis of the kinetic theory of gases, temperature is defined as the average kinetic energy of the molecules. According with the first law of thermodynamics, the units of heat and work are the same:

$$\oint dQ = \oint dW$$

According with this definition, the dimensions of heat are force and length [FL] equivalent to:

$$Heat\ (Q) = Work = [FL] = [LMT^{-2}][L] = [ML^2T^{-2}].$$

Table 3.5 shows the dimensions of some common variables involving heat transfer.

Gravitational force: The gravitational constant is a constant but in dimensional analysis is considered a variable.

Gravity is the force that attracts two bodies toward each other and is the product of mass times the acceleration of the gravitational constant on earth (g). The units of g are m/s^2 or equivalently in N/kg. The value of g is not constant on earth but

Table 3.4 Physical quantities expressed in terms of their dimensions

Physical quantity	Symbol	LMT	LFT	LMTθ	LFTθ
Velocity	U	LT^{-1}	LT^{-1}	LT^{-1}	LT^{-1}
Acceleration	a	LT^{-2}	LT^{-2}	LT^{-2}	LT^{-2}
Density	ρ	ML^{-3}	$FT^2 L^{-4}$	ML^{-3}	$FT^2 L^{-4}$
Energy	E	ML^2T^{-2}	FL	ML^2T^{-2}	FL
Power	P	ML^2T^{-3}	$FT^{-1}L$	ML^2T^{-3}	$FT^{-1}L$
Force (gravity, etc.)	F	LMT^{-2}	F	LMT^{-2}	F
Mass	M	M	$L^{-1}FT^2$	M	$L^{-1}FT^2$
Thermal conductivity	k	–	–	$MLT^{-3}\theta^{-1}$	$FT^{-1}\theta^{-1}$
Diffusion coefficient	D	L^2T^{-1}	L^2T^{-1}		L^2T^{-1}
Surface tension	σ	MT^{-2}			
Dynamic viscosity	μ	$ML^{-1} T^{-1}$	FTL^{-2}	$ML^{-1} T^{-1}$	FTL^{-2}
Kinematic viscosity	ν	L^2T^{-1}	L^2T^{-1}	L^2T^{-1}	L^2T^{-1}
Heat transfer coefficient	h	–	–	$MT^{-3}\theta^{-1}$	$FT^{-1}L^{-1}\theta^{-1}$
Pressure	p	$ML^{-1} T^{-2}$	FL^{-2}	$ML^{-1} T^{-2}$	FL^{-2}
Specific heat capacity	C	–	–	$L^2T^{-2}\theta^{-1}$	$(FLM^{-1}\theta^{-1})$
Angle	–	LL^{-1}	LL^{-1}	LL^{-1}	LL^{-1}

Table 3.5 Dimensions of variables involving heat transfer

Physical quantity	Symbol	LMTθ	LFTθ	LQTθ
Heat	Q	$ML^2 T^{-2}$	FL	Q
Rate of heat transfer	$\dot{Q}$	ML^2T^{-3}	FLT^{-1}	QT^{-1}
Entropy	S	$ML^2T^{-2}\theta^{-1}$	$LF\theta^{-1}$	$Q\theta^{-1}$
Thermal conductivity	k	$MLT^{-3}\theta^{-1}$	$FT^{-1}\theta^{-1}$	$QL^{-1} T^{-1}\theta^{-1}$
Specific heat capacity	C	$L^2T^{-2}\theta^{-1}$	$FLM^{-1}\theta^{-1}$	

changes with location from about 9.780 m/s^2 at the Equator to about 9.832 m/s^2 at the poles, so an object will weigh approximately 0.5% more at the poles than at the Equator [15]. In 1901, at the third General Conference on Weights and Measures it was defined a standard value equal to 9.80665 m/s^2 based on measurements done at the Pavillon de Breteuil near Paris in 1888.

Dimensions g	Symbol	LMT
Gravity of earth	g	LT^{-2}

3.3 Principle of Dimensional Homogeneity

Any physical equation contains terms on both sides, *The principle of dimensional homogeneity is based on the fact that dimensions and units in every term of one equation have to be the same in order to obey the equality.* If this condition is satisfied the equation is dimensionally homogeneous. This principle, which is familiar to any student today and could be considered part of the common sense because, from an algebraic view point, only oranges can be added with oranges, it was formally defined until 1822 by Fourier.

For example, consider the Bernoulli equation; This equation represents an energy balance per unit volume and is defined as follows:

$$P_1 + \frac{1}{2}\rho U_1^2 + \rho g h_1 = P_2 + \frac{1}{2}\rho U_2^2 + \rho g h_2$$

Let's confirm if this equation is dimensionally homogeneous: the dimensions on both sides of the equation should be the same:

$$\text{Pressure} = \left[\frac{\text{Force}}{\text{Area}}\right] = \left[\frac{\text{mass} \times \text{acceleration}}{\text{Area}}\right] = \left[\frac{\text{mass}}{\text{Area}}\frac{\text{velocity}}{\text{time}}\right] = \left[\frac{M}{L^2}\frac{\frac{L}{T}}{T}\right] = ML^{-1}T^{-2}$$
$$\frac{1}{2}\rho U^2 = \left[\frac{M}{L^3}\left(\frac{L}{T}\right)^2\right] = ML^{-1}T^{-2}$$
$$\rho g h = \left[\frac{M}{L^3}\frac{L}{T^2}L\right] = ML^{-1}T^{-2}$$

The dimensions are the same in both sides of the equation, therefore, it is confirmed that the equation is dimensionally homogeneous. If any equation doesn't follow the principle of dimensional homogeneity, it is incorrect. In terms of dimensions, it is to be noted that the previous equation can be described in terms of a product of dimensions, each one elevated to an exponent, called monomial power law.

$$\left[ML^{-1}T^{-2}\right] + \left[ML^{-1}T^{-2}\right] + \left[ML^{-1}T^{-2}\right] = \left[ML^{-1}T^{-2}\right] + \left[ML^{-1}T^{-2}\right] + \left[ML^{-1}T^{-2}\right]$$

In a general way, in any experiment we can have n number of physical variables (Y, X) that are related by an expression dimensionally homogeneous, as follows:

$$f(Y, X_1, X_2, \ldots, X_n) = 0$$

If Y is the dependent variable:

$$Y = f(X_1, X_2, \ldots, X_n)$$

where; X_1, X_2, ..., X_n represent the independent variables and assuming that all variables correspond to that of a mechanical system, described by the dimensions

MLT, then the previous relationship can be defined in terms of the principle of dimensional homogeneity, as follows:

$$\left[M^a \cdot L^b \cdot T^c\right] = f\left[M^{a_1} \cdot L^{b_1} \cdot T^{c_1}\right]\left[M^{a_2} \cdot L^{b_2} \cdot T^{c_2}\right]\left[M^{a_3} \cdot L^{b_3} \cdot T^{c_3}\right]\cdots\left[M^{a_n} \cdot L^{b_n} \cdot T^{c_n}\right]$$

This equation defines the principle of dimensional homogeneity. The main point here is to observe that each physical variable is defined as a monomial containing the product of their dimensions each one elevated to a power and is always this way and no other.

3.3.1 Derivation of the Principle of Dimensional Homogeneity [16, 17]

A physical quantity is measured and expressed with a number and units:

$$X = xU = x'U'$$

where x and x' represent numbers, U and U′ represent different type of units. The values x and x' depend on the type of units. One example is 1 m = 100 cm. The value 1 m, mathematically described is the product of two numbers, the number 1 and the number 1 m, the first number is a scalar and the second makes reference to the relative scale of units to measure length in this case. The absolute value of the physical variable is independent of the type of units. One way to cancel units is by defining the physical variable as a ratio.

In oder to make the physical quantity dimensionless and therefore independent of the type of units, it has to be expressed as a ratio of units.

$$X = \frac{x'U'}{xU}$$

X is dimensionless but depends on terms that depend on the set of dimensions. To illustrate this relationship, it is expressed as follows:

$$X = f(L, M, T)$$

This equation is called the *dimensional function which is independent of the type of units.* On the contrary x and x' depend on the type of units chosen, for example if we define two types of units called 1 and 2, their values are defined as follows:

$$x = f(L_1, M_1, T_1)$$
$$x' = f(L_2, M_2, T_2)$$

In order to define the value of the dimensional function, f (L, M, T) which is independent of the type of units it is necessary to use a ratio of units. For example, the area of a land measuring 4 m by 2 m is the same as that measuring 157.48 inches by 78.74 inches, in both cases the ratio is two: $4/2 = 2$ and $157.48/78.74 = 2$. This property of a ratio of units, for the same type of dimension, suggests its use in the final form of the dimensional function. A ratio of units, considering the two types of units (1 and 2) within the same dimension is given below:

$$\frac{x'}{x} = \frac{f(L_2, M_2, T_2)}{f(L_1, M_1, T_1)} = f\left(\frac{L_2}{L_1}, \frac{M_2}{M_1}, \frac{T_2}{T_1}\right)$$

This expression is called a *functional equation*:

$$\frac{f(L_2, M_2, T_2)}{f(L_1, M_1, T_1)} = f\left(\frac{L_2}{L_1}, \frac{M_2}{M_1}, \frac{T_2}{T_1}\right)$$

Since selection of units is arbitrary, an alternative choice is:

$$\frac{f(L, M, T)}{f(L_1, M_1, T_1)} = f\left(\frac{L}{L_1}, \frac{M}{M_1}, \frac{T}{T_1}\right)$$

Making $L_2 = L_1 = L$, $M_2 = M_1 = M$ and $T_3 = T_2 = T$:

$$\frac{f(L, M, T)}{f(L, M, T)} = f\left(\frac{L}{L}, \frac{M}{M}, \frac{T}{T}\right)$$

What follows is the analysis to proof that a power law monomial satisfies a functional equation. This process involves three steps; Partial derivatives with respect to L, M and T.

Partial derivatives with respect to L:

$$\frac{\frac{\partial f}{\partial L}(L,M,T)}{f(L,M,T)} = \frac{1}{L}\frac{\partial f}{\partial \frac{L}{L}}(1, 1, 1)$$
$$\frac{1}{L}\frac{\partial f}{\partial \frac{L}{L}}(1, 1, 1) = a$$
$$\frac{d f(L,M,T)}{f(L,M,T)} = a\frac{dL}{L}$$

Integrating the previous differential equation:

$$\ln\{f(L, M, T)\} = a \ln L + \ln C_1(M, T)$$
$$f(L, M, T) = L^a C_1(M, T)$$

Notice that the new function, C_1, depends only on M and T.

$$\frac{C_1(M_2, T_2)}{C_1(M_1, T_1)} = C_1\left(\frac{M_2}{M_1}, \frac{T_2}{T_1}\right)$$

Continuing the derivation of the new function C_1 with respect to M.

$$\frac{\frac{\partial C_1}{\partial M}(M,T)}{f(M,T)} = \frac{1}{M}\frac{\partial C_1}{\partial \frac{M}{M}}(1, 1, 1)$$

$$\frac{1}{\frac{M}{M}}\frac{\partial C_1}{\partial \frac{M}{M}}(1, 1, 1) = b$$

$$\frac{d\mathring{C}_1(M,T)}{C_1(M,T)} = b\frac{dM}{M}$$

$$C_2(T) = L^a M^b C_2(T)$$

The process is repeated for with the new function C_2:

$$C_2(T) = T^c C_3$$

Then,

$$f(L, M, T) = k\, L^a M^b T^c$$

k is a constant equal to 1 since $L = M = T = 1$ indicating that the value of the quantity X remains unchanged.

$$f(L, M, T) = L^a M^b T^c$$

This equation defines the principle of dimensional homogeneity for any variable and is the basis of all methods of dimensional analysis. The result proves that only a power law monomial satisfies a functional equation. The same result is obtained applying the approximation theory, as shown by Szirtes [18]. It was reported by Maxwell in 1871 [19] and Rayleigh in 1877 [17]. As a concept, in dimensional analysis a physical quantity is treated only in terms of their dimensions, more specifically as the product of these dimensions elevated to a power. Since dimensional analysis is not concerned with units, only with dimensions, there is drastic difference how variables in one equation are defined using the conventional approach and using dimensional analysis.

The typical relationship between two variables in the conventional way is:

$$Y = f(X_1)$$

Applying the principle of dimensional homogeneity, it was found that a dimensional function is expressed as a function of power monomial. Assuming a case where the variables depend on the dimensions MLT, the previous equation using dimensional analysis has the following form:

$$\left[M^a \cdot L^b \cdot T^c\right] = f\left[M^{a_1} \cdot L^{b_1} \cdot T^{c_1}\right]$$

In the case with one independent variable elevated to a power, k_1:

$$Y = f(X_1)^{k_1}$$

$$[M^a \cdot L^b \cdot T^c] = f[M^{a_1} \cdot L^{b_1} \cdot T^{c_1}]^{k_1}$$

Dimensions will be the same in both sides of the equation if the following is satisfied:

$$a = a_1 k_1$$
$$b = b_1 k_1$$
$$c = c_1 k_1$$

An even more general expression considering n independent variables

$$Y = f(X_1^{e_1}, X_2^{e_2}, \ldots, X_n^{e_n})$$

On application of the principle of dimensional homogeneity, the previous equation becomes:

$$[M^a \cdot L^b \cdot T^c] = f[M^{a_1} \cdot L^{b_1} \cdot T^{c_1}]^{e_1} [M^{a_2} \cdot L^{b_2} \cdot T^{c_2}]^{e_2} [M^{a_3} \cdot L^{b_3} \cdot T^{c_3}]^{e_3} \ldots$$
$$[M^{a_n} \cdot L^{b_n} \cdot T^{c_n}]^{e_n}$$

One of the most important consequences of the principle of dimensional homogeneity is that the variables can be grouped in terms that are dimensionless also called dimensionless numbers. In this case the principle of dimensional homogeneity is satisfied if the following balances of exponents is satisfied:

$$a = a_1 e_1 + a_2 e_2 + \cdots + a_n e_n$$
$$b = b_1 e_1 + b_2 e_2 + \cdots + b_n e_n$$
$$c = c_1 e_1 + c_2 e_2 + \cdots + c_n e_n$$

An equivalent form is the following

$$[M^0 \cdot L^0 \cdot T^0] = f[M^a \cdot L^b \cdot T^c][M^{a_1} \cdot L^{b_1} \cdot T^{c_1}]^{e_1} [M^{a_2} \cdot L^{b_2} \cdot T^{c_2}]^{e_2} [M^{a_3} \cdot L^{b_3} \cdot T^{c_3}]^{e_3} \ldots$$
$$[M^{a_n} \cdot L^{b_n} \cdot T^{c_n}]^{e_n}$$

$$0 = a + a_1 e_1 + a_2 e_2 + \cdots + a_n e_n$$
$$0 = b + b_1 e_1 + b_2 e_2 + \cdots + b_n e_n$$
$$0 = c + c_1 e_1 + c_2 e_2 + \cdots + c_n e_n$$

A general relationship in terms of dimensionless numbers (or dimensionless groups) is obtained applying the principle of dimensional homogeneity and one method of dimensional analysis.

References

1. D.C. Ipsen, *Units, Dimensions, and Dimensionless Numbers* (McGraw-Hill, USA, 1960).
2. J.H. Williams, *Defining and Measuring Nature.* (Morgan & Claypool, 2014), pp. 1–9
3. M.H. Stone, The cubit: a history and measurement commentary. J. Anthropol. 1–11 (2014)
4. H. Estrada, J. Ruiz, J. Triana, The origin of meter and the trust of mathematics. Matemáticas: enseñanza universitaria **19**, 89–101 (2011)
5. F. Cardarelli, *Encyclopaedia of Scientific Units, Weights and Measures. Their SI Equivalences and Origins* (Springer, London, 2006).
6. W. Thopmson, *Electrical Units of Measurement* (Institution of Civil Engineers, London, 1884), p. 149
7. P. Tunbridge, *Lord Kelvin: His Influence on Electric Measurements and Units* (Peter Peregrinus Ltd., London UK, 1992), p. 60
8. S.V. Gupta, *Units of Measurement; Past, Present and Future International System of Units* (Springer, London UK, 2010), pp. 117–126
9. I. Mills, P. Mohr, T. Quinn, B. Taylor, E. Williams, Redefinition of the kilogram, ampere, kelvin and mole: a proposed approach to implementing CIPM recommendation 1(CI-2005). Metrologia **43**, 227–246 (2006)
10. I. Mills, P. Mohr, T. Quinn, B. Taylor, E. Williams, Adapting the international system of units to the twenty-first century. Phil. Trans. R. Soc. A **369**, 3907–3924 (2011)
11. H. Andres, H. Haerri, B. Niederhauser, S. Wunderli, U. Feller, No Rationale for a redefinition of the mole. Chimia **63**, 616–618 (2009)
12. https://www.npl.co.uk/news/international-system-of-units-overhauled-in-historic-vote. Accessed 2 Dec 2018)
13. P.W. Bridgman, *Dimensional Analysis* (Yale University Press, USA, 1963), p. 20
14. https://physics.nist.gov/cuu/Units/SIdiagram.html. Accessed 22 Nov 2020)
15. https://en.wikipedia.org/wiki/Gravity_of_Earth
16. G.I. Barenblatt, *Scaling, Self-Similarity, and Intermediate Asymptotics* (Cambridge University Press, USA, 1996), pp. 34–36
17. J. Worstell, *Dimensional Analysis Practical Guides in Chemical Engineering* (Elsevier, UK, 2014), pp. 45–48
18. T. Szirtes, *Applied Dimensional Analysis and Modeling* (Elsevier, USA, 2007), p. 1333
19. Q.-M. Tan, *Dimensional Analysis: With Case Studies in Mechanics* (Springer, Berlin, 2011), p. 8

Chapter 4
Dimensionless Numbers

4.1 Concept of a Dimensionless Number

This chapter provides information on some relevant dimensionless numbers, giving a description of the forces involved and a short biography of the people bearing the names of those dimensionless numbers.

Some variables are dimensionless by nature, for example an angle. An angle is a ratio of the arc length to the radius, measured in radians. The ratio of the arc length with respect to the full circumference gives the angle in revolutions and if the angle is the ratio of the arc length with respect to 1/360 of its full circumference it value is reported in degrees. The number 360 probably derives from a first approximation of the number of days in a year. In any ratio of similar units, the variable is dimensionless.

The importance of a dimensionless number is that it clearly defines the ratio of forces, energies, diffusion rates, etc., involved. When the value of a dimensionless number is obtained it is pretty useful to know if this value is small or large because in these two cases, one of the forces can be neglected and therefore the analysis is simplified.

Some of these forces are indicated below:

Inertial forces: are always present in all fluids in motion.

$$F = ma = \rho L^3 \frac{dU}{dt} = \rho L^3 \frac{dU}{dx}\frac{dx}{dt} = \rho U^2 L^2$$

Gravitational forces: Gravity controls the movement of a free surface or interface. Only confined flows are not subject to gravitational forces.

$$F = mg = \rho g L^3$$

Buoyancy forces (gas–liquid system): Due to bubble rising in a liquid

$$F = \left(\rho_l - \rho_g\right)gL^3 = \Delta\rho g L^3$$

© The Author(s), under exclusive license to Springer Nature Singapore Pte Ltd. 2021
A. N. Conejo, *Fundamentals of Dimensional Analysis*,
https://doi.org/10.1007/978-981-16-1602-0_4

Buoyancy forces (pure liquid): Due to thermal gradients

$$F = \rho_1 g L^3 \beta \Delta T$$

where: β is the coefficient of volume expansion,

$$\beta = -\frac{1}{\rho}\frac{\partial \rho}{\partial T}$$

Applying Boussinesq approximation, after Joseph Boussinesq (1842-1929), valid if the temperature gradients are small and fluid's density remains almost constant:

$$\rho = \rho_0 - \rho_0 \beta \Delta T$$
$$\rho_1 g = (\rho_0 - \rho_0 \beta \Delta T)g$$

Viscous forces:

$$F = \mu L^2 \frac{dU}{dx} = \mu UL$$

Pressure forces:

$$F = p L^2$$

Surface tension forces:

$$F = \sigma L$$

A dimensionless number is a ratio of dimensional variables in both numerator and denominator and the ratio of dimensions becomes dimensionless. The ratio of variables is also called dimensionless group. Bezuglyi et al. [1] has classified two types of dimensionless numbers based on the number of forces involved. A general case involves dimensionless numbers with two types of forces but there are also dimensionless numbers with more than two forces.

A large amount of dimensionless numbers has been assigned names. Some dimensionless numbers have variations in the variables, a given dimensionless number can be described as a ratio or product with other dimensionless numbers and furthermore, some dimensionless numbers can be identified with different names, for example the Euler number is the same as the Newton number. The book by Kunes [2] in 2012, reports 1200 dimensionless numbers in eight fields. A total of 239 numbers in the field of fluid mechanics, 270 in thermomechanics and 98 in electromagnetism.

4.2 Common Dimensionless Numbers

Reynolds number: [3–9]

The numerator defines the inertial (dynamic) forces and the denominator the viscous forces. It is used to describe if the flow is laminar or turbulent. During laminar flow the flow follows the same direction and during turbulent flow, some regions of the flow change direction and oppose to the flow, creating eddies. The ratio of forces was first described by George Stokes in 1851, however the name of the number was given by Arnold Sommerfeld in 1908 in honor of Osborn Reynolds.

$$\mathrm{Re} = \frac{\rho U^2 L^2}{\mu UL} = \frac{\rho UL}{\mu} = \frac{UL}{\nu}$$

where: U is the velocity, ρ is the density, μ is the dynamic viscosity, ν is the kinematic viscosity and L represents a characteristic length.

The Re number should always be considered when the viscosity of the fluid plays a significant role.

At low Reynolds numbers (below 2000) the flow is laminar and at high values (above 4000) the flow is turbulent. Low values are an indication that viscous forces predominate and consequently the elementary volumes of fluid cannot be easily deformed. On the other hand, at high values the velocity of the fluid predominates and the flow becomes unstable.

The Re, Pe and Sc numbers are related as follows,

$$\mathrm{Re} = \frac{\mathrm{Pe}}{\mathrm{Sc}}$$

In 1883 Reynolds carried out his famous experiments on instability of tube flow. A large glass tube was placed inside a tank of water. Water could flow from the tank into the tube when a valve at the other end of the tube was opened. When the water was flowing a dye would be 'pulled' inside the tube, working as a tracer of the fluid motion. He measured the maximum velocity in the glass pipe, continuously increasing the flow rate until the flow became turbulent. He found that for one liquid, the product of the transition velocity times the radius, was constant. Changing the fluid or changing its temperature observed that the previous ratio divided by the kinematic viscosity of the fluid was also constant. Reynolds submitted his work for publication in the proceedings of the Royal Society. The two reviewers, George Stokes and Lord Rayleigh recommended its publication. Further theoretical aspects reported to the Royal Society in 1894 were not fully understood by Stokes who also reviewed the paper. In this paper Reynolds introduced the concepts of mean and fluctuating components of the flow, which constitute the first attempts to describe the nature of turbulence. It is mentioned that Stokes first developed the ratio of inertial to viscous forces to control fluid flow, in a paper published 35 years earlier [10]. It is surprising that he doesn't mention anything about that in his review.

The Reynolds number of microscopic particles, for example non-metallic inclusions in steel, is insignificant and are thus mainly driven by the viscous forces of the fluid. For such objects, a fluid would feel significantly more rigid, i.e., it would be difficult for an inclusion to force a path through a moving fluid not following the streamlines. As the objects grow larger, their inertia starts to dominate over the viscous forces. For most fish, the Reynolds number is in the range of 1×10^5, for a human it is in the range of 1×10^6. At higher Reynolds numbers, an object is able to force its way through a flow field even across the streamlines. An example is a ship (with Reynolds numbers in the range of 1×10^9) compared to a paper boat: the large vessel can force its way through the current and the waves, whereas the light paper boat would not be able to do so. Rather it has to stay with the streamline and will be dragged along [9].

Osborne Reynolds (23 August 1842–21 February 1912) was born in Belfast, Ireland. He was the second of three children. His mother died after the birth of his younger brother. Reynolds' family was British, from East Anglia where they had extensive agricultural states. After Reynolds' birth the family returned to Essex where he attended Dedham grammar school.

His father, with the same name, was a mathematician but after graduation was trained to become a priest following his family tradition for three generations. At the age of 19, he took employment for a year at the workshop of Edward Hayes, a mechanical engineer of Stony Stratford, before he went to Queens' College in Cambridge to study mathematics, in 1863 and then graduated in 1867. Got married with Charlotte Chadwick in 1868. His first wife died a year later, twelve days after the birth of their first son.

Once he graduated, he got a job with civil engineering consultants in London. Owens college (currently the University of Manchester) was funded by a grant from John Owens, a wealthy textile merchant and opened in 1851. A chair of civil and mechanical engineering was opened in 1868 and Reynolds applied for that chair on January 18th. The salary was criticized at an engineering journal and the committee decided to raise it to £500 per year, also, at one time Prof. W. Rankine, a recognized scientist holding the chair of engineering at the University of Glasgow considered to apply. Reynolds was finally chosen from a list of 27 candidates and delivered his introductory address on October 5th, 1868. He was the second full time professor of engineering in England and was only 25 years old. The new department lacked research laboratories and for 5 years he had to do some research work at home on natural phenomena such as the formation of hailstones and snowflakes. In 1873 a new campus was built and began to study heat transfer. In 1881 he married Annie Wilkinson, 22 years old with whom he had four more children. In 1877 he was named fellow of the Royal Society. Reynolds is better known for two papers published in 1883 and 1895 on fluid mechanics, which represent the definition of laminar and turbulent flow as well as the first attempt to explore the complex nature of turbulence, however he made significant contributions in other areas, for example developing the theory of thin-film lubrication. A detailed description of his research work is given by Jackson [6]. His contributions on turbulence were ahead of his time.

By 1905 Reynolds' health deteriorated and had to retire, having being working for 37 years. In 1908 he moved to St Decuman's, a village community on the hill just above Watchet in Somerset, a small coastal port. There, on February 21, 1912 died from influenza, at the age of 69. Reynolds' remains are buried in the churchyard.

Froude number: [11–13]

The Froude number represents the ratio of inertial forces to gravitational or uplift forces or alternatively as the ratio of the kinetic energy to the potential energy.

$$Fr = \frac{\rho U_0^2 L^2}{\rho g L^3} = \frac{U_0^2}{gL}$$

where: U_0 is a characteristic flow velocity, g is the gravitational field constant and L is a characteristic length. Equivalent expressions are the square root of the previous ratio,

$$Fr = \frac{U_0}{\sqrt{gL}}$$

and in terms of the gas flow rate. $Q = UA$, where U is the superficial velocity and A is the area of the injection device.

$$Fr = \frac{U_s^2}{gd_n} = \frac{Q^2}{gd_n^5}$$

The modified Froude Number is a ratio of inertial and buoyancy forces. The inertial forces are caused by the motion of the bubble and the buoyancy forces result from the density difference between the liquid and the gas phase.

$$Fr_m = \left[\frac{\rho_g}{(\rho_1 - \rho_g) gd_n} \frac{U_0^2}{} \right] \cong \left[\frac{\rho_g}{\rho_1} \frac{U_0^2}{gd_n} \right]$$

The inverse value of the modified Froude number is called the Tundish Richardson number, *Tu*. This name was given by Damle and Sahai [14]. It represents a ratio of buoyancy to inertial forces. In this case the buoyancy forces are due to density differences in the fluid.

$$Tu = \left[\frac{gL}{U_0^2}\beta\Delta T\right]$$

where β is the coefficient of thermal expansion.

William Froude (28 November 1810–4 May 1879) was a British civil and naval engineer. He was the fourth son of Archdeacon Richard Hurrell Froude. William was born in Dartington, Devon County, west of Torquay harbor, in southwest England. Attended school at Buckfastleigh and then entered Oriel College, Oxford University, in 1828, graduating as a civil engineer in 1832, at 22 years old. Got married in 1838 and got 5 children.

He worked as civil engineer for 14 years, working for the Bristol and Exeter Railway. Ship design was a subject that attracted his attention as early as 1833. It is important to recall that the Royal Navy played a key role in establishing the British Empire as the main world power during the 19th and first part of the twentieth centuries.

He retired to take care of his father ill's health in 1846. Lived with him for 13 years. A major event occurred in 1857. The largest ship in the world, the *Great Eastern* was built, however its speed was so low that it could not cover its own fuel expenses. Froude then started experiments in fluid dynamics. At Dartington in 1850 he made screw-propeller tests on the River Dart. He built two scale models called Raven and Swan which helped him to demonstrate that there was no ideal form and that performance varied with speed. In 1861 reported his first paper at the first meeting of the Institution of Naval Architects (INA). He conducted research on fluid flow for the Torquay waterworks from 1863 to 1866. In the early 1870s got government funding to build a towing tank in spite some opposition from members of the INA. The drag force acting on the models was measured with a custom-designed dynamometer. He investigated the most efficient hull shape using scale models. In a hydraulic tank built in 1871 he developed the law of comparison, now called dynamic similarity principle. He built a 1/16 geometric scale of the ship *Greyhound*, towed at speeds from 3 to 12.5 knots (5.5 -23 km/hr) and measured the relationship between resistance and the speed of the ship. The towing tank was filled with water on March 21, 1871 and the first experiment was conducted on March 3, 1872. The financial support requested was 717 lb for materials and 500 lb/year for labor (his son Robert Edmund Froude worked as his assistant).

In a paper published in 1874 he reported the law of comparison. He assumed that the total resistance was the sum of wave formation and skin friction and that each could be scaled independently. He observed that geometrically similar hulls produced different wave patterns, however above a ratio proportional to the velocity squared/hull length, the wave patterns were nearly identical or similar. This ratio can also be expressed as a ratio velocity/square root of their dimensions. One consequence of this law is that the wave effects would be similar in model and prototype if the velocity were reduced in proportion to the square root of the length. His results represent the foundation of similarity and scaling laws. In 1870 was elected fellow of the Royal Society. In 1876 received an honorary Doctor of Law degree from the University of Glasgow and a medal from the Royal Society for "his researches both theoretical and experimental on the behavior of ships, their oscillation, their resistance, and their propulsion".

At the time of his death was vice-president of INA. In 1878 he went on vacations to South Africa and got ill from Dysentery, dying the following year at an age of 69 years. His grave remains at the Naval Cemetery, Simons Town, South Africa. His son R. Edmund continued his work until 1924. Their collected work was reported in 34 papers in the Transaction of INA. Froude's idea about science is summarized in his quote; "Our sacred duty [is] to doubt each and every proposition put to us including our own".

Weber number:

The number describes the ratio of inertial forces to surface tension forces. It is useful to analyze the formation of bubbles and droplets. The force due to surface tension is tangential to the surface and has the same magnitude perpendicular to any line element along the surface.

$$\text{We} = \frac{\rho U^2 L^2}{\sigma L} = \frac{\rho U^2 L}{\sigma}$$

If the Weber number is high, the kinetic energy dominates (surface tension forces can be neglected). With high We values a droplet has tendency for splashing when the kinetic energy is released at the impact on a surface.

The number was named by Franz Eisner in honor of Moritz Gustav Weber (18 July 1871–10 June 1951). He was born in Leipzig and raised up in Hannover by his grandfather Moritz Rühlmann. His tutor was the mathematician Felix Klein, then assistant professor at the University of Göttingen. He first worked in Berlin as a government architect for the railways. He was in charge of the electrification and construction projects of the Charlottenburg power station. In 1904 he became professor of mechanics at the Gottfried Wilhelm Leibniz University in Hannover and, from 1913 professor of ship and marine engineering at the Technical University of Berlin Charlottenburg (then Technischen Hochschule Berlin). He was particularly interested in the mechanics of similarity and argued about its importance on many fields. In 1930 he published the paper "The general principle of similarity in physics and its connection with the theory of dimensions and model science". Got married to Margarethe Leyn and had two children. He retired in 1936. He is responsible for naming the Reynolds and Froude numbers [15]. He died in Neuendettelsau. He is buried in the Evangelische Kirchhof Nikolassee in Berlin.

The Bond number (also called Eötvös number, Eo) is similar to the Weber number. It is a ratio of body forces (gravitational or buoyancy) to surface tension forces.

$$Bo = \frac{\rho g L^3}{\sigma L} = \frac{\rho g L^2}{\sigma}$$

The name was given after the physicist Loránd Eötvös, a Hungarian baron. He studied the gravitational gradient on the Earth's surface. He died in Budapest in 1919 at the age of 70 years. Is buried in the Kerepesi Cemetery in Budapest.

Sherwood number: [16]

The Sherwood number is equivalent to the Nusselt number for heat transfer. It represents the ratio of the convective mass transfer to the rate of diffusive mass transport.

$$Sh = \frac{k_m L}{D}$$

where: k_m is the convective mass transfer in ms^{-1}, L is a characteristic length (m), D is mass diffusivity ($m^2 s^{-1}$).

There are several semiempirical relationships between Sherwood, Reynolds and Schmidt numbers which are very useful to estimate the mass transfer coefficient. One of those classic relationships is the Ranz- Marshall equation, valid for low values of Re number (2 < Re < 200):

$$Sh = 2 + 0.6 Re^{\frac{1}{2}} Sc^{\frac{1}{3}}$$

Clift et al. proposed another relationship for higher values of the Re and Sc numbers (Sc > 200, 100 < Re < 2000):

$$Sh = 1 + 0.724 Re^{0.48} Sc^{\frac{1}{3}}$$

Thomas Kilgore Sherwood (July 25, 1903–January 14, 1976) was a USA chemical engineer, born in Columbus, Ohio. In 1923 received his B.S. from McGill University, and entered the Massachusetts Institute of Technology (MIT) where he was a student of W. K. Lewis. He completed his Ph.D. in 1929. In 1930 he entered MIT as an assistant professor and became full professor in 1941. He retired from MIT in 1969 to become professor of chemical engineering at the University of California, Berkeley.

During WWII acted as consultant for the war department.

His main work was related to mass transfer theory. He published a book in 1937 on absorption and extraction and later on was rewritten with Pigford and Wilke in 1974 under the title "Mass Transfer", with a large influence in the academic community. He married twice. His second wife made him into a mountaineer.

Schmidt number: [13]

The Schmidt number is employed in mass transfer. Represents a ratio of kinematic viscosity or momentum transfer by internal friction to molecular diffusivity.

$$Sc = \frac{\mu}{\rho D} = \frac{\nu}{D}$$

The Schmidt number depends only on the material properties and does not depend on the flow conditions.

	T, °C	ρ, kg/m^3	$\mu \times 10^{-3}$, N·s/m^2	$\nu \times 10^{-6}$, m^2·s^{-1}	$D \times 10^{-6}$, m^2·s^{-1}	Sc
Liquid steel	1585	6900	4.5	652.2	❧2810	4.3
Water	20	1000	1.003	1.003	25.8 @ 25 °C	25.8
Air	20	1.225	0.0182	14.857		
Argon	20	1.6228	0.02125	13.095		

❧Ref. [17]

The number was named in honor of Ernst Heinrich Wilhelm Schmidt (February 11, 1892- January 22, 1975). He was born in Vögelsen, Northwestern Germany. During WWI interrupted his studies and served in the German Army. He got a degree in electrical engineering in 1919. From 1918–1922 got an assistant position and studied at the Institute for Applied Physics at Munich University. From 1922–1925 he carried research on heat economy at a research institute, developing a device to measure heat losses. In 1925 became Professor and Director of the Engine laboratory at Danzig Institute of Technology, replacing Rudolf Plank who had moved to Karlsruhe. In 1936 became Professor at Dresden Institute of Technology but after one year had to move to an Aeronautical Research Lab. in Braunschweig due to opposition from the Nazi party. After WWII moved for a short time to England and then back to Braunschweig Institute of Technology. Later on, he moved to Munich Institute of Technology to occupy the chair for Thermodynamics as a successor of Wilhelm Nusselt, until his retirement in 1960 to become Emeritus Professor.

He obtained the first patent to use aluminum foil as a thermal insulator. In 1929 published a paper that described the similarities between mass and heat transfer. He carried out new measurements of thermal properties of materials thanks to his invented instruments. In 1936 published a book on engineering thermodynamics.

Prandtl number: [18]

The Prandtl number is a ratio between the momentum diffusivity (or kinematic viscosity), ν, and the thermal diffusivity (α). In other words, a *ratio of momentum transport to heat transport*. It doesn't include a length scale and only depends on the properties of the fluid. For small values (<<1) the thermal diffusivity dominates and for large values (>>1) the momentum diffusivity dominates. Liquids metals have low values indicating that heat diffuses very quickly. As viscosity increases, the momentum transport dominates over the heat transport.

$$Pr = \frac{\nu}{\alpha} = \frac{\left(\frac{\mu}{\rho}\right)}{\left(\frac{k}{\rho C_p}\right)} = \frac{\mu C_p}{k}$$

Ceotto [19] defined a practical approach to compute the thermal diffusivity of steel in m^2/s. Requires the calculation of density, thermal conductivity and heat capacity of steel.

$$\alpha = \frac{k}{\rho C_p}$$

The density of steel, in kg/m^3, can be calculated with the following expression (After Miettinen):

$$\rho = 8319.49 - 0.835T + (-83.19 + 0.00835T)x_c$$

where: T is the temperature in °C and x_c is wt%C.

The heat capacity (C_p) in J/kg·K, is computed from free energy data for Fe–C alloys (After Turkevich), using the regular solution model, as follows:

$$C_p = -T\frac{\partial^2 G_L}{\partial T^2}$$
$$G_L = x_{Fe}G_{Fe}^L + x_C G_C^L + RT(x_{Fe}\ln x_{Fe} + x_C \ln x_C) + x_{Fe}x_C L_{FeC}^L$$
$$G_{Fe}^L = 13265 + 117.576T - 23.5143T\ln T - 4.39752 \times 10^{-3}T^2$$
$$- 5.89269 \times 10^{-8}T^3 - 3.6751 \times 10^{-21}T^7 + 77358.5 \times T^{-1}$$
$$G_C^L = 10^5 + 146.1T - 24.3T\ln T - 0.4723 \times 10^{-3}T^2 + 2562600 \times T^{-1}$$
$$- 2.6439 \times 10^8 T^{-2} + 1.2 \times 10^{10}T^{-3}$$
$$L_{FeC}^L = -124320 + 28.5T + 19300(x_C - x_{Fe}) + (49260 - 19T)(x_C - x_{Fe})^2$$

The thermal conductivity, in W/m·K, is computed from a relationship between thermal conductivity and electric resistivity (After Nishi):

$$k = \frac{2.445 \times 10^{-8}T}{\rho_e}$$

The electric resistivity for pure iron, in $\mu\Omega$cm, is calculated After Kita et al.

$$\rho_e = 0.0154T + 112.3$$

where: T is temperature in °C.

Combining the expressions for ρ, C_p and k, the final expression developed by Ceotto [19] is the following:

$$\alpha \times 10^{-6} = -3.2629 + 0.0102T - 0.2476x_C - 2.8698 \times 10^{-6}T^2$$
$$+ 8.0664 \times 10^{-5}Tx_C$$

The Prandtl number also results from the ratio of Peclet and Reynolds numbers:

$$Pr = \frac{Pe}{Re} = \frac{\left(\frac{\rho C_p UL}{k}\right)}{\left(\frac{\rho UL}{\mu}\right)} = \frac{\mu C_p}{k}$$

Because of the connection among the Pe, Re and Pr numbers, a flow involving Re and Pe numbers can alternatively be handled by Re and Pr numbers.

Since the Pr number is a ratio of momentum diffusivity to thermal diffusivity and also a ratio between Pe and Re numbers, it can be inferred that the Pe number is an inverse measure of the tendency for thermal energy to diffuse and the Re number is an inverse measure of the tendency for momentum to diffuse.

The Prandtl number is a dimensionless number that describes an intrinsic property of a fluid. Liquids with a low Prandtl number are good heat conducting liquids. The thermal conductivity of metals is up to 100 times higher than water. In the case of steel, it has a thermal conductivity of 0.07 cal/cm·s·°C, 50 times higher than that of water with a value of 0.0014 cal/cm·s·°C.

	T, °C	k, W/m·K	$\mu \times 10^{-3}$, Pa·s	C_p, J/kg·K	Pr
Liquid steel	1600	35[§]	5.23[†]	762[*]	0.11
Water	25	0.5948	1.002	4.182	7.04
Air	30	0.026	0.019	1010	0.72
Argon	30	0.018	22.90	18	22.77

[§]Ref. [20] [†]Ref. [21] [*]Ref. [22]. Others CRC Handbook of chemistry and physics

The Schmidt number is the mass transfer analog of the Prandtl number. The thermal diffusion rate is replaced by the mass diffusion rate (D).

Ludwig Prandtl (4 February 1875–15 August 1953), was born in Freising, near Munich. The only surviving son of three children. His father Alexander carried out research on milk production. Her mother became mentally ill due to the death of several children and miscarriages. In 1894 he spent three months in Nuremberg gaining practical experience working in a foundry. Graduated from the Technische Hochschule in Munich in 1898 and was awarded a Ph.D. in Physics in 1900. When Prandtl was 23 years old he worked as Föppl's assistant from 1898 to 1899.

In 1901 he became professor of fluid mechanics at the Technische Hochschule University of Hannover. He became the youngest professor in Prussia. Head of the Institut fur Technische Physik at the University of Gottingen in 1904. In a paper published the same year, "On the motion of fluids in very little friction", he described the concept of the boundary layer and its importance for drag in aerodynamics that granted him a leading position at the University of Gottingen.

In 1908, Prandtl and his student T. Meyer developed the first theories of supersonic shock waves and flow. Theodore von Kármán, a student of Prandtl achieved the full development of this theory. In 1909, Prandtl married Gertrud Föppl in Munich. He was a pragmatic supporter of the Third Reich. He was made head of the Kaiser Wilhelm Institute for fluid Mechanics in 1925 to counteract offers to move to Hannover. He was elected to the Royal Society of London in 1928, having been awarded their Gold Medal the previous year, and received an honorary degree from the University of Cambridge in 1936. His health began to deteriorate from 1950.

Blasius shape of boundary layer: [23]

The Blasius boundary layer shows that the skin friction decreases as the Reynolds number increases.

$$\frac{\delta}{x} = \frac{4.9}{\sqrt{Re_x}}$$

where: δ is the boundary layer thickness and Re_x is the local Reynolds number.

Paul Richard Heinrich Blasius was born on August 9, 1883 in Berlin. Blasius was Prandtl's first Ph.D. student at Gottingen University. His Ph.D. work in 1907 provided a solution for the calculation of the boundary layer theory. Was research assistant at the hydraulics laboratory of Berlin Technical University from 1908 until 1912, then moved to the technical college of Hamburg as lecturer. His research activities lasted only six years. He stayed at the mechanical engineering department from 1912 until 1950. Head of the department from 1945–1950. Published several books; Heat transfer in 1931, Mechanics in 1934 and a book on lubrication in 1961 when he was almost 78 years old.

He died on April 24, 1970 in Hamburg.

Peclet Number: [24]

The Peclet number has two definitions depending if it is applied to heat or mass transfer.

Thermal Peclet number:

The thermal Pe number is important to define thermal similarity. Is a ratio of heat flow by bulk motion to conductive heat transfer

$$Pe = \frac{UL}{\alpha} = \left(\frac{\rho C_p UL}{k} \right)$$

An increment in Pe number is an indication that heat conduction decreases and the heat transfer by convection increases.

The thermal Peclet number is related to the Prandtl and Reynolds numbers:

$$Pe = Pr \cdot Re = \frac{\text{kinematic viscosity}}{\text{thermal diffusivity}} \times \frac{\text{inertia}}{\text{viscous}} = \left(\frac{\mu C_p}{k} \right) \left(\frac{\rho UL}{\mu} \right) = \left(\frac{\rho C_p UL}{k} \right)$$

Mass Peclet number is a ratio of convective transport to diffusive transport.

$$Pe = \frac{UL}{D}$$

Jean Claude Eugène Péclet (February 10, 1793–December 6,1857) was a French physicist, born in Besançon, during the French Revolution. He went to the École Normale in Paris in 1812. Gay-Lussac and Dulong were his professors.

From 1816 until 1827 was a faculty member at the Collège de Marseille, then returned as a Professor of Physics and was co-founder of the École Centrale des Arts et Manufactures in Paris, where he received support from Ampere.

He was a physicist interested in the study and application of heat in industry. Conducted studies in ventilation, fresh air requirements and removal of water vapor and odors, recommending a fresh air supply rate of 10 m^3/hr per person. He studied the effect of water vapor on ventilation. Wrote several books: Chemistry and Physics in 1823, Illumination in 1827 and Treatise on Heat in 1830. He married Cécile Henriette Coriolis, sister of Gaspard-Gustav de Coriolis. In 1840, Péclet became inspecteur général de l'instruction publique. He retired in 1852 and died in Paris.

Biot number: [25]

Represents a ratio of heat transfer by convection to heat transfer by conduction. At low values, heat transfer by conduction dominates the process reducing the temperature gradient in the solid. At large values, convection dominates the process of heat transfer. For small values of the Biot number, less than 0.1, it is valid to assume a uniform temperature along the solid. It can also be viewed as a ratio of two resistances; resistance in the solid to the resistance in the fluid.

$$Bi = \frac{\frac{L}{kA}}{\frac{1}{hA}} = \frac{hL}{k}$$

where h is the convective heat transfer coefficient in $Wm^{-2} K^{-1}$, k is the thermal conductivity of the solid in $Wm^{-1} K^{-1}$ and L is a characteristic length in m.

The Biot Number and the Nusselt Number have the same group of physical parameters. The Biot number is used the characterize the heat transfer resistance "inside" a solid body; h represents the heat transfer coefficient from the "surface of the solid body" to the surrounding fluid and k is the thermal conductivity of the solid body. On the other hand, the Nusselt Number is used to characterize the heat flux from a solid surface to a fluid; k represents the thermal conductivity of the fluid.

It was named after the French physicist Jean-Baptiste Biot (April 21,1774–February 3,1862). French mathematician who worked in astronomy, elasticity, electricity and magnetism, heat, optics and geometry. Born in Paris.

He joined the French army in September 1792, and served in the artillery at the Battle of Hondschoote in September 1793. After the battle Biot, suffering from an illness, decided to leave the army. In Paris was arrested as a deserter (he was still in uniform) but a stranger intervened and he was set free. He studied at the École Polytechnique, founded in 1794. One of his classmates was Siméon-Denis Poisson. He married the sister of his school friend Barnabé Brisson, Gabrielle in 1797. He graduated in the same year. He became Professor of Mathematics at the École Centrale de l'Oise at Beauvais in 1797. Supported by Laplace he was appointed Professor of Mathematical Physics at the Collège de France in 1800. Biot and Gay-Lussac developed the first balloon for scientific purposes, to measure magnetic, electrical, and chemical properties of the atmosphere at various heights, in 1804.

The first balloon ascent made for scientific purposes, departed from the garden of the Conservatoire des Arts on 24 August 1804. In 1808, he carried out measurements of the length of the pendulum at different points on the meridian, in particular at Bordeaux and at Dunkirk. In 1809 Biot was appointed Professor of Physical Astronomy at the Faculty of Sciences. He held this position for over fifty years.

Biot studied a wide range of mathematical topics, mostly on the applied mathematics side. He made advances in astronomy, elasticity, electricity and magnetism, heat and optics on the applied side while, in pure mathematics, he also did important work in geometry. Biot, together with Felix Savart, discovered that the intensity of the magnetic field set up by a current flowing through a wire varies inversely with the distance from the wire. This is now known as Biot-Savart's Law. He tried twice for the post of Secretary to the Académie des Sciences but he lost to Fourier in 1822 and Arago in 1830. His name is inscribed in the Eiffel Tower.

Fourier number: [26]

Represents a ratio of diffusive or conductive transport rate to the storage rate, where the storage rate can be either heat or mass.

$$Fo_h = \frac{\alpha t}{L^2}$$

$$F_{O_m} = \frac{Dt}{L^2}$$

where: α is the thermal diffusivity and D is the mass diffusivity.

Joseph Fourier (March 21st, 1768 Auxerre–May 16th, 1830 Paris) was the nineth son of his father's second marriage. Was left an orphan at the age of 10 but was lucky to be recommended to the bishop of Auxerre and got education at the Ecole Royale Militaire. He wished to pursue a military career but was cancelled because of the lack of nobility origins, then he moved in 1787 with the Benedictines to become a priest. In 1789 all religious orders were cancelled. The Ecole Normale was opened in 1795 and he was selected as one of their faculty members together with Lagrange, Laplace and Berthollet. The school was closed months later and he moved to the Ecole Polytechnique. In 1798 he was selected as engineer to form Napoleon's Egyptian campaign. He conducted studies on Egypt, made studies on algebraic equations and returned in 1801. In 1802 Napoleon appointed him as prefect of Isere (Grenoble region). He is also considered an excellent administrator who made important changes in roads and other improvements in the land. He wrote the book *Analytical Theory of Heat* in 1807. When Napoleon was defeated and exiled in 1814, he avoided him during his pass through the south of France. He kept his position but one year later during the 100 days period he was initially loyal to the King. Once Napoleon took control Fourier met him, then he was named prefect of Rhone but later on resigned in disagreement with Lazare Carnot's policies. Once Napoleon was finally defeated in 1815, Fourier went to Paris and got a position at the Statistical Bureau. In 1822 he published his book; *The analytical theory of heat*. Fourier never married. In his last five years he suffered chronic rheumatism, working very hard enclosed in a box-like chair.

Guthrie number: [27]

The Guthrie number is a ratio of two time scales; the residence time of liquid steel in the tundish (volume tundish/flow rate of liquid steel) to the flotation time of inclusions. The number was named by Chattopadhyay and Isac in honor of Prof. Roderick Guthrie.

$$Gu = \frac{\tau_{tundish}}{\tau_{NMI}}$$

Roderick Guthrie (England 1941–) is a professor of metallurgy at McGill University since 1967 after he graduated from Imperial College. His work has been focused on the application of transport phenomena to the processing of metals. He developed an in-situ method to remove inclusion from liquid aluminum and has made important contributions on strip casting. He was a pioneer in studies on ferroalloy melting. He published the book Thermophysical Properties of Metallic Liquids in 1988 and the book Engineering in Process Metallurgy in 1989. He has many outstanding former students like Prof. Gordon Irons (1978) and Dipak Mazumdar (1985).

Archimedes Number (Ar):

The Archimedes number define the motion of fluids due to density differences and involves a ratio between the product of gravitational forces and buoyancy forces to the square of the viscous forces. Characterizes the flotation of particles, drops and bubbles.

$$Ar = \frac{\rho_f(\rho - \rho_f)gL^3}{\mu^2} = \frac{gL^3}{\nu^2}\frac{(\rho - \rho_f)}{\rho_f}$$

ρ_f is the density of the fluid, ρ is the density of the body, μ is the dynamic viscosity, L is the characteristic length.

Example: [28] Information provided: Fluid flow phenomena due to argon injection in metallurgical ladles is studied using water models and injecting air instead of argon. Calculate the velocity of an air bubble of 5 mm in diameter rising through water at 20 °C.

Properties: $\rho_f = 999\,kg/m^3$, $\rho_p = 1.2047\,kg/m^3$, $\mu = 1001 \times 10^{-6}\,kg/m \cdot s$.

Solution: The motion of sphere immersed in an infinite fluid defines a relationship between the Reynolds and Arquimedes numbers. The Reynolds number is associated with the terminal velocity as follows:

$$Re = \frac{\rho_f U_t D_p}{\mu}$$

Therefore, first we calculate the Arquimedes number and then the Reynolds number.

$$Ar = \frac{\rho_f(\rho - \rho_f)gL^3}{\mu^2} = \frac{999(1.2047 - 999)(9.81)(5 \times 10^{-3})^3}{(100110^{-6})^2} = -1.219 \times 10^6$$

The negative sign is only an indication that the bubble motion is in a direction opposite to gravity, i.e., it is rising.

The force balance for fluid flow past a single sphere yields an implicit expression between Re and Ar numbers. To simplify the calculation the following explicit relationship is employed: [29]

$$Re = \frac{Ar}{18}\left(1 + 0.0579 Ar^{0.412}\right)^{-1.214}$$
$$Re = \frac{1.219 \times 10^6}{18}\left(1 + 0.0579(1.219 \times 10^6)^{0.412}\right)^{-1.214} = 1825$$
$$U_t = \frac{\mu Re}{\rho_f D_p} = \frac{(1001 \times 10^{-6})(1825)}{(999)(5 \times 10^{-3})} = 0.37\,m/s$$

Archimedes [30] (c. 287–c. 212 BC) was born in Syracuse in the island of Sicily, a Greek city-state. Hiero II seized power in 275 BC, inaugurating a period of 50 years of peace and prosperity.

Archimedes studied in Alexandria, Egypt, learning Euclid geometry. He is considered the greater mathematician of antiquity. He anticipated modern calculus and analysis by applying concepts of infinitesimals and the method of exhaustion to derive and rigorously prove a range of geometrical theorems, including: the area of a circle; the surface area and volume of a sphere; deriving an accurate approximation of pi (approx. 3.1408–3.1429); founder of hydrostatics, gave an explanation of the principle of the lever. His knowledge of the lever is expressed in his famous quote: "Give me a place to stand on, and I can move the earth". He is credited with designing innovative machines, such as his screw pump for irrigation, and defensive war machines (catapults, a burning mirror and a system of pulleys) to protect his native Syracuse from Roman invasion in 214 BC.

Hiero's successor, the young Hieronymus (ruled from 215 BC), broke the alliance with the Romans after their defeat at the Battle of Cannae and accepted Carthage's support. The Romans, led by consul Marcus Claudius Marcellus, besieged the city in 214 BC for three years, finally defeating the Greeks in 212 BC. Archimedes died during the Siege of Syracuse, where he was killed by a Roman soldier at the age of 75. The place of his tomb is unknown.

Morton number: [31]

The Morton number (Mo) represents a ratio of viscous forces to surface tension forces. It is constant for a given fluid and is used to characterize the shape of bubbles or drops moving in a surrounding fluid. The bubble shape in a fluid directly depends on the values of the Eotvos (or Bond) and Reynolds numbers. The Morton number is defined from Eo and Re numbers.

$$\text{Mo} = \left(\frac{\mu^4 g \Delta \rho}{\rho^2 \sigma^3} \right)$$

where: g is the acceleration of gravity, $\Delta \rho$ is the density difference between the particle and the fluid, ρ is the density of the fluid and σ is the surface tension.

If the density of the particle is negligible, for example a gas bubble, the expression is reduced to:

$$\text{Mo} = \left(\frac{\mu^4 g}{\rho \sigma^3} \right) = \frac{\text{We}^3}{\text{Fr Re}^4}$$

It was named after Rose Morton who described it with W. Haberman in 1953. They conducted experiments on the drag coefficient of rising air bubbles in different fluids at different temperatures.

Clift et al. first called the dimensionless group, the Morton number. The group has been used before it was published by Haberman and Morton.

Rose Katherine Morton was born on December 3, 1925 in Albemarle NC and went to the University of North Carolina to become a mathematician, graduating in 1948. She worked for the Navy doing research on fluid mechanics of liquids and bubbles with Haberman. She married C. Sayre in 1953 and continued with her work until 1958. She died on November 1999 in Washington DC.

Grashof number: [32]

The Grashof number is a ratio of three forces: inertial forces, buoyancy forces and viscous forces. When the buoyancy forces are due concentration differences or due to a bubble rising in a liquid, it is called Grashof mass transfer number.

$$Gr = \frac{\left(\rho_1 U^2 L^2\right)\left(\Delta \rho g L^3\right)}{(\mu UL)^2} = \frac{\rho_1 \Delta \rho g L^3}{\mu^2}$$

If the buoyancy forces are due to thermal gradients in the fluid, is called Grashof heat transfer number.

$$Gr = \frac{\left(\rho_1 U^2 L^2\right)\left(\rho_1 g L^3 \beta \Delta T\right)}{(\mu UL)^2} = \frac{\rho^2 g L^3 \beta \Delta T}{\mu^2} = \frac{g L^3}{v^2}\beta \Delta T = Ga\,\beta \Delta T$$

where: Ga is the Galilei number.

The Gr number describes non-isothermal convection of the fluid duye to density differences caused by temperature gradients.

It was named in honor of Franz Grashof (July 7, 1826–October 26, 1893). He was born in Düsseldorf, Prussia. He studied at the Berlin Industrial Academy from 1844 to 1847. Took part in the German revolution of 1848. Named director of weights and measures in Berlin. He founded the association of German engineers (VDI) in 1856. In 1863 became a professor at Karlsruhe Polytechnic.

He reported the dimensionless group that bears his name in 1875.

He suffered a stroke in 1882 and once again in 1891. He died in Karlsruhe, German Empire.

In the previous expressions, it can be observed that it is possible to report a dimensionless number in different ways. This is valid and can be done according to the convenience of the researcher. For example, the Reynolds number:

$$\text{Re} = \frac{\rho U^2 L^2}{\mu UL} = \left[\frac{\rho UL}{\mu}\right] = \left[\frac{UL}{\nu}\right] = \frac{\text{Pe}}{\text{Sc}} = \frac{UL^2}{\nu L} = \frac{\frac{L^2}{\nu}}{\frac{L}{U}}$$

In the last term of the previous equation, it can be noticed that both terms on the numerator and denominator have units of time, on this basis Chattopadhyay and Isac[5] point out that many dimensionless numbers, in particular from fluid mechanics, can be described in terms of two time scales.

References

1. B.A. Bezuglyi, N.A. Ivanova, L.V. Sizova, Transport phenomena and dimensionless numbers: towards a new methodological approach. Eur. J. Phys. **38**, 1–25 (2017)
2. J. Kunes, *Dimensionless Physical Quantities in Science and Engineering* (Elsevier USA, 2012)
3. D. Jackson, B. Launder, Osborne Reynolds and the publication of his papers on turbulent flow. Annu. Rev. Fluid Mech. **39**, 19–35 (2007)
4. D. Zuck, Osborne Reynolds, 1842–1912, and the flow of fluids through tubes. Br. J. Anaesth. **43**, 1175–1182 (1971)
5. A. Cameron, Osborne Reynolds. Tribol. Ser. **11**, 3–13 (1987)
6. J.D. Jackson, Osborne Reynolds: scientist, engineer and pioneer. Proc. R. Soc. London. Ser. A Math. Phys. Sci. **451**, 49–86 (1995)
7. B.E. Launder, Horace Lamb & Osborne Reynolds: remarkable mancunians … and their interactions. J. Phys. Conf. Ser. **530**, 1–17 (2014)
8. B. Launder, D. Jackson, Osborne Reynolds: a turbulent life. Chap. 1 in a voyage through turbulence, ed. by P. Davidson, Y. Kaneda, K. Moffatt, K. Sreenivasan (Cambridge University Press, Cambridge UK, 2011), pp. 1–39
9. B. Rapp, *Microfluidics: Modeling, Mechanics, and Mathematics* (Elsevier, Cambridge USA, 2017), p. 261
10. G.G. Stokes, On the effect of the internal friction of fluids on the motion of pendulums. Trans. Camb. Philos. Soc. **9**(10), 8–106 (1856). (Read 9 Dec 1850)
11. W. Froude, (1810–1979). Nature 90–91 (1933)
12. C.L. Vaughan, M.J. O'Malley, Froude and the contribution of naval architecture to our understanding of bipedal locomotion. Gait Posture **21**, 350–362 (2005)
13. W.H. Hager, O. Castro-Orgaz, William Froude and the Froude number. J. Hydraul. Eng. **143**, 02516005 (2017)
14. C. Damle, Y. Sahai, A criterion for water modeling of non-isothermal melt flows in continuous casting tundishes. ISIJ Int. **36**(6), 681–689 (1996)
15. R. Bird, Who was who in transport phenomena. Chem. Eng. Educ. **35**(4), 256–265 (2001)
16. H. Hottel, Thomas Kilgore Sherwood: 1903–1976 (Biographical memoir, National Academy of Sciences, Washington DC, 1994)
17. A. Meyer, L. Hennig, F. Kargl, T. Unruh, Iron self-diffusion in liquid pure iron and iron-carbon alloys. J. Phys. Condens. Matter **31**, 1–5 (2019)
18. A. Busemann, Ludwig Prandtl: 1875–1953. Biogr. Mem. Fellows R. Soc. **5**, 193–205 (1960)
19. D. Ceotto, Thermal diffusivity, viscosity and Prandtl number for molten iron and low carbon steel. Hight Temp. **51**(1), 131–134 (2013)

20. T. Nishi, H. Shibata, H. Ohta, Y. Waseda, Thermal conductivities of molten iron, cobalt, and nickel by laser flash method. Metall. Matls. Trans. A **34**, 2801–2807 (2003)
21. M. Assael, K. Kakosimos, R. Banish, J. Brillo, I. Egry, R. Brooks, P. Quested, K. Mills, A. Nagashima, Y. Sato, W. Wakeham, Reference data for the density and viscosity of liquid aluminum and liquid iron. J. Phys. Chem. Ref. Data **35**(1), 285–300 (2006)
22. J. Valencia, P. Quested, Thermophysical properties, in *ASM Handbook*, vol. 15, (Casting: ASM International, USA, 2006), p. 470
23. W. Hager, Blasius: a life in research and education. Exp. Fluids **34**, 566–571 (2003)
24. E. Roberts, Jean Claude Eugene Peclet. CIBSE Heritage Group. https://www.hevac-heritage.org/built_environment/biographies/surnames_M-R/peclet/P1-PECLET.pdf. Accessed 1 Aug 2020
25. https://mathshistory.st-andrews.ac.uk/Biographies/Biot/. Accessed 2 Aug 2020
26. E. Prestin, Chapter 1: Joseph Fourier: the man and the mathematician, in *The Evolution of Applied Harmonic Analysis Models of the Real World*, (Springer, 2016), pp. 1–24
27. K. Chattopadhyay, M. Isac, Dimensionless numbers for tundish modelling and the Guthrie number (Gu). IM & SM **39**(4), 278–283 (2012)
28. I. Tosun, *Modeling in Transport Phenomena: A Conceptual Approach*. 2nd edn. (Elsevier, 2007), pp. 66–69
29. R. Turton, N. Clark, An explicit relationship to predict spherical particle terminal velocity. Powder Technol. **53**, 127 (1987)
30. T. Heath, *The Works of Archimedes* (Cambridge University Press, London UK, 1897).
31. M. Pfister, W. Hager, History and significance of the Morton number in hydraulic engineering. J. Hydraulic Eng. **140**(5), 02514001 (2014)
32. C.J. Sanders, J.P. Holman, Franz Grashof and the Grashof number. Int. J. Heat Moss Transf. **15**, 562–563 (1972)
33. E. Eckert, To Ernst Schmidt on his 70th birthday. Int. J. Heat Mass Transf. **5**, 113–115 (1962)

Chapter 5
Rayleigh's Method

5.1 John William Strutt [1–4]

This is the first method of dimensional analysis, also known as method of indices. It was developed by John William Strutt. A brilliant researcher who had it all, except good health. His grandfather was the Coronel Joseph Strutt who fought in the war against Napoleon and the King George IV made him baron Rayleigh in 1821. His father, the second baron Rayleigh died in 1873 and he became the third baron Rayleigh inheriting a large state at Terling Essex.

John William Strutt was born on November 12, 1842 in Langford Grove near Maldon Essex and died on June 30, 1919 in Terling Essex at the age of 78 years.

In 1853 attended Eton college for a short time where he contracted whooping cough, then moved to Wimbledon and then Harrow until 1856, there he contracted a chest infection, then due to bad health he got private education. In 1861 at the age of 19 he became student at Trinity College in Cambridge where he studied mathematics. E. Routh was his Math tutor and attended lectures from G. Stokes. He graduated in 1865 at the age of 23. While at Trinity he met A. Balfour and her two sisters, Evelyn and

Leonor. Upon graduation as mathematician, he accomplished the maximum honors, winning the first Wrangler and Smith's prize. He was elected fellow at Trinity college from 1866 until 1871 when he married Evelyn Balfour. The fellowship position was only for bachelors. In 1873 he got elected member of the Royal Society and the same year his father passed away. He became third baron Rayleigh and returned to Terling to take care of his state. At Terling he built a laboratory on the west wing of his house. At Terling he got sick due to a rheumatic fever. To avoid the winter, he moved on December 1872 to live on a houseboat on the river Nile and returned on May 1873. From 1871 until 1879 he worked on the diffraction of light. In 1877 he published his book *"The theory of sound"*. In 1879 the chair of experimental physics at Cambridge was vacant due to the death of James Maxwell and he was awarded his position until 1884. After Cambridge he became professor at the Royal Institution of Great Britain, from 1887 to 1905. He served as president of the Royal Society from 1905 to 1908. The he became chancellor at Cambridge university from 1908 until 1919.

The contributions made by Rayleigh to science are enormous. He published 446 papers, mostly as the single author. His contributions cover some of the following fields: optics, acoustics, the discovery of argon, standardization of the ohm, heat transfer, capillarity, elasticity, hydrodynamics, etc. He was part of the committee that founded the National Physical Laboratory (NPL).

In 1904 he was awarded the Nobel prize in physics for the discovery of argon. He donated the money of the award to Cambridge University.

As the author of the first method of dimensional analysis he applied the principle of dynamic similarity as early as 1871 to prove that light is scattered by particles which are very small and the ratio of the amplitudes of the vibrations of the scattered incident light varies inversely as the inverse of the square of the wave-length. His publication from 1915 on "the principle of similitude" is a detailed description of his method [4].

Before Rayleigh died, he kept working intensively. On December 27, 1918 dined with the king but on January 21, 1919 he fainted while travelling to attend a meeting at the NPL. On May 14, 1919 he delivered an address to the Society of Physical Research. On June 30, 1919 he died of a heart attack. He was 77 years old.

5.2 Rayleigh's Method

This method involves the following steps:

1. Define the true independent variables that affect the dependent variable.
2. Define the functional relationship, write the equation as a product of variables raised to arbitrary exponents and define each variable in terms of their dimensions.
3. Applying the principle of dimensional homogeneity, equate the exponents on the two sides of the equation to obtain a set of simultaneous equations and solve

these equations to obtain the value of the exponents. If the number of unknowns is higher than the number of equations, fix values to obtain one solution.

4. Substitute the values of the exponents in the main equation and form the non-dimensional terms by grouping the variables with like exponents.

Two important rules for all methods:

- Once the dimensionless groups have been defined, make sure there is only one dependent variable
- Make sure the dependent variable appears in only one dimensionless group.

In the general case, when one dependent variable is a function of n independent variables and applying the principle of dimensional homogeneity

$$Y = f(X_1)^{k_1} (X_2)^{k_2} \cdots (X_n)^{k_n}$$

$$[M^a \cdot L^b \cdot T^c] = f[M^{a_1} \cdot L^{b_1} \cdot T^{c_1}]^{k_1} [M^{a_2} \cdot L^{b_2} \cdot T^{c_2}]^{k_2} \cdots [M^{a_n} \cdot L^{b_n} \cdot T^{c_n}]^{k_n}$$

Equating the exponential terms for the same dimension gives a system of three equations.

$$a = a_1 k_1 + a_2 k_2 + \cdots + a_n k_n$$
$$b = b_1 k_1 + b_2 k_2 + \cdots + b_n k_n$$
$$c = c_1 k_1 + c_2 k_2 + \cdots + c_n k_n$$

There is a unique solution only when the number of unknowns is three, if this number is higher it is necessary to arbitrarily assign values.

An alternative approach is to include all the variables in one side of the equation

$$f(X_1)^{k_1} (X_2)^{k_2} \cdots (X_n)^{k_n} = 0$$

In this case, X_1 represents the dependent variable. All the exponents on the right side are zero. For example, if the dimensions are [MLT]:

$$f[M^{a_1} \cdot L^{b_1} \cdot T^{c_1}]^{k_1} [M^{a_2} \cdot L^{b_2} \cdot T^{c_2}]^{k_2} \cdots [M^{a_n} \cdot L^{b_n} \cdot T^{c_n}]^{k_n} = [M^0 \cdot L^0 \cdot T^0]$$

Equating the exponents on both sides of the equation, the following system of equations is obtained:

$$a_1 k_1 + a_2 k_2 + \cdots + a_n k_n = 0$$
$$b_1 k_1 + b_2 k_2 + \cdots + b_n k_n = 0$$
$$c_1 k_1 + c_2 k_2 + \cdots + c_n k_n = 0$$

Rayleigh's method has some drawbacks:

- It cannot anticipate the number of dimensionless groups
- It becomes complex when the number of variables increases.

5.3 Examples of Application of Rayleigh's Method

The period of a pendulum: One of the simplest and most common examples of dimensional analysis is the pendulum. This example is also used to illustrate the problems associated with the correct choice of independent variables. A pendulum is a mass (bob) suspended from a fixed point which, when pulled back and released, swings back and forth. The time for one complete cycle, a left swing and a right swing, is called the period. The amplitude is the width of the pendulum's swing.

During the Han dynasty the pendulum was employed to detect earthquakes. Galileo Galilei in 1602 discovered the property isochronism of a pendulum which means that for small swings the period is independent of amplitude. The famous Big Ben is based on the pendulum.

In 1656 the Christiaan Huygens, from the Netherlands, built the first pendulum clock. Huygens proposed the motion of the pendulum to define the meter. The meter was defined as the length required to give a period of 2 s. On the basis of this definition $g = \pi^2 = 9.87$ m/s^2. The idea was rejected because the force of gravity affects different depending on the altitude.

Leon Foucault built in 1851 a pendulum to measure the motion of the earth. The motion of the earth is 465 m/s. This motion creates a tangential force but is only 0.3% in comparison with the force of gravity.

The equation that defines the period of a pendulum can be derived from a force balance, applying Newton's second law. Figure 5.1 shows the decomposition of forces, F_T represents the tension of the string and (mg cosθ) is the force that balances the tension of the string. The component (mg sinθ) is the component of the force that opposes the movement of the bob in the tangential direction and is called the restoring force of the pendulum. The minus sign is shown to indicate that opposes to the movement.

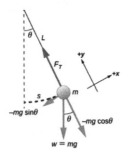

Consider an ideal pendulum (one with no friction, a massless string and a miniscule bob).

$$F = -mg \sin \theta = ma_t$$

a_t represents the tangential acceleration and is equal to $l\alpha = l\left(\frac{d^2\theta}{dt^2}\right)$ where α is the angular acceleration. Replacing this value yields a differential equation that describes the sinusoidal motion of the pendulum.

$$-mg \sin \theta = ml\left(\frac{d^2\theta}{dt^2}\right)$$

$$\frac{d^2\theta}{dt^2} + \frac{g}{l} \sin \theta = 0$$

To simplify the analysis, two terms can be modified. First, If the amplitude is small, for values of θ less than $20°$, $\sin \theta \approx \theta$, then (mg sin$\theta$) reduces to mg$\theta$ and second, the tangential acceleration is defined in a similar way as the centripetal acceleration ($a_c = \omega^2 l \approx \omega^2 x$). Furthermore, the arc length, named as x is defined as $x = l\theta$, hence $\theta = x/l$.

$$F = -mg \sin \theta = -mg\theta = -mg\left(\frac{x}{l}\right) = ma_t = -m\omega^2 x$$

$$-mg\left(\frac{x}{l}\right) = -m\omega^2 x$$

The angular velocity is then mass independent:

$$\omega = \sqrt{\frac{g}{l}}$$

The period, T, is the time to swing a distance 2π over the angular velocity ω:

$$T = \frac{2\pi}{\omega} = 2\pi\sqrt{\frac{l}{g}}$$

This period of a pendulum only depends on the length of the string and the force of gravity and not on the mass of the bob or the amplitude.

Assuming that we ignore the previous analysis and we want to define the equation for the period of a pendulum just on the basis of simple observation. Observing the motion of a pendulum it seems obvious to include the length of the string, the force of gravity, the mass of the bob and the amplitude.

First step: Define the variables involved.

$$T = f(l, g, m, \theta)$$

Second step: Apply the principle of dimensional homogeneity and define the dimensions of each variable.

$$T = f(l)^a(g)^b(m)^c(\theta)^d$$

$$T = f(L)^a(LT^{-2})^b(M)^c(-)^d$$

In this case, T is used to describe the variable period and also the dimension of time.

The amplitude is already a dimensionless number and could have been ignored from the beginning because it will not play a role in the analysis.

Third step: Set the system of equations based on the principle of dimensional homogeneity and solve for the unknowns.

$$L:\ 0 = a + b$$
$$T:\ 1 = -2b$$
$$M:0 = c$$

Then; $b = -1/2$ and $a = \frac{1}{2}$

Important note: in some cases, the system of equations can involve more terms. It is always suggested to confirm that the solution satisfies all equations, otherwise a simple step can mislead to wrong results.

Fourth step: Replace the value of the unknowns.

$$T = f(l)^{\frac{1}{2}}(g)^{-\frac{1}{2}}(m)^0$$

$$T = f\sqrt{\frac{l}{g}}$$

This is the general function that defines the period of a pendulum. It should be noticed that the dimensional analysis ends with a general equation that defines the dimensionless number(s) that describe a phenomenon. In this case the dimensionless number can be expressed as follows:

$$T\sqrt{\frac{g}{l}} = \text{constant}$$

To check it is dimensionless, we substitute units:

$$\sec\sqrt{\frac{m}{\sec^2 m}} \equiv -$$

A useful lab practice is available [5]. This practice can be employed to estimate the value of the gravitational constant. Additionally, a set of experimental data is available in ref [6].

If, for any reason, the variables are wrongly chosen, and even if the dimensional analysis gives a result, it will not match the experimental data. Ruzicka [7] gives the outcome when a different set of variables is chosen. For example, if only the length of the string is considered. The result is that the equation of dimensional homogeneity is inconsistent because on both sides of the equation are different types of units which leads to a contradiction like $1 = 0$. Another case is when the properties of the fluid are taken into account. In this case, if the dynamic viscosity is included, the method of dimensional analysis works but the final equation will not match the experimental data. The equation is obtained as follows:

$$T = f(l, \mu)$$

$$T = f(l)^a(\mu)^b$$

$$L : 0 = a + b$$
$$T : 1 = -2b$$

$$b = -1 \text{ and } a = 2$$

$$T = f\left(\frac{l^2}{\mu}\right)$$

The Rayleigh–Riabouchinsky controversy: This example illustrates the combination of two variables and treated as one variable in dimensional analysis. The problem involves a metallic sphere immersed in a stream of fluid in motion and heat transfer in steady state conditions. Boussinesq originally described this problem. A dimensional analysis was reported by Rayleigh in 1915 [4]. He reported two dimensionless groups. Later, in the same year, Riabouchinsky [8] from a different perspective did the same analysis reporting three dimensionless groups.

Rayleigh analysis:

First step: Rayleigh defined the problem with six variables involving four dimensions; heat, temperature, length and time.

$$\dot{Q} = f(k, l, U, \theta, \rho C_V)$$

The dependent variable is the rate of heat transfer $\left(\dot{Q}\right)$. The independent variables are: k represents the thermal conductivity; l is a characteristic length (diameter), θ is the temperature gradient for heat transfer, U is the velocity of the fluid and the product of variables, ρC_V, is a heat capacity term based on volume rather than mass. The product ρC_V is involved in the general heat conduction equation.

Second step: Define system of dimensions (LMT θ) and apply principle of dimensional homogeneity.

$$\dot{Q} = f(k, l, U, \theta, \rho C_V)$$

$$\dot{Q} = f(k)^a (l)^b (U)^c (\theta)^d (\rho C_V)^e$$

$$\left[\frac{ML^2}{T^3}\right] = f\left(\frac{ML}{T^3 \theta}\right)^a (L)^b \left(\frac{L}{T}\right)^c (\theta)^d \left(\frac{M}{LT^2 \theta}\right)^e$$

Third step: Set system of equations and solve

$$
\begin{aligned}
L: \quad & 2 = a + b + c - e \\
M: \quad & 1 = a + e \\
T: \quad & -3 = -3a - c - 2e \\
\theta: \quad & 0 = -a + d - e
\end{aligned}
$$

There are five unknowns and four equations. One unknown should be fixed and define the others in terms of that one. The final results change and different dimensionless groups can be obtained depending on this selection, however are also correct. If the exponent e is fixed, it is necessary to define all of the other terms as a function of e, as follows:

$$
\begin{aligned}
a &= 1 - e \\
b &= 1 + e \\
c &= e \\
d &= 1 \\
e &= e
\end{aligned}
$$

Fourth step: The exponents are replaced and the dimensionless groups are formed with like exponents.

$$\dot{Q} = f(k)^{1-e} (l)^{1+e} (U)^e (\theta)^1 (\rho C_V)^e$$

$$\dot{Q} = f k l \theta \left(\frac{\rho C_V U l}{k}\right)^e$$

$$\frac{\dot{Q}}{k l \theta} = f\left(\frac{\rho C_V U l}{k}\right)^e$$

In words of Rayleigh; "Since e is undetermined, any number of terms of this form may be combined, and all that we can conclude is that:"

$$\frac{\dot{Q}}{kl\theta} = f'\left(\frac{\rho C_V Ul}{k}\right)$$

Confirmation that the two groups are dimensionless:

$$\frac{\dot{Q}}{kl\theta} \equiv \frac{ML^2}{T^3}\frac{T^3\theta}{ML}\frac{1}{L}\frac{1}{\theta} \equiv -$$

$$\frac{\rho C_V Ul}{k} \equiv \frac{M}{LT^2\theta}\frac{L}{T}\frac{L\,T^3\theta}{ML} \equiv -$$

If the exponent a is fixed, the result is also two dimensionless groups but the dimensionless group from the left side is different:

$$\frac{\dot{Q}}{\rho C_V l^2 U\theta} = f\left(\frac{\rho C_V Ul}{k}\right)^{-a}$$

It can be reduced to:

$$\frac{\dot{Q}}{\rho C_V l^2 U\theta} = f'\left(\frac{\rho C_V Ul}{k}\right)^{b}$$

However, it can be shown[11] that this expression can be reduced to the previous two dimensionless numbers, after some algebraic manipulation:

$$\frac{\dot{Q}}{l\theta}\frac{1}{\rho C_V Ul}\frac{k}{k} = f'\left(\frac{\rho C_V Ul}{k}\right)$$

$$\frac{\dot{Q}}{kl\theta}\frac{k}{\rho C_V Ul} = f'\left(\frac{\rho C_V Ul}{k}\right)$$

$$\frac{\dot{Q}}{kl\theta} = f'\left(\frac{\rho C_V Ul}{k}\right)\left(\frac{\rho C_V Ul}{k}\right)$$

$$\frac{\dot{Q}}{kl\theta} = f''\left(\frac{\rho C_V Ul}{k}\right)$$

In the original analysis from Reynolds in 1915 he employed the system of dimensions [QLT θ]. The procedure applying Reynold's method with this system of dimensions is shown below:

$$f\left(\dot{Q}, k, l, U, \theta, \rho C_V\right) = 0$$

$$f\left(\dot{Q}\right)^a (k)^b (l)^c (U)^d (\theta)^e (\rho C_V)^f = 0$$

Replacing dimensions in each variable:

$$f\left[QT^{-1}\right]^a \left(\frac{Q}{LT\theta}\right)^b (L)^c \left(\frac{L}{T}\right)^d (\theta)^e \left(\frac{Q}{L^3\theta}\right)^f = \left[Q^0 L^0 T^0 \theta^0\right]$$

This is equivalent to:

$$f\left[QT^{-1}\right]^a \left(QL^{-1}T^{-1}\theta^{-1}\right)^b (L)^c \left(LT^{-1}\right)^d (\theta)^e \left(QL^{-3}\theta^{-1}\right)^f = \left[Q^0 L^0 T^0 \theta^0\right]$$

Equating exponents on both sides:

$$
\begin{aligned}
Q^0 &: a + b + f & = 0 \\
T^0 &: -a - b - d & = 0 \\
L^0 &: -b + c + d - 3f &= 0 \\
\theta^0 &: -b + e - f & = 0
\end{aligned}
$$

The system contains six unknowns and four equations. Two unknowns should be fixed. If a and b are fixed, after some algebraic manipulations, the other variables can be defined.

$$
\begin{aligned}
a &= a \\
b &= b \\
c &= -2 - b \\
d &= -a - b \\
e &= -a \\
f &= d
\end{aligned}
$$

After replacing the exponents, the final equation with two dimensionless numbers is:

$$f\left(\frac{\dot{Q}}{\rho C_V l^2 U\theta}\right)^a \left(\frac{k}{\rho C_V U l}\right)^b = 0$$

Which is equivalent to:

$$\dot{Q} = f k l \theta \left(\frac{\rho C_V U l}{k}\right)$$

This example has been reviewed in the past by many researchers [9, 10]. In this form, it clearly shows that the heat loss from the sphere is proportional to the fluid's thermal conductivity, the diameter of the sphere, the temperature gradient, fluid's velocity and the volumetric heat capacity.

Riabouchinsky analysis:

Riabouchinsky [8] proposed a different solution to eliminate temperature as a preliminary variable. He replaced temperature by energy because by definition, temperature is the average kinetic energy of a substance. Using this definition for temperature, its dimensions are:

$$\theta = m\bar{c}^2 \equiv M\frac{L^2}{T^2}$$

The dimensions of θ are replaced in the thermal conductivity and volumetric heat capacity. The new equation with the new set of dimension can be expressed as follows [11]:

$$\dot{Q} = f(k)^a (l)^b (U)^c (m\bar{c}^2)^d (\rho C_V)^e$$

$$\left[\frac{ML^2}{T^3}\right] = f\left(\frac{1}{LT}\right)^a (L)^b \left(\frac{L}{T}\right)^c \left(\frac{ML^2}{T^2}\right)^d \left(\frac{1}{L^3}\right)^e$$

Which is equivalent to:

$$\left[ML^2T^{-3}\right] = f\left(L^{-1}T^{-1}\right)^a (L)^b \left(LT^{-1}\right)^c \left(ML^2T^{-2}\right)^d \left(L^{-3}\right)^e$$

Equating exponents on both sides:

$$
\begin{aligned}
L: \quad & 2 = -a + b + c + 2d - 3e \\
M: \quad & 1 = d \\
T: \quad & -3 = -a - c - 2d
\end{aligned}
$$

There are 5 unknowns and only three equations. In order to get the same equation as Riabouchinsky, noticing that k and l appear in two groups which means that should not be fixed, we choose to fix the values of c and e:

$$
\begin{aligned}
a &= 1 - c \\
b &= b - 2c + 3e \\
c &= c \\
d &= 1 \\
e &= e
\end{aligned}
$$

After replacing the exponents:

$$\dot{Q} = f(k)^a (l)^b (U)^c \left(m\bar{c}^2\right)^d (\rho C_V)^e$$

$$\dot{Q} = f\left(m\bar{c}^2\right) kl \left(\frac{U}{kl^2}\right)^c \left(\rho C_V l^3\right)^e$$

$$\frac{\dot{Q}}{kl\left(m\bar{c}^2\right)} = f\left(\frac{U}{kl^2}\right)^c \left(\rho C_V l^3\right)^e$$

The result is correct to define the heat transfer from a sphere immersed in a fluid, however the analysis made by Rayleigh is more simple and allows to conclude that a change in velocity is equivalent to the same change in the volumetric heat capacity. With the product of two terms in the equation reported by Riabouchinsky, it is not possible to make a precise statement. Rayleigh's reply [12] to Riabouchinsky was that he did the analysis based on the energy equation:

"The question raised by Dr. Riabouchinsky belongs rather to the logic than the use of the principle of similitude, with which I was mainly concerned. It would be well worthy of further discussion. The conclusion that I gave follows on the basis of the usual Fourier equations for the conduction of heat, in which temperature and heat are regarded as sui generis. It would indeed be a paradox if the further knowledge of the nature of heat afforded us by molecular theory put us in a worse position than before in dealing with a particular problem. *The solution would seem to be that the Fourier equations embody something as to the nature of heat and temperature which is ignored in the alternative argument of Dr. Riabouchinsky* [12].

The Rayleigh-Riabouchinsky controversy exposes the problem of defining the "truly" fundamental units. Because it is not only the definition of temperature in its dimensions of energy but also ignoring other possible variables [9], such as liquid viscosity, compressibility, density, thermal expansivity, absolute temperature or the gravitational constant. In this analysis it is clearly found the essential prerequisite to have of a sound experience on the nature of the problem.

References

1. P.N.T. Wells, Lord Rayleigh: John William Strutt, third Baron Rayleigh. IEEE Trans. Ultrason. Ferroelectr. Freq. Control **54**, 591–596 (2007)
2. B.R. Masters, Lord Rayleigh: a scientific life. Opt. Photonics News **20**, 36 (2009)
3. A.T. Humphrey, Lord Rayleigh—the last of the great victorian polymaths. Gen. Electr. Co. GEC Rev. **7**(3), 167. https://www.trevorwright.com/GEC/Journals/GEC_Review/v7n3/p167/GECReviewv7n3p167.htm
4. Rayleigh (Lord), The principle of similitude. Nature **95**, 66–68 (1915).
5. https://www.phys.utk.edu/labs/SimplePendulum.pdf. Accessed 1 Nov 2018
6. https://www.physicsclassroom.com/class/waves/Lesson-0/Pendulum-Motion. Accessed 25 Nov 2020
7. M.C. Ruzicka, On dimensionless numbers. Chem. Eng. Res. Design **86**, 835–868 (2008)
8. D. Riabouchinsky, The principle of similitude. Nature **95**, 591 (1915)

9. P.W. Bridgman, *Dimensional Analysis* 1st edn. (Yale University Press, New Haven and London) (USA, 1922), pp. 9–12
10. R.I.L. Guthrie, *Engineering in Process Metallurgy* (Oxford University Press, USA, 1993), pp. 154–155
11. J.C. Gibbins, *Dimensional Analysis* (Springer, London, 2011), pp. 136–138
12. Rayleigh (Lord), Reply note. Nature **95**, 644 (1915)

Chapter 6
Buckingham's π Theorem

6.1 Edgar Buckingham

Buckingham's π-theorem was formally reported by E. Buckingham in 1914 [1] who also extensively promoted its application in subsequent publications [2–4]. Buckingham defined the dimensionless groups with the symbol π and later on, Bridgman [5] defined the method as π-theorem. The theorem arrives to the definition of the number of dimensionless groups given the number of independent variables (n) and dimensions (k). It indicates that the number of dimensionless groups is always lower than the number of dimensional variables.

Edgar Buckingham [6, 7] was born on July 8, 1867 in Philadelphia. His father was a linguist. This photo is from the times of his graduation. He graduated from Harvard in 1885 with a degree in physics. He got a position as graduate assistant at

© The Author(s), under exclusive license to Springer Nature Singapore Pte Ltd. 2021   69
A. N. Conejo, *Fundamentals of Dimensional Analysis*,
https://doi.org/10.1007/978-981-16-1602-0_6

the University of Strasbourg and the University of Leipzig. He obtained his Ph.D. from this last university in 1893. He was student of W. Ostwald. In 1893 he became lecturer at Bryn Mawr College in Pennsylvania and left in 1899 as an associate professor. Between 1899 and 1900 he worked for a mining company. In 1901 he got married to Elizabeth Holstein, who met at Bryn Mawr College. During the year 1901 he worked at the University of Wisconsin. During 5 years, from 1902 to 1906 he worked for the USDA Bureau of Soils (BOS). The BOS's director was a stubborn man and created a poor working atmosphere. His immediate boss resigned and he did that later on. During this time he made important contribution to soil physics. He developed quasi-empirical formulas with broad application that helped to improve agriculture. After leaving the BOS he worked for the National Bureau of Standards (NBS) and worked there from 1907 until 1937. From 1918–1919 he was scientific associate at the US embassy in Rome.

Buckingham died on April 29, 1940 in Washington DC, at the age of 72.

6.2 General π-Theorem

There are several ways to define the general π-theorem [1, 8–10]. The following approach is based on David and Nolle [9]. First, a reminder about three concepts needed in this development:

(i)   Any mathematical function can be expressed as follows:

$$Y = f(X_1, X_2, \ldots, X_k, X_{k+1}, \ldots X_n)$$

An alternate form is:

$$f(X_1, X_2, \ldots, X_k, X_{k+1}, \ldots X_n) = 0$$

where: k is a symbol to define the maximum number of dimensions (A), n defines the total number of variables (X), X_1 is the dependent variable and X_2 to X_n are the independent variables. In the general case, the number of variables is larger than the number of dimensions.

(ii)   According with the principle of dimensional homogeneity, a variable is related with its dimensions by the following equation:

$$X = (A_1)^{m_1} (A_2)^{m_2} \cdots (A_k)^{m_k}$$

(iii)   Is possible to make a change of units in a variable using a conversion factor:

$$X_i' = \gamma_i X_i$$

Since the number of variables X_1 to X_k is based on the maximum number of dimensions and those dimensions are not be the same in each variable, variables 1 to k are used as primary variables, each one with one independent dimension. In addition to this, because these dimensions form the system of units, its exponent is the unity. The rest of variables is defined in the normal way for any general variable, as follows:

$$X_1 = (A_1)$$
$$X_2 = (A_2)$$
$$\cdots\cdots\cdots$$
$$X_k = (A_k)$$
$$X_{k+1} = (A_1)^{m_1}(A_2)^{m_2}\cdots(A_k)^{m_k}$$
$$\cdots\cdots\cdots$$
$$X_n = (A_1)^{p_1}(A_2)^{p_2}\cdots(A_k)^{p_k}$$

The dimensions for the variables from $k+1$ to n are the total number of dimensions defined previously from 1 to k. Now, in order to define dimensionless variables, a ratio can be defined using two different systems of units and γ as a conversion factor:

$$X'_1 = \gamma_1 X_1 = \gamma_1 A_1$$
$$X'_2 = \gamma_2 X_2 = \gamma_2 A_2$$
$$\cdots\cdots\cdots$$
$$X'_k = \gamma_k X_k = \gamma_k A_k$$
$$X'_{k+1} = (\gamma_1 A_1)^{m_1}(\gamma_2 A_2)^{m_2}\cdots(\gamma_k A_k)^{m_k}$$
$$= \left[(\gamma_1)^{m_1}(\gamma_2)^{m_2}\cdots(\gamma_k)^{m_k}\right]\left[(A_1)^{m_1}(A_2)^{m_2}\ldots(A_k)^{m_k}\right]$$
$$= \left[(\gamma_1)^{m_1}(\gamma_2)^{m_2}\cdots(\gamma_k)^{m_k}\right]X_{k+1}$$
$$\cdots\cdots\cdots$$
$$X'_n = (\gamma_1 A_1)^{p_1}(\gamma_2 A_2)^{p_2}\cdots(\gamma_k A_k)^{p_k}$$
$$= \left[(\gamma_1)^{p_1}(\gamma_2)^{p_2}\cdots(\gamma_k)^{p_k}\right]\left[(A_1)^{p_1}(A_2)^{p_2}\ldots(A_k)^{p_k}\right]$$
$$= \left[(\gamma_1)^{p_1}(\gamma_2)^{p_2}\cdots(\gamma_k)^{p_k}\right]X_n$$

Since the functional equation doesn't change with the change of units:

$$f\left(X'_1, X'_2, \ldots, X'_k, X'_{k+1}, \ldots X'_n\right) = 0$$

Substituting the values of X'_i :

$$f\left\{\gamma_1 X_1, \gamma_2 X_2, \ldots \gamma_k X_k, (\gamma_1)^{m_1}(\gamma_2)^{m_2}\cdots(\gamma_k)^{m_k} X_{k+1}, \ldots (\gamma_1)^{p_1}(\gamma_2)^{p_2}\cdots(\gamma_k)^{p_k} X_n\right\} = 0$$

Since units are arbitrary as well as their conversion factors, the values for the conversion factors is conveniently defines as follows:

$$\gamma_1 = \frac{1}{X_1}; \gamma_2 = \frac{1}{X_2}, \ldots \gamma_k = \frac{1}{X_k}$$

This change makes the first k terms unity and the rest is also affected, for example:

$$(\gamma_1)^{m_1}(\gamma_2)^{m_2} \cdots (\gamma_k)^{m_k}X_{k+1} = \left(\frac{1}{X_1}\right)^{m_1}\left(\frac{1}{X_2}\right)^{m_2} \cdots \left(\frac{1}{X_k}\right)^{m_k}X_{k+1}$$

$$= \frac{X_{k+1}}{(X_1)^{m_1}(X_2)^{m_2} \cdots (X_k)^{m_k}}$$

The new expression contains n-k terms that are dimensionless.

$$f\left\{1, 1, \ldots 1, \frac{X_{k+1}}{(X_1)^{m_1}(X_2)^{m_2} \cdots (X_k)^{m_k}}, \ldots \frac{X_n}{(X_1)^{p_1}(X_2)^{p_2} \cdots (X_k)^{p_k}}\right\} = 0$$

Each term is called a π-group:

$$\Pi = \frac{X_{k+1}}{(X_1)^{m_1}(X_2)^{m_2} \cdots (X_k)^{m_k}}$$

Substituting the original dimensions of these variables it is proved they are dimensionless and therefore independent of the original system of units:

$$\Pi = \frac{X_{k+1}}{(X_1)^{m_1}(X_2)^{m_2} \cdots (X_k)^{m_k}} = \frac{(A_1)^{m_1}(A_2)^{m_2} \cdots (A_k)^{m_k}}{(A_1)^{m_1}(A_2)^{m_2} \cdots (A_k)^{m_k}} = 1$$

Up to this point it has been found that the original equation containing n dimensional variables, such as:

$$f(X_1, X_2, \ldots, X_k, X_{k+1}, \ldots X_n) = 0$$

Can also be expressed as a function of n-k dimensionless terms (Π):

$$f(\Pi_{k+1}, \Pi_{k+2}, \ldots \Pi_n) = 0$$

$$f(\Pi_1, \Pi_2, \ldots \Pi_{n-k}) = 0$$

where:

$$\Pi_1 = \frac{X_{k+1}}{(X_1)^{m_1}(X_2)^{m_2} \cdots (X_k)^{m_k}}$$

$$\cdots \cdots \cdots$$

$$\Pi_{n-k} = \frac{X_n}{(X_1)^{m_1}(X_2)^{m_2} \cdots (X_k)^{m_k}}$$

It is important to notice that each term on the numerator of each dimensionless term is elevated to the power of one and a product of variables elevated to a given power. The denominator is the same for all the π-groups. This feature is the reason the method is also called the method of repeating variables.

Based on the definition of the π-groups, the π-method is reduced to the following procedure:

- The number of dimensionless groups defined by the π-theorem = n – k.
- The number of repeated variables is equal to the number of dimensions (k)
- The non-repeating variables have their dimensions elevated to the first power.

It should be taken into account that from the total number of dimensionless groups (n-k) not all are mutually independent. Therefore, k, the number of dimensions is only a first approximation but still needs to be defined more precisely.

One example of Π dimensionless group can be defined as follows:

$$\Pi = X_1 (X_2)^{a_1} (X_3)^{a_2} \cdots (X_n)^{a_n}$$

The π-theorem, based on the previous analysis, can be summarized as follows [8]: A dimensionally homogeneous equation between n variables can be reduced to a relationship among a set of (n-k) dimensionless products.

The conventional method using the π-theorem can be described in the following steps:

1. *Define all the main variables that affect a physical quantity*: These variables should describe the system's geometry, relevant properties of all materials involved and the external forces. The variables should be mutually independent. The number of variables is defined as n.
2. *Define the set of dimensions for each variable.* This number is defined as k.
3. *Number of dimensionless groups*: The number of dimensionless groups or π-terms is the difference between the number of variables and the number of dimensions (n-k).
4. *Choose the number of repeating variables*: This number is equal to the number of dimensions. Each repeating variable will be raised to an exponent.
    If the dependent variable is chosen as a repeating variable it can appear in more than one dimensionless group. To avoid this, the dependent variable should not be chosen as a repeating variable.
    Also, if the dimensions of the repeating variables are less than the number of unknowns the system will not yield a unique solution. To prevent this chose a combinations of variables that includes all the dimensions for that system.
5. *Form the dimensionless groups*: A dimensionless group is formed multiplying one non-repeating variable with the set of repeating variables. The repeating variables include the unknown exponents. There is one equation involved for each non-repeating variable.
6. *System of equations*: A system of equations is defined applying the principle of dimensional homogeneity. Solve the system of equations and substitute the results in the dimensionless groups. Check that the terms are dimensionless.

6.3 Examples of Application of π-Theorem

Sphere heating immersed in a fluid: A metallic sphere is immersed in a stream of fluid in motion and heat transfer occurs in steady state conditions.
First step: Main independent variables. n = 6

$$\dot{Q} = f(k, l, U, \theta, \rho C_V)$$

Second step: Let's choose the following dimensions [11]: QLT θ. k = 4

$$f\left(\dot{Q}, k, l, U, \theta, \rho C_V\right) = 0$$

$$\dot{Q} \equiv \left[QT^{-1}\right]$$

$$k \equiv \left[QL^{-1}T^{-1}\theta^{-1}\right]$$

$$l \equiv [L]$$

$$U \equiv \left[LT^{-1}\right]$$

$$\theta \equiv [\theta]$$

$$\rho C_V \equiv \left[QL^{-3}\theta^{-1}\right]$$

Third step: Number of dimensionless groups. n $-$ k = 6 $-$ 4 = 2. Two dimensionless groups (π_1 and π_2).

Fourth step: Choose the fixed variables (two). One has to be the heating rate $\left(\dot{Q}\right)$ and the second one can be the volumetric heat capacity (ρC_V). The others become the repeating variables.

Fifth step: Form the dimensionless groups.

$$\pi_1 = \dot{Q}(k)^{a_1} (l)^{b_1} (U)^{c_1} (\theta)^{d_1}$$

$$\pi_2 = \rho C_V (k)^{a_1} (l)^{b_1} (U)^{c_1} (\theta)^{d_1}$$

$$\pi_1 = f(\pi_2)$$

Sixth step: System of equations.
First π group.

$$\pi_1 = \left[QT^{-1}\right]\left(QL^{-1}T^{-1}\theta^{-1}\right)^{a_1}(L)^{b_1}\left(LT^{-1}\right)^{c_1}(\theta)^{d_1} = Q^0T^0L^0\theta^0$$

$$\begin{array}{ll}
\text{Q}: & 1+a_1 = 0 \\
\text{T}: & -1-a_1-c_1 = 0 \\
\text{L}: & -a_1+b_1+c_1 = 0 \\
\theta: & -a_1+d_1 = 0
\end{array}
\qquad
\begin{array}{l}
\text{Solution}: \\
a_1 = -1 \\
b_1 = -1 \\
c_1 = 0 \\
d_1 = -1
\end{array}$$

Therefore, first dimensionless group becomes:

$$\pi_1 = \dot{Q}(k)^{-1}(l)^{-1}(U)^0(\theta)^{-1}$$

$$\pi_1 = \frac{\dot{Q}}{kl\theta}$$

Second π group.

$$\pi_2 = \rho C_V(k)^{a_2}(l)^{b_2}(U)^{c_2}(\theta)^{d_2}$$

$$\pi_2 = \left[QL^{-3}\theta^{-1}\right]\left(QL^{-1}T^{-1}\theta^{-1}\right)^{a_2}(L)^{b_2}\left(LT^{-1}\right)^{c_2}(\theta)^{d_2} = Q^0T^0L^0\theta^0$$

$$\begin{array}{ll}
\text{Q}: & 1+a_2 = 0 \\
\text{T}: & -a_2-c_2 = 0 \\
\text{L}: & -3-a_2+b_2+c_2 = 0 \\
\theta: & -1-a_2+d_2 = 0
\end{array}
\qquad
\begin{array}{l}
\text{Solution}: \\
a_2 = -1 \\
b_2 = 1 \\
c_2 = 1 \\
d_2 = 0
\end{array}$$

$$\pi_2 = \rho C_V(k)^{-1}(l)^1(U)^{-1}(\theta)^0$$

$$\pi_2 = \left(\frac{\rho C_V Ul}{k}\right)$$

Then, the final result is:

$$\pi_1 = f(\pi_2)$$

$$\frac{\dot{Q}}{kl\theta} = f\left(\frac{\rho C_V Ul}{k}\right)$$

Particular solution:

This note applies to all methods of dimensional analysis. Once the general solution is defined, for example, in the previous case the general relationship has the form

$$\pi_1 = i(\pi_2)^j$$

where: i is a constant and j is an exponent. The values of i and j should be obtained with statistical analysis of experimental data.

All methods define the same functional relationship. The only difference is that using the π-method the groups are called π-groups. For a more general relationship:

$$\pi_1 = i(\pi_2)^{j_2}(\pi_3)^{j_3}..(\pi_n)^{j_n}$$

The equation is transformed into log space:

$$\log \pi_1 = \log(i) + j_2 \log(\pi_2) + j_3 \log(\pi_3) + \ldots.. + j_n \log(\pi_n)$$

The values of the constant i and exponents j_n, are obtained by multiple regression analysis.

References

1. E. Buckingham, On physically similar systems; illustrations of the use of dimensional equations. Phys. Rev. **4**, 345–376 (1914)
2. E. Buckingham, The principle of similitude. Nature **96**, 396–397 (1915)
3. E. Buckingham, Notes on the method of dimensions. Phil. Magazine Series **6**(42), 696–719 (1921)
4. E. Buckingham, Dimensional analysis. Phil. Magazine Series **6**(48), 141–145 (1924)
5. P.W. Bridgman, *Dimensional Analysis* (Yale University Press, New Haven USA, 1931).
6. J.R. Nimmo, E.R. Landa, The soil physics contributions of Edgar Buckingham. Soil Sci. Soc. Am. J. **69**, 328–342 (2005)
7. T.N. Narasimhan, Buckingham, 1907: an appreciation. Vadose Zo. J. **4**, 434–441 (2005)
8. L.I. Sedov, *Similarity and Dimensional Methods in Mechanics* (Academic Press, USA, 1959), pp. 16–19
9. F.W. David, H. Nolle, *Experimental Modelling in Engineering* (Buttterworths, London, 1982), pp. 15–18
10. A. Flaga, Basic principles and theorems of dimensional analysis and the theory of model similarity of physical phenomena. Tech. Trans. Civil Eng. 241–271 (2015)
11. R.I.L. Guthrie, *Engineering in Process Metallurgy* (Oxford University Press, USA, 1993), pp. 156–157

Chapter 7
Step by Step Method (Ipsen's Method)

David C. Ipsen an associate research engineer and lecturer in mechanical engineering the University of California-Berkeley published in 1960 the book "Units, Dimensions and dimensionless numbers" which defines in simple and practical terms dimensional analysis [1]. In this book the author proposed a new method called the "Step by step approach". This method is simple; its objective is to eliminate all dimensions by simple algebraic operations.

7.1 Step by Step Method

The method can be described as follows:

1. *Define all the main variables that affect a physical quantity and set the dimensions for each variable.*: These variables should describe the system´s geometry, relevant properties of all materials involved and the external forces. The variables should be mutually independent. Put the variables and dimensions on a table. The variables in the first row and dimensions in the second row. The dependent variable in the first cell of the table.
2. *Elimination of dimensions of all variables*: What follows are the rules to make the dimensions of all variables the unity:

   2.1 One dimension is eliminated in each step reducing the number of variables by one. In some particular cases, two dimensions can be eliminated at a time. Elimination is carried out by simple algebraic operations, however, when one dimension is eliminated from the dependent variable, only the independent variables with the same dimension are taken into account and the dimensions of the other variables remain the same.

© The Author(s), under exclusive license to Springer Nature Singapore Pte Ltd. 2021
A. N. Conejo, *Fundamentals of Dimensional Analysis*,
https://doi.org/10.1007/978-981-16-1602-0_7

2.2   When one dimension is going to be eliminated in the dependent variable, identify the independent variable(s) that contain the same dimension. Choose one of the independent variables, called "reference", and multiply by its own dimensions in such a way that the result is a dimensionless independent variable and removal of one dimension from the dependent variable. Since the dimensions of the independent variable employed in this step have vanished, this variable can be eliminated from the table.

2.3   The number of terms in the new variables is modified in each step. Its content changes due to division or multiplication with other variables.

2.4   The elimination of one dimension in all variables is based on the exponent of that dimension in each variable. For example, if the dimension to eliminate is L and the table contains several variables with that dimension, say Q_1, is ML^2 and that of the reference variable, Q_2, is $1/L$, we need to multiply $Q_1Q_2^{-2}$ to eliminate L from Q_1. The form of the new variable is then $Q_1Q_2^{-2}$ and its new dimension is M, in this example.

2.5   Update the new dimensions in the table

2.6   Continue the process of elimination of all dimensions in all variables. The groups of variables remaining are the dimensionless groups.

2.7   The order in which the dimensions of the dependent variable are eliminated can affect the final form of the dimensionless groups.

7.2   Examples of Application of the Step by Step Method

Heating of sphere submerged in a fluid

This is the example analyzed by Rayleigh, which involves six variables: Rate of heat transfer ($\dot{Q}$), k represents the thermal conductivity; l is a characteristic length (diameter), θ is the temperature gradient for heat transfer, U is the velocity of the fluid and ρC_V, is a volumetric heat capacity. The following table summarizes all variables and its dimensions in the system MLTθ.

$\dot{Q}$	k	l	U	θ	ρC_V
$\left[\frac{ML^2}{T^3}\right]$	$\left[\frac{ML}{T^3\theta}\right]$	$[L]$	$\left[\frac{L}{T}\right]$	$[\theta]$	$\left[\frac{M}{LT^2\theta}\right]$

In this example it should first be noticed that the dependent variable doesn´t contain the dimension for temperature difference, therefore, this dimension should first be eliminated from the independent variables.

Elimination of θ:

To disappear the variable and its dimension, divide the variable by itself but to disappear this dimension in the other independent variables it is a multiplication:

$\dot{Q}$	$k\theta$	l	U	$\frac{\theta}{\theta}$	$\rho C_V \theta$
$\left[\frac{ML^2}{T^3}\right]$	$\left[\frac{ML}{T^3}\right]$	$[L]$	$\left[\frac{L}{T}\right]$	$[-]$	$\left[\frac{M}{LT^2}\right]$

We can eliminate the variable θ because it will not play any role in the subsequent steps.

$\dot{Q}$	$k\theta$	l	U	$\rho C_V \theta$
$\left[\frac{ML^2}{T^3}\right]$	$\left[\frac{ML}{T^3}\right]$	$[L]$	$\left[\frac{L}{T}\right]$	$\left[\frac{M}{LT^2}\right]$

Elimination of L:

Choose the variable l as the reference. Division by l^2 will eliminate L in the dependent variable. Elimination of L in the independent variables is made according with the exponent of L in their terms, as follows

$\frac{\dot{Q}}{l^2}$	$\frac{k\theta}{l}$	$\frac{l}{l}$	$\frac{U}{l}$	$\rho C_V l \theta$
$\left[\frac{M}{T^3}\right]$	$\left[\frac{M}{T^3}\right]$	$[-]$	$\left[\frac{1}{T}\right]$	$\left[\frac{M}{T^2}\right]$

We can eliminate the variable l because it will not play any role in the subsequent steps.

$\frac{\dot{Q}}{l^2}$	$\frac{k\theta}{l}$	$\frac{U}{l}$	$\rho C_V l \theta$
$\left[\frac{M}{T^3}\right]$	$\left[\frac{M}{T^3}\right]$	$\left[\frac{1}{T}\right]$	$\left[\frac{M}{T^2}\right]$

Elimination of M:

Taking as a reference variable; $\frac{k\theta}{l}$, if the term containing the dependent variable is multiplied by the reciprocal of the reference term $\frac{k\theta}{l}$, then simultaneously two dimensions are eliminated, M and T. In order to continue, we need to manipulate the two remaining dimensional groups that still contain M and T, using the reference variable from this step. This can be done in several ways; multiplying $\frac{U}{l}$ by $\frac{l}{k\theta}$ introduces the dimensions M and T as shown below:

$\frac{\dot{Q}}{l^2}\frac{l}{k\theta}$	$\frac{k\theta}{l}\frac{l}{k\theta}$	$\frac{U}{l}\frac{l}{k\theta} = \frac{U}{k\theta}$	$\rho C_V l \theta$
$[-]$	$[-]$	$\left[\frac{T^2}{M}\right]$	$\left[\frac{M}{T^2}\right]$

Using the term $\frac{U}{k\theta}$ as the final reference, the dimensions can be eliminated:

$\frac{\dot{Q}}{l^2}\frac{l}{k\theta}$	$\frac{U}{k\theta}\frac{k\theta}{U}=1$	$\rho C_V l\theta\frac{U}{k\theta}=\frac{\rho C_V lU}{k}$
[–]	[–]	[–]

Or

$\frac{\dot{Q}}{kl\theta}$	$\frac{\rho C_V lU}{k}$
[–]	[–]

The other way is first to eliminate M in the last group and leave its dimensions only as a function of T.

$\frac{\dot{Q}}{l^2}\frac{l}{k\theta}$	$\frac{U}{l}$	$\rho C_V l\theta\frac{l}{k\theta}=\frac{\rho C_V l^2}{k}$
[–]	$[\frac{1}{T}]$	[T]

Using the term $\frac{U}{l}$ as the final reference, the dimensions can be eliminated:

$\frac{\dot{Q}}{kl\theta}$	$\frac{U}{l}\frac{l}{U}=1$	$\frac{\rho C_V l^2}{k}\frac{U}{l}=\frac{\rho C_V lU}{k}$
[–]	[–]	[–]

Therefore, the final result indicates two dimensionless groups_

$\frac{\dot{Q}}{kl\theta}$	$\frac{\rho C_V lU}{k}$
[–]	[–]

Which is equivalent to:

$$\dot{Q} = f kl\theta\left(\frac{\rho C_V Ul}{k}\right)$$

Energy released by an atomic bomb

The estimated energy released before the first nuclear test, the Trinity test, on July 1945 ranged from 0.3–10 kilotons of TNT [2]. Pictures of the explosion were released in TIME magazine in 1950 showing the size of the explosion as a function of time. Dimensional analysis can be employed to define the dimensionless numbers that describe this phenomenon. Deakin [3] discussed the possible variables involved in an explosion and how the resulting dimensionless numbers can be simplified to one, which is the one reported in many publications. G. Taylor, member of the Manhattan project, calculated the energy released. The information available is the radius of the explosion as a function of time since detonation. The amount of energy released (Q) is defined as a function of the following variables [3]: The radius of the explosion (R), elapsed time (t) and air density (ρ):

Q	R	t	ρ
$\left[\frac{ML^2}{T^2}\right]$	[L]	[T]	$\left[\frac{M}{L^3}\right]$

Elimination of L:

Use R as the reference group.

$\frac{Q}{R^2}$	$\frac{R}{R}$	t	ρR^3
$\left[\frac{M}{T^2}\right]$	[–]	[T]	[M]

$\frac{Q}{R^2}$	t	ρR^3
$\left[\frac{M}{T^2}\right]$	[T]	[M]

Elimination of M:

Use ρR^3 as the reference group.

$\frac{Q}{R^2}\frac{1}{\rho R^3} = \frac{Q}{\rho R^5}$	t	$\frac{\rho R^3}{\rho R^3} = 1$
$\left[\frac{1}{T^2}\right]$	[T]	[–]

Elimination of T:

$\frac{Q}{\rho R^5}t^2$	$\frac{t}{t} = 1$
[–]	[–]

Therefore, the final results is:

$$\frac{Qt^2}{\rho R^5} = \text{constant}$$

or:

$$Q = \rho C R^5 t^2$$

References

1. D. C. Ipsen, *Units, Dimensions and Dimensionless Numbers* (Mc Graw-Hill, New York, USA, 1970), pp. 168–170
2. https://en.wikipedia.org/wiki/Trinity_(nuclear_test)
3. M. A. Deakin, G.I. Taylor and the trinity test. Int. J. Math. Edu. Sci. Tech. **42**(8), 1069–1079 (2011)

Chapter 8
Matrix Method

The original Buckingham π-theorem states that the number of dimensionless groups is the difference between the number of variables and the number of dimensions. This method can result in dimensionless groups that are not mutually independent. Van Driest [1] in 1946 found a solution to this limitation, defining the following rule:

> The number of dimensionless products in a complete set is equal to the total number of variables minus the maximum number of these variables that will not form a dimensionless product.

The maximum number of those variables that will not form a dimensionless product correspond to those that are mutually independent. The variables and its dimensions can be arranged in the form of a matrix, called dimensional matrix and using matrix algebra to define the maximum number of columns or rows that cannot be converted into zeros using elementary Gaussian elimination. This number is the rank of a matrix. Therefore, the maximum number of mutually independent dimensionless products is the difference between the number of variables and the rank of the dimensional matrix.

To know not only the rank of a matrix but also the algebraic operations to define the dimensionless numbers requires some basic knowledge about matrix algebra. This information is provided in the following section.

8.1 Matrix Algebra

A matrix is an array of numbers ordered by m rows and n columns, called m x n matrix with its elements a_{ij}, as shown below:

$$\begin{pmatrix} a_{11} & a_{12} & \cdots & a_{1n} \\ a_{21} & a_{22} & \cdots & a_{2n} \\ \vdots & \vdots & \vdots & \vdots \\ a_{m1} & a_{m2} & \cdots & a_{mn} \end{pmatrix}$$

Dimensional matrix: The dimensional matrix results from the principle of dimensional homogeneity. The principle of dimensional homogeneity is a balance of dimensions considering all variables. In its general form, for n variables and three dimensions (MLT), is expressed as follows:

$$\left[M^{b_1} \cdot L^{b_2} \cdot T^{b_3} \right] = f \left[M^{a_{11}} \cdot L^{a_{21}} \cdot T^{a_{31}} \right]^{k_1} \left[M^{a_{12}} \cdot L^{a_{22}} \cdot T^{a_{32}} \right]^{k_2} \left[M^{a_{13}} \cdot L^{a_{23}} \cdot T^{a_{33}} \right]^{k_3} \cdots$$
$$\left[M^{a_{1n}} \cdot L^{a_{2n}} \cdot T^{a_{3n}} \right]^{k_n}$$

The balance of dimensions can be expressed in the form of a set of algebraic simultaneous equations:

$$a_{11}k_1 + a_{12}k_2 + a_{13}k_3 + \cdots + a_{1n}k_n = b_1$$
$$a_{21}k_1 + a_{22}k_2 + a_{23}k_3 + \cdots + a_{2n}k_n = b_2$$
$$a_{31}k_1 + a_{32}k_2 + a_{33}k_3 + \cdots + a_{3n}k_n = b_3$$

In matrix form is expressed as:

$$DK = B$$

where D is the dimensionless matrix, K is the vector of unknowns and B is the vector of constants.

$$D = \begin{pmatrix} a_{11} & a_{12} & \cdots & a_{1n} \\ a_{21} & a_{22} & \cdots & a_{2n} \\ \vdots & \vdots & \vdots & \vdots \\ a_{m1} & a_{m2} & \cdots & a_{mn} \end{pmatrix}$$

$$K = \begin{pmatrix} k_1 \\ k_2 \\ \vdots \\ k_n \end{pmatrix} \text{ and } B = \begin{pmatrix} b_1 \\ b_2 \\ b_3 \end{pmatrix}$$

The constants are zero when the system is dimensionless. This gives an *homogeneous system of equations* represented as:

$$DK = 0$$

It is important to notice that columns are associated with the variables and rows are associated with dimensions. Each element a_{ij} is a number that represents the exponent of each dimension for a corresponding variable (Q).

$$
\begin{array}{cccc}
 & Q_1 & Q_2 & \cdots & Q_n \\
D_1 & a_{11} & a_{12} & \ldots & a_{1n} \\
D_2 & a_{21} & a_{22} & \ldots & a_{2n} \\
 & \vdots & \vdots & \vdots & \vdots & \vdots \\
D_m & a_{m1} & a_{m2} & \ldots & a_{mn}
\end{array}
$$

Square matrix: is a matrix when the number of rows is equal to the number of columns. The main diagonal in a square matrix are the elements with subscripts *ii*.

$$
\begin{pmatrix}
a_{11} & a_{12} & a_{13} \\
a_{21} & a_{22} & a_{23} \\
a_{31} & a_{32} & a_{33}
\end{pmatrix}
$$

Identity o unit matrix (I): Is a diagonal matrix whose nonzero elements are all "1".

$$
I = \begin{pmatrix}
1 & 0 & 0 \\
0 & 1 & 0 \\
0 & 0 & 1
\end{pmatrix}
$$

Augmented matrix: Results from the combination of two matrices (for example A and B) and is represented as (A|B). Depending on the matrices, it can result in the addition of columns or rows.

$$
(A|B) = \begin{pmatrix}
a_{11} & a_{12} & a_{13} & b_1 \\
a_{21} & a_{22} & a_{23} & b_2 \\
a_{31} & a_{32} & a_{33} & b_3
\end{pmatrix}
$$

Singular matrix: The determinant of the matrix is zero, $|A|=0$, otherwise is non-singular.

Transpose of a matrix: It is obtained by interchanging the rows and columns of a matrix.

$$
K = \begin{pmatrix}
k_1 \\
k_2 \\
\vdots \\
k_n
\end{pmatrix}
\quad K^T = \begin{pmatrix} k_1 & k_2 & \cdots & k_n \end{pmatrix}
$$

Null matrix: A zero matrix or null matrix is a matrix in which all the elements are equal to 0.

Inverse matrix: The inverse of a matrix (A^{-1}) is that matrix which, when multiplied by the original matrix, gives an identity matrix.

$$AA^{-1} = I$$
$$AA^T = I$$

Orthogonal matrix: The inverse of a matrix is equal to its transpose.

Vectors: A matrix with one column m × 1 is called column vector and with one row 1 × n is called row vector.

Partitioned matrix: A matrix can be split into several sub-matrices by means of horizontal and vertical lines.

$$A = \begin{pmatrix} a_{11} \ a_{12} & a_{13} \ a_{14} \ a_{15} \\ a_{21} \ a_{22} & a_{23} \ a_{24} \ a_{25} \\ a_{31} \ a_{32} & a_{33} \ a_{34} \ a_{35} \\ a_{41} \ a_{42} & a_{43} \ a_{44} \ a_{45} \end{pmatrix} = \begin{pmatrix} A_{11} & A_{12} \\ A_{21} & A_{22} \end{pmatrix}$$

Addition/Subtraction of matrices: $R = A \pm B$, implies that $r_{ij} = a_{ij} \pm b_{ij}$.

Product of two matrices (AB): it is allowed only if they are compatible. It occurs if the number of columns of A equals the number of rows of B, for example, if $A = m \times r$ and $B = r \times n$

$$A \cdot B = \left(\sum_{k=1}^{r} a_{ik} \cdot b_{kj} \right), i = 1, 2, \ldots m; j = 1, 2, \ldots n$$

Determinant of a matrix: The determinant of a matrix is a number (scalar) and is denoted as |A| or det (A). Matrices with large determinants denote variables that are independent of one another.

Rank of a matrix: The definition depends on the method employed to compute the rank of a matrix. For our purpose, the rank of a matrix is the maximum number of variables that will not form a dimensionless product and therefore, instead of subtracting variables minus dimensions to define the number of dimensionless groups the correct way is the number of variables minus the rank of the dimensional matrix. Rosza [2] describes a method called "minimal dyadic decomposition" which is summarized in the following steps:

Step 1: set $n = 1$.
Step 2: Select a pivot "p" in dimensional matrix, A_n.
Step 3: Generate matrix A_{n+1}:

$$A_{n+1} = A_n - \frac{1}{p} \underbrace{\begin{bmatrix} \ \end{bmatrix}}_{\text{column of p}} \cdot \overbrace{\begin{bmatrix} \ \end{bmatrix}}^{\text{row of p}}$$

Step 4: If $A_{n+1} = 0$, the rank of the matrix is n, otherwise set $n = n + 1$, and go to step 2 and keep the procedure. Stop when the last step gives a null matrix. The rank of the matrix is the last value of n.

8.2 Matrix Method

Once the Buckingham´s π-theorem was improved with a proper definition of the maximum number of variables that will not form a dimensionless product, different approaches have been proposed to define its value. Langhaar [3] was one of the first to use the rank of a matrix. Subsequently more researchers [4] have reported alternatives of solution. The method explained below is based on the book by Worstell [5]. Tzirtes book [6] provides about 200 examples of application of the matrix method.

Previously, the principle of dimensional homogeneity was defined in matrix form as follows:

$$DK = B$$

where D is the dimensional matrix and K the unknowns vector. It was also pointed out to observe that in this matrix the columns are associated with the variables and row with the dimensions.

$$
\begin{array}{ccccc}
 & Q_1 & Q_2 & \cdots & Q_n \\
D_1 & a_{11} & a_{12} & \ldots & a_{1n} \\
D_2 & a_{21} & a_{22} & \ldots & a_{2n} \\
\vdots & \vdots & \vdots & \vdots & \vdots \\
D_m & a_{m1} & a_{m2} & \ldots & a_{mn}
\end{array}
$$

In a general case, the number of unknowns is higher than the number of equations. For example, assuming a case of 3 equations and 5 unknowns:

$$a_{11}k_1 + a_{12}k_2 + a_{13}k_3 + a_{14}k_4 + a_{15}k_5 = b_1$$
$$a_{21}k_1 + a_{22}k_2 + a_{23}k_3 + a_{24}k_4 + a_{25}k_5 = b_2$$
$$a_{31}k_1 + a_{32}k_2 + a_{33}k_3 + a_{34}k_4 + a_{35}k_5 = b_3$$

In this case, to get a particular solution, the values of two unknowns should be fixed. This condition defines the two additional equations needed to form a square matrix:

$$k_1 + 0 + 0 + 0 + 0 = k_1$$
$$0 + k_2 + 0 + 0 + 0 = k_2$$
$$a_{11}k_1 + a_{12}k_2 + a_{13}k_3 + a_{14}k_4 + a_{15}k_5 = b_1$$

$$a_{21}k_1 + a_{22}k_2 + a_{23}k_3 + a_{24}k_4 + a_{25}k_5 = b_2$$
$$a_{31}k_1 + a_{32}k_2 + a_{33}k_3 + a_{34}k_4 + a_{35}k_5 = b_3$$

In matrix form:

$$\begin{pmatrix} 1 & 0 & 0 & 0 & 0 \\ 0 & 1 & 0 & 0 & 0 \\ a_{11} & a_{12} & a_{13} & a_{14} & a_{15} \\ a_{21} & a_{22} & a_{23} & a_{24} & a_{25} \\ a_{31} & a_{32} & a_{33} & a_{34} & a_{35} \end{pmatrix} \begin{pmatrix} k_1 \\ k_2 \\ k_3 \\ k_4 \\ k_5 \end{pmatrix} = \begin{pmatrix} k_1 \\ k_2 \\ b_1 \\ b_2 \\ b_3 \end{pmatrix}$$

The solution to this system of equation is by obtaining the inverse of the augmented matrix:

$$\begin{pmatrix} k_1 \\ k_2 \\ k_3 \\ k_4 \\ k_5 \end{pmatrix} = \begin{pmatrix} 1 & 0 & 0 & 0 & 0 \\ 0 & 1 & 0 & 0 & 0 \\ a_{11} & a_{12} & a_{13} & a_{14} & a_{15} \\ a_{21} & a_{22} & a_{23} & a_{24} & a_{25} \\ a_{31} & a_{32} & a_{33} & a_{34} & a_{35} \end{pmatrix}^{-1} \begin{pmatrix} k_1 \\ k_2 \\ b_1 \\ b_2 \\ b_3 \end{pmatrix}$$

The solution to this system is obtained when the elements of the vector of constants are zero. This condition defines a dimensionless system. It can be observed that the augmented matrix also includes the identity matrix (I) and the null (0) matrix. In addition to this, the solution reported by Tzirtes [6] involves partitioning of the dimensional matrix into two: S and R. S is a square matrix, taking elements from the right side of the dimensional matrix and R is he remaining matrix. The size of the square matrix is the same as the rank of the dimensional matrix.

$$S = \begin{pmatrix} a_{13} & a_{14} & a_{15} \\ a_{23} & a_{24} & a_{25} \\ a_{33} & a_{34} & a_{35} \end{pmatrix}$$

$$R = \begin{pmatrix} a_{11} & a_{12} \\ a_{21} & a_{22} \\ a_{32} & a_{32} \end{pmatrix}$$

Substituting this values, the system reduced to:

$$\begin{pmatrix} k_1 \\ k_2 \\ k_3 \\ k_4 \\ k_5 \end{pmatrix} = \begin{pmatrix} I & 0 \\ R & S \end{pmatrix}^{-1} \begin{pmatrix} k_1 \\ k_2 \\ b_1 \\ b_2 \\ b_3 \end{pmatrix}$$

The solution is expressed as follows:

$$
\begin{pmatrix} k_1 \\ k_2 \\ k_3 \\ k_4 \\ k_5 \end{pmatrix} = \begin{pmatrix} I & 0 \\ -S^{-1}R & S^{-1} \end{pmatrix} \begin{pmatrix} k_1 \\ k_2 \\ 0 \\ 0 \\ 0 \end{pmatrix}
$$

The matrix just right to the equal sign is called the total matrix [5]. The number and form of the dimensionless groups is taken from this matrix:

– The number of dimensionless groups correspond to the number of columns of the identity matrix. The example shows two columns of the identity matrix, corresponding to two dimensionless groups.

$$
\begin{array}{c c c c c c}
 & \pi_1 & \pi_2 & & & \\
Q_1 & 1 & 0 & 0 & 0 & 0 \\
Q_2 & 0 & 1 & 0 & 0 & 0 \\
Q_3 & k_{11} & k_{12} & k_{13} & k_{14} & k_{15} \\
Q_4 & k_{21} & k_{22} & k_{23} & k_{24} & k_{25} \\
Q_5 & k_{31} & k_{32} & k_{33} & k_{34} & k_{35}
\end{array}
$$

In this example, the values k_{ij} represent the results of the matrix operations involving the inverse of matrix R.

– The exponents of the variables for each dimensionless group is read from each column below the identity matrix.

$$
\pi_1 = \left(Q_1^1 Q_2^0 Q_3^{k_{11}} Q_4^{k_{21}} Q_5^{k_{31}} \right)
$$

$$
\pi_2 = \left(Q_1^0 Q_2^1 Q_3^{k_{12}} Q_4^{k_{22}} Q_5^{k_{32}} \right)
$$

Taking into account that the matrix $-S^{-1}R$ includes the final value of exponents in the dimensionless matrix, the presence of all zeros in any column of its transpose would be an indication that a variable is irrelevant.

$$
-S^{-1}R = \begin{pmatrix} k_{11} & k_{12} \\ k_{21} & k_{22} \\ k_{31} & k_{32} \end{pmatrix}
$$

$$
\left(-S^{-1}R \right)^T = \begin{pmatrix} k_{11} & k_{21} & k_{31} \\ k_{12} & k_{22} & k_{32} \end{pmatrix}
$$

8.3   Examples of Application of Matrix Method

<u>Sphere heating immersed in a fluid</u>: A metallic sphere is immersed in a stream of fluid in motion and heat transfer occurs in steady state conditions.

Similar to previous methods the first step is to define variables.

Variables. n = 6

$$\dot{Q} = f(k, l, U, \theta, \rho C_V)$$

Second step: Dimensional matrix. MLT θ.

$$
\begin{array}{c c c c c c c}
 & \dot{Q} & k & l & U & \theta & \rho C_V \\
M & 1 & 1 & 0 & 0 & 0 & 1 \\
L & 2 & 1 & 1 & 1 & 0 & -1 \\
T & -3 & -3 & 0 & -1 & 0 & -2 \\
\theta & 0 & 0 & -1 & 0 & 1 & -1
\end{array}
$$

$$
D = \begin{pmatrix}
1 & 1 & 0 & 0 & 0 & 1 \\
2 & 1 & 1 & 1 & 0 & -1 \\
-3 & -3 & 0 & -1 & 0 & -2 \\
0 & -1 & 0 & 0 & 1 & -1
\end{pmatrix}
$$

Third step: Define matrices S and R.

Compute the rank of the square matrix. Make the augmented matrix with an identity matrix equal to the number of variables minus the rank of the square matrix and add the corresponding null matrix.

Employing internet tools [7] it is found the rank to be 4. Then the size of the identity matrix is; 6–4 = 2. The augmented matrix becomes:

$$
\begin{pmatrix}
1 & 0 & 0 & 0 & 0 & 0 \\
0 & 1 & 0 & 0 & 0 & 0 \\
1 & 1 & 0 & 0 & 0 & 1 \\
2 & 1 & 1 & 1 & 0 & -1 \\
-3 & -3 & 0 & -1 & 0 & -2 \\
0 & -1 & 0 & 0 & 1 & -1
\end{pmatrix}
$$

$$
S = \begin{pmatrix}
0 & 0 & 0 & 1 \\
1 & 1 & 0 & -1 \\
0 & -1 & 0 & -2 \\
0 & 0 & 1 & -1
\end{pmatrix}
$$

$$R = \begin{pmatrix} 1 & 1 \\ 2 & 1 \\ -3 & -3 \\ 0 & -1 \end{pmatrix}$$

Fourth step: Define the value of the following product of matrices $-S^{-1}R$:

$$S^{-1} = \begin{pmatrix} 0 & 0 & 0 & 1 \\ 1 & 1 & 0 & -1 \\ 0 & -1 & 0 & -2 \\ 0 & 0 & 1 & -1 \end{pmatrix}^{-1} = \begin{pmatrix} 3 & 1 & 1 & 0 \\ -2 & 0 & -1 & 0 \\ 1 & 0 & 0 & 1 \\ 1 & 0 & 0 & 0 \end{pmatrix}$$

$$-S^{-1} = \begin{pmatrix} -3 & -1 & -1 & 0 \\ 2 & 0 & 1 & 0 \\ -1 & 0 & 0 & -1 \\ -1 & 0 & 0 & 0 \end{pmatrix}$$

$$-S^{-1}R = \begin{pmatrix} -3 & -1 & -1 & 0 \\ 2 & 0 & 1 & 0 \\ -1 & 0 & 0 & -1 \\ -1 & 0 & 0 & 0 \end{pmatrix} \begin{pmatrix} 1 & 1 \\ 2 & 1 \\ -3 & -3 \\ 0 & -1 \end{pmatrix} = \begin{pmatrix} -2 & -1 \\ -1 & -1 \\ -1 & 0 \\ -1 & -1 \end{pmatrix}$$

Since the identity matrix has two columns, each column corresponds to one dimensionless group. In this example the variables, in this order are: $\dot{Q}$, k, l, $\dot{U}$, θ, ρC_V. The value of the exponent foe each of these variables is in the same order of each column.

$$\begin{pmatrix} k_1 \\ k_2 \\ k_3 \\ k_4 \\ k_5 \end{pmatrix} = \begin{pmatrix} 1 & 0 & 0 & 0 & 0 & 0 \\ 0 & 1 & 0 & 0 & 0 & 0 \\ -2 & -1 & 0 & 0 & 0 & 1 \\ -1 & -1 & 1 & 1 & 0 & -1 \\ -1 & 0 & 0 & -1 & 0 & -2 \\ -1 & -1 & 0 & 0 & 1 & -1 \end{pmatrix} \begin{pmatrix} k_1 \\ k_2 \\ 0 \\ 0 \\ 0 \\ 0 \end{pmatrix}$$

Another way is to use the transpose to read by rows, as follows:

$$\pi_1 \equiv \begin{array}{cccccc} \dot{Q} & k & l & \dot{U} & \theta & \rho C_V \\ 1 & 0 & -2 & -1 & -1 & -1 \end{array}$$

The first group is:

$$\pi_1 = \frac{\dot{Q}}{l^2 \dot{U}\theta\rho C_V}$$

Similarly, for the second dimensionless group:

$$\pi_2 \equiv \begin{array}{cccccc} \dot{Q} & k & l & U & \theta & \rho C_V \\ 0 & 1 & -1 & -1 & 0 & -1 \end{array}$$

$$\pi_2 = \frac{k}{lU\rho C_V}$$

It is important to check that both groups are indeed dimensionless. The two groups are dimensionless. However, their form is not identical to the groups reported by Rayleigh.

$$\pi_1 = f(\pi_2)$$

$$\frac{\dot{Q}}{l^2 U\theta \rho C_V} = f\left(\frac{k}{lU\rho C_V}\right)$$

It is possible to obtain the same results as Rayleigh, after performing some manipulations. On one hand, the inverse of the second pi-group and on the other hand, multiplication of $\pi_1 \pi_2^{-1} = \pi_3$

$$\pi_3 = \left(\frac{\dot{Q}}{l^2 U\theta \rho C_V}\right)\left(\frac{lU\rho C_V}{k}\right) = \left(\frac{\dot{Q}}{kl\theta}\right)$$

The new form of the two dimensionless groups is now:

$$\frac{\dot{Q}}{kl\theta} = f'\left(\frac{lU\rho C_V k}{lU\rho C_V}\right)$$

The final form of the dimensionless groups will change depending on the order in which the variables are placed in the dimensional matrix. Choosing this sequence; l, θ, U, $\dot{Q}$, ρC_V, k, the following groups are obtained:

$$\frac{k^2 \theta}{U\dot{Q}\rho C_V} = f\left(\frac{lU\rho C_V}{k}\right)$$

By choosing the sequence: $\dot{Q}$, U, ρC_V, k, l, θ, the form of the two groups reported by Rayleigh is obtained directly. In all cases, if the groups are indeed dimensionless are all valid.

The example illustrates the advantages of dimensional analysis to reduce the number of experiments. The problem involves five independent variables. Applying the dimensionless equation, it is enough five experiments to cover all variables and change the value of one independent dimensionless variable. On the contrary, with the equation with dimensional variables, there are five variables that need to be changed and evaluate the effect of each variable on the others, would lead to a large amount of experiments; $5^5 = 3125$.

References

1. E.R. Van Driest, On dimensional analysis and the presentation of data in fluid flow problems. J. Appl. Mech. **13**(1), A-34 (1946)
2. P. Rosza, *Chapter 1 in T. Zirtes: Applied Dimensional Analysis and Modeling*. 2nd edn. (Elsevier-BH, UK, 2007), pp. 14–16
3. H.L. Langhaar, *Dimensional Analysis and Theory of Models* (John Wiley, USA, 1951), pp. 30–46
4. A.D. Sloan, W.W. Happ, Matrix methods of dimensional analysis. NASA Rept. ERCICQD 66–623, (1968)
5. T. Zirtes, *Applied Dimensional Analysis and Modeling*, 2nd edn. (UK, Elsevier-BH, 2007), pp. 14–16
6. J. Worstell, *Dimensional Analysis Practical Guides in Chemical Engineering* (Elsevier, UK, 2014), p. 53
7. https://www.emathhelp.net/calculators/linear-algebra/

Chapter 9
Inspection Method

9.1 Inspection Method

This method is quite simple if the number of dimensionless groups is small, particularly up to two. It is based on three principles [1]:

- The number of dimensionless groups is known from the π-theorem.
- One dimensionless group contains the dependent variable.
- The dimensionless groups should be independent.

The method seeks to make dimensionless one variable using the dimensions of the other variables. Once the first group is formed, the second group is built starting with the variables not employed in the previous step.

9.2 Examples of Application of the Inspection Method

Let's continue using the Bousinessq problem on heat transfer.

First step: Variables, dimensions and dimensionless groups.

Six variables ($\dot{Q}$, k, l, U,θ, ρC_V), four dimensions (MLTθ) and two dimensionless groups, according with the π-theorem.

Second step: Dimensional groups manipulating the dimensions in a free form.

First dimensional group should contain the dependent variable and make it dimensionless using the other variables available.

$$\left[\frac{\dot{Q}}{k}\right] = \left[\frac{ML^2}{T^3}\right]\left[\frac{T^3\theta}{ML}\right] = [L\theta]$$

$$\left[\frac{\dot{Q}}{kl\theta}\right] = \left[\frac{ML^2}{T^3}\right]\left[\frac{T^3\theta}{ML}\right] = \left[\frac{L\theta}{L\theta}\right] = 1$$

Second dimensional group using starting with the variables not employed previously.

$$[\rho C_V U] = \left[\frac{M}{LT^2\theta}\right]\left[\frac{L}{T}\right] = \left[\frac{M}{T^3\theta}\right]$$

$$\left[\frac{\rho C_V U}{k}\right] = \left[\frac{M}{T^3\theta}\right]\left[\frac{T^3\theta}{ML}\right] = \left[\frac{1}{L}\right]$$

$$\left[\frac{\rho C_V Ul}{k}\right] = \left[\frac{1}{L}\right][L] = 1$$

In fewer steps the two dimensionless groups are defined:

$$\frac{\dot{Q}}{kl\theta} = f\left(\frac{lU\rho C_V k}{lU\rho C_V}\right)$$

Ladle shroud: In continuous casting the transfer of liquid steel from the ladle to the tundish is carried out by opening a slide-gate at the bottom of the ladle. In this way liquid steel is discharged. If the stream of liquid steel is not protected during this transfer operation, oxygen and nitrogen from the surrounding atmosphere interact with the stream of liquid steel. In many applications this reoxidation should be fully prevented. This is done using a shroud partially immersed in the tundish and injecting a protective gas inside the shroud. The protective gas surrounds a jet of liquid steel once is discharged from the ladle. The height of the free jet has been defined by the following set of variables:

$$H_m = f(Q_g, Q_L, D_s, g)$$

where: H_m is the height of the free jet, Q_g is the gas flow rate, Q_L is the liquid's flow rate, D_s is the diameter of the shroud and g is the acceleration gravity constant.
Dimensional analysis: Five variables and two dimensions:

H_m	Q_g	Q_L	D_s	g
[L]	$[L^3T^{-1}]$	$[L^3T^{-1}]$	[L]	$[L\,T^{-2}]$

H_m is made dimensionless using D_s:

$$\left[\frac{H_m}{D_s}\right] = \left[\frac{L}{L}\right] = 1$$

Q_g is made dimensionless using Q_L:

$$\left[\frac{Q_g}{Q_L}\right] = \left[\frac{L^3 T^{-1}}{L^3 T^{-1}}\right] = 1$$

The remaining variable is g, which is made dimensionless using the variables available:

$$\left[\frac{g}{Q_L^2}\right] = \left[\frac{LT^{-2}}{L^6 T^{-2}}\right] = \left[L^{-5}\right]$$

$$\left[\frac{D_s^5 g}{Q_L^2}\right] = \left[L^5\right]\left[L^{-5}\right] = 1$$

Then, the final relationship becomes:

$$\frac{H_m}{D_s} = C\left(\frac{Q_g}{Q_L}\right)\left(\frac{D_s^5 g}{Q_L^2}\right)$$

The inspection method is quite simple and provides a fast solution.

Reference

1. D. F. Young, B. R. Munson, T. H. Okiishi, W. W. Huebsch, *A Brief Introduction to Fluid Mechanics*, 5th edn. (Wiley, 2011), pp. 248–249

Chapter 10
Engineering Problems with and without Experimental Data

10.1 General Engineering Problems with and without Experimental Data

10.1.1 Bubble Shape Dynamics Rising in a Liquid

Gas injection in liquid steel is a common practice in order to improve mixing conditions and removal of non-metallic inclusions. Bubble rising in an infinite medium (dimension of the fluid is much larger than the dimensions of the bubbles) is affected by the following eight variables:

– Densities of surrounding fluid and bubble (ρ_f, ρ)
– Viscosities of surrounding fluid and bubble (μ_f, μ)
– Gravitational force (g)
– Surface tension (σ)
– Bubble diameter (D)
– Bubble velocity (U)

Dimensions:

Variable	Symbol	Units	Dimensions
Bubble diameter	D	m	$[L]$
Density	ρ	kg/m^3	$[ML^{-3}]$
Viscosity	μ	kg/(m × s)	$[ML^{-1}T^{-1}]$
Surface tension	Σ	N/m	$[MT^{-2}]$
Gravitational acceleration	g	m/s^2	$[LT^{-2}]$
Velocity	U	m/s	$[LT^{-1}]$

© The Author(s), under exclusive license to Springer Nature Singapore Pte Ltd. 2021　　　99
A. N. Conejo, *Fundamentals of Dimensional Analysis*,
https://doi.org/10.1007/978-981-16-1602-0_10

π-method of repeating variables:

There are 8 variables and 3 dimensions. According with the π-theorem this yields five dimensionless groups. As non-repeating variables we choose; ρ_f, μ_f and D. Next, we define the values for the five dimensionless groups as follows:

$$\pi_1 = U(\rho_f)^{a_1}(\mu_f)^{b_1}(D)^{c_1}$$

$$\pi_2 = \mu(\rho_f)^{a_2}(\mu_f)^{b_2}(D)^{c_2}$$

$$\pi_3 = \rho(\rho_f)^{a_3}(\mu_f)^{b_3}(D)^{c_3}$$

$$\pi_4 = g(\rho_f)^{a_4}(\mu_f)^{b_4}(D)^{c_4}$$

$$\pi_5 = \sigma(\rho_f)^{a_5}(\mu_f)^{b_5}(D)^{c_5}$$

First π-group:

$$\pi_1 = U(\rho_f)^{a_1}(\mu_f)^{b_1}(D)^{c_1}$$

$$\pi_1 = (LT^{-1})(ML^{-3})^{a_1}(ML^{-1}T^{-1})^{b_1}(L)^{c_1} = M^0T^0L^0$$

Balance:

$$M: \quad a_1 + b_1 = 0$$
$$L: \quad 1 - 3a_1 - b_1 + c_1 = 0$$
$$T: \quad -1 - b_1 = 0$$

Solution:

$$b_1 = -1$$
$$a_1 = -b_1 = 1$$
$$c_1 = 3a_1 + b_1 - 1 = 1$$

Substituting values:

$$\pi_1 = \frac{\rho_f U D}{\mu_f}$$

Second π-group:

$$\pi_2 = \mu(\rho_f)^{a_2}(\mu_f)^{b_2}(D)^{c_2}$$

$$\pi_2 = \left(ML^{-1}T^{-1}\right)\left(ML^{-3}\right)^{a_2}\left(ML^{-1}T^{-1}\right)^{b_2}(L)^{c_2} = M^0T^0L^0$$

Balance:

$$M: \quad 1 + a_2 + b_2 = 0$$
$$L: \quad -1 - 3a_2 - b_2 + c_2 = 0$$
$$T: \quad -1 - b_2 = 0$$

Solution:

$$b_2 = -1$$
$$a_2 = -b_2 - 1 = 0$$
$$c_2 = 3a_2 + b_2 + 1 = 0$$

Substituting values:

$$\pi_2 = \frac{\mu}{\mu_f}$$

Third π-group:

$$\pi_3 = \rho(\rho_f)^{a_3}(\mu_f)^{b_3}(D)^{c_3}$$

$$\pi_3 = \left(ML^{-3}\right)\left(ML^{-3}\right)^{a_3}\left(ML^{-1}T^{-1}\right)^{b_3}(L)^{c_3} = M^0T^0L^0$$

Balance:

$$M: \quad 1 + a_3 + b_3 = 0$$
$$L: \quad -3 - 3a_3 - b_3 + c_3 = 0$$
$$T: \quad -b_3 = 0$$

Solution:

$$b_3 = 0$$
$$a_3 = -b_3 - 1 = -1$$
$$c_3 = 3 + 3a_3 + b_3 = 0$$

Substituting values:

$$\pi_3 = \frac{\rho}{\rho_f}$$

Fourth π-group:

$$\pi_4 = g(\rho_f)^{a_4}(\mu_f)^{b_4}(D)^{c_4}$$

$$\pi_4 = \left(LT^{-2}\right)\left(ML^{-3}\right)^{a_4}\left(ML^{-1}T^{-1}\right)^{b_4}(L)^{c_4} = M^0T^0L^0$$

Balance:

$$M : a_4 + b_4 = 0$$
$$L : 1 - 3a_4 - b_4 + c_4 = 0$$
$$T : -2 - b_4 = 0$$

Solution:

$$b_4 = -2$$
$$a_4 = -b_4 = 2$$
$$c_4 = -1 + 3a_4 + b_4 = -1 + 3a_4 - a_4 = -1 + 2a_4 = 3$$

Substituting values:

$$\pi_4 = \frac{gD^3\rho_f^2}{\mu_f^2}$$

Fifth π-group:

$$\pi_5 = \sigma(\rho_f)^{a_5}(\mu_f)^{b_5}(D)^{c_5}$$

$$\pi_5 = \left(MT^{-2}\right)\left(ML^{-3}\right)^{a_5}\left(ML^{-1}T^{-1}\right)^{b_5}(L)^{c_5} = M^0T^0L^0$$

Balance:

$$M : 1 + a_5 + b_5 = 0$$
$$L : -3a_5 - b_5 + c_5 = 0$$
$$T : -2 - b_5 = 0$$

Solution:

$$b_5 = -2$$
$$a_5 = -b_5 - 1 = 1$$
$$c_2 = -1 + 3a_5 + b_5 = 0$$

Substituting values:

$$\pi_5 = \frac{\sigma\,\rho_f}{\mu_f^2}$$

Substituting the previous results:

$$\pi_1 = f(\pi_2)(\pi_3)(\pi_4)(\pi_5)$$

$$\frac{\rho_f U D}{\mu_f} = f\left(\frac{\mu}{\mu_f}\right)\left(\frac{\rho}{\rho_f}\right)\left(\frac{gD^3\rho_f^2}{\mu_f^2}\right)\left(\frac{\sigma\,\rho_f}{\mu_f^2}\right)$$

In order to define the particular solution, the exponents in the following relationship should be defined:

$$\frac{\rho_f U D}{\mu_f} = f\left(\frac{\mu}{\mu_f}\right)^{x_1}\left(\frac{\rho}{\rho_f}\right)^{x_2}\left(\frac{gD^3\rho_f^2}{\mu_f^2}\right)^{x_3}\left(\frac{\sigma\,\rho_f}{\mu_f^2}\right)^{x_4}$$

In the previous analysis it recognized that π_1 is the Reynolds number.

$$\mathrm{Re} = \frac{\rho_f U D}{\mu_f}$$

Bubble motion and bubble dynamics have been extensively investigated in the past, concluding that these phenomena are affected not only by the Reynolds number but also by the Eotvos and Morton numbers, therefore in order to obtain these dimensionless groups it is noticed that ρ_f, σ and g are repeating variables. The process is repeated using the following arrangement of variables:

$$\pi_1 = U(g)^{a_1}(\sigma)^{b_1}(\rho_f)^{c_2}$$

$$\pi_2 = \rho(g)^{a_2}(\sigma)^{b_2}(\rho_f)^{c_2}$$

$$\pi_3 = \mu(g)^{a_3}(\sigma)^{b_3}(\rho_f)^{c_3}$$

$$\pi_4 = \mu_f(g)^{a_4}(\sigma)^{b_4}(\rho_f)^{c_4}$$

$$\pi_5 = D(g)^{a_5}(\sigma)^{b_5}(\rho_f)^{c_5}$$

First π-group:

$$\pi_1 = U(g)^{a_1}(\sigma)^{b_1}(\rho_f)^{c_2}$$

$$\pi_1 = \left(LT^{-1}\right)\left(LT^{-2}\right)^{a_1}\left(MT^{-2}\right)^{b_1}\left(ML^{-3}\right)^{c_1} = M^0T^0L^0$$

Balance:

$$M : a_1 + c_1 = 0$$
$$L : \ 1 + a_1 - 3c_1 = 0$$
$$T : \ -1 - 2a_1 - 2b_1 = 0$$

Solution:

Subtracting the first two equations : $c_1 = 1/4$

$$a_1 = -c_1 = -1/4$$
$$2b_1 = -1 - 2a_1 \quad \therefore \quad b_1 = -1/4$$

Substituting values:

$$\pi_1 = \frac{\rho_f^{1/4} \, U}{g^{1/4} \sigma^{1/4}} = \frac{\rho_f U^4}{g \sigma}$$

Second π-group:

$$\pi_2 = \rho(g)^{a_2} (\sigma)^{b_2} (\rho_f)^{c_2}$$

$$\pi_2 = \left(ML^{-3}\right)\left(LT^{-2}\right)^{a_2} \left(MT^{-2}\right)^{b_2} \left(ML^{-3}\right)^{c_2} = M^0 T^0 L^0$$

Balance:

$$M : 1 + b_2 + c_2 = 0$$
$$L : \ -3 + a_2 - 3c_2 = 0$$
$$T : \ -2a_2 - 2b_2 = 0 \quad \therefore \quad b_2 = -a_2$$

Solution:

$$-a_2 + c_2 = -1 \quad \therefore a_2 = c_2 + 1$$
$$a_2 - 3c_2 = 3 \quad \therefore \quad c_2 = -1, a_2 = 0$$
$$\therefore b_2 = -a_2 = 0$$

Substituting values:

$$\pi_2 = \frac{\rho}{\rho_f}$$

Third π-group:

$$\pi_3 = \mu(g)^{a_3} (\sigma)^{b_3} (\rho_f)^{c_3}$$

$$\pi_3 = \left(ML^{-1}T^{-1}\right)\left(LT^{-2}\right)^{a_3}\left(MT^{-2}\right)^{b_3}\left(ML^{-3}\right)^{c_3} = M^0T^0L^0$$

Balance:

$$
\begin{aligned}
M: &\quad 1 + b_3 + c_3 = 0 \quad \therefore \quad b_3 = 1 + c_3 \\
L: &\quad -1 + a_3 - 3c_3 = 0 \\
T: &\quad -1 - 2a_3 - 2b_3 = 0
\end{aligned}
$$

Solution:

Rearranging equations 2 and 3 and adding $L_2 + L_3$:

$$2a_3 - 6c_3 = 2$$
$$-2a_3 + 2c_3 = -1 \quad \therefore \quad c_3 = -1/4, b_3 = -3/4, a_3 = 1/4$$

Substituting values:

$$\pi_3 = \frac{\mu g^{1/4}}{\rho_f^{1/4}\sigma^{3/4}} = \frac{\mu^4 g}{\rho_f \sigma^3}$$

Fourth π-group:

$$\pi_4 = \mu_f(g)^{a_4}(\sigma)^{b_4}(\rho_f)^{c_4}$$

$$\pi_4 = \left(ML^{-1}T^{-1}\right)\left(LT^{-2}\right)^{a_4}\left(MT^{-2}\right)^{b_4}\left(ML^{-3}\right)^{c_4} = M^0T^0L^0$$

This group is similar to the previous one, except that viscosity is the fluid's viscosity

$$\pi_4 = \frac{\mu_f g^{1/4}}{\rho_f^{1/4}\sigma^{3/4}} = \frac{g\mu_f^4}{\rho_f \sigma^3} = Mo$$

Fifth π-group:

$$\pi_5 = D(g)^{a_5}(\sigma)^{b_5}(\rho_f)^{c_5}$$

$$\pi_5 = (L)\left(LT^{-2}\right)^{a_5}\left(MT^{-2}\right)^{b_5}\left(ML^{-3}\right)^{c_5} = M^0T^0L^0$$

Balance:

$$
\begin{aligned}
M: &\quad b_5 + c_5 = 0 \quad \therefore \quad c_5 = -b_5 \\
L: &\quad 1 + a_5 - 3c_5 = 0 \\
T: &\quad 2a_5 - 2b_5 = 0 \quad \therefore \quad a_5 = b_5
\end{aligned}
$$

Solution:

Rearranging L_2 and L_3

$$2a_5 - 6c_5 = -2$$
$$- 2a_5 + 2c_5 = 0 \therefore c_5 = 1/2, b_5 = -1/2, a_5 = 1/2$$

Substituting values:

$$\pi_5 = \frac{Dg^{1/2}\rho_f^{1/2}}{\sigma^{1/2}} = \frac{gD^2\rho_f}{\sigma} = Eo$$

Substituting into the general expression:

$$\pi_1 = f(\pi_2)(\pi_3)(\pi_4)(\pi_5)$$

$$\frac{\rho_f U^4}{g\sigma} = f\left(\frac{\rho}{\rho_f}\right)\left(\frac{g\mu^4}{\rho_f\sigma^3}\right)\left(\frac{g\mu_f^4}{\rho_f\sigma^3}\right)\left(\frac{gD^2\rho_f}{\sigma}\right)$$

$$\frac{\rho_f U^4}{g\sigma} = f\left(\frac{\rho}{\rho_f}\right)(Mo')(Mo)(Eo)$$

However, the three dimensionless numbers Re, Mo and Eo are not obtained together in this particular analysis of the variables.

Smolianski et al. [1] reported a similar analysis, loosely describing space and time as two additional variables and reporting in total 6 dimensionless numbers but choosing 5 of them. Their dimensional analysis report Re, Mo and Eo numbers as well as a dimensionless time.

The shape of the bubble can be described based on the values of the Re and Eo numbers. The weber number can be used to replace the Eo number. The Morton number results from a ratio between Eo and Re numbers, as follows:

$$Mo = \frac{Eo^3}{Re^4}$$

Figure 10.1 describes the shape of the bubbles as a function of the previous dimensionless numbers.

10.1.2 Motion of Non-Metallic Inclusions in Liquid Metals

Flotation of non-metallic inclusions in a stationary liquid occurs due to buoyancy forces. The density of non-metallic inclusions is similar to slags, for example silica inclusions have a density around 2800 kg/m^3. Assuming the inclusion has the shape

Fig. 10.1 Bubble shape as a function of Re, Eo and We numbers (After Clift et al. 1978)

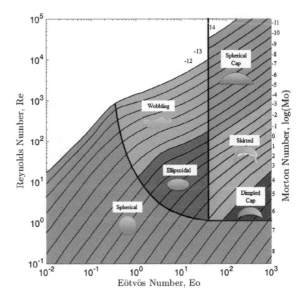

of a sphere, its terminal velocity is reached when the system of forces reaches equilibrium. The buoyancy force acts upwards and the weight of the particle acts downwards. The buoyancy force results from the liquid displaced. A third force is the drag force which acts in a direction opposite to the motion of the particle. The drag force represents all normal, tangential and shear forces imposed by the liquid. The force balance is expressed as follows,

$$\rho_l g \left(\frac{4}{3}\pi R^3 \right) + 6\pi\mu R U_\infty = \rho_s g \left(\frac{4}{3}\pi R^3 \right)$$

The terminal velocity (U_∞) is separated and defined as follows;

$$U_\infty = \frac{4}{18}\frac{(\rho_l - \rho_s)gR^2}{\mu} = \frac{2}{9}\frac{(\rho_l - \rho_s)gR^2}{\mu} = \frac{(\rho_l - \rho_s)gd^2}{18\mu}$$

This equation is known as Stokes law. Guthrie [2] provided some data on the flotation of Bakelite particles on liquid bromoform and Mazumdar and Evans [3] reported the resulting dimensionless groups that describe the flotation of inclusions. With information from a physical model on the flotation rate for one size particle it is needed to define the rising velocity in a metallic system.

The flotation rate (U) of inclusions depends on four variables; its weight inside the liquid (W), its particle size (d) and the properties of the liquid, density (ρ_l) and viscosity (μ).

$$U = f(W, d, \mu, \rho)$$

Dimensional analysis: Let's choose the MLT system of dimensions and the π-theorem.

Variable	Symbol	Units	Dimensions
Rising velocity	U	m/s	$[LT^{-1}]$
Particle weight	W	kg m/s^2	$[MLT^{-2}]$
Particle diameter	D	M	$[L]$
Fluid's density	ρ	kg/m^3	$[ML^{-3}]$
Fluid's viscosity	μ	kg m/s	$[ML^{-1}T^{-1}]$

The π-theorem defines two dimensionless groups because there are 5 variables—3 dimensions $= 2\pi$ groups. As non-repeating variables we choose; d, μ and ρ. Now we define the values for the two dimensionless groups as follows:

$$\pi_1 = U(d)^{a_1}(\rho)^{b_1}(\mu)^{c_1}$$

$$\pi_2 = W(d)^{a_2}(\rho)^{b_2}(\mu)^{c_2}$$

$$\pi_1 = f(\pi_2)$$

First π-group:

$$\pi_1 = U(d)^{a_1}(\rho)^{b_1}(\mu)^{c_1}$$

$$\pi_1 = \left(LT^{-1}\right)(L)^{a_1}\left(ML^{-3}\right)^{b_1}\left(ML^{-1}T^{-1}\right)^{c_1} = M^0T^0L^0$$

Balance:

$$M: \quad b_1 + c_1 = 0$$
$$L: \quad 1 + a_1 - 3b_1 - c_1 = 0$$
$$T: \quad -1 - c_1 = 0$$

Solution:

$$b_1 = -c_1$$
$$c_1 = -1 \therefore b_1 = 1$$
$$a_1 = -1 + 3b_1 + c_1 \therefore a_1 = 1$$

$$a_1 = 1, b_1 = 1, c_1 = -1$$

Substitution of these values on the π_1 -group yields:

$$\pi_1 = \frac{\rho U d}{\mu}$$

Second π-group:

$$\pi_2 = W(d)^{a_2}(\rho)^{b_2}(\mu)^{c_2}$$

$$\pi_2 = (MLT^{-2})(L)^{a_2}(ML^{-3})^{b_2}(ML^{-1}T^{-1})^{c_2} = M^0T^0L^0$$

Balance:

$$
\begin{aligned}
M: \quad & 1 + b_2 + c_2 = 0 \\
L: \quad & 1 + a_2 - 3b_2 - c_2 = 0 \\
T: \quad & -2 - c_2 = 0
\end{aligned}
$$

Solution:

$$
\begin{aligned}
L_1 : b_2 = -c_2 - 1 \\
L_3 : c_2 = -2 \therefore b_2 = 1
\end{aligned}
$$

$$L_2 : a_2 = -1 + 3b_2 + c_2 \therefore a_2 = 0$$

$$a_2 = 0, b_2 = 1, c_2 = -2$$

Substitution of these values on the π_2-group yields:

$$\pi_2 = \frac{W\rho}{\mu^2}$$

The final relationship is as follows,

$$\frac{\rho U d}{\mu} = f\left(\frac{W\rho}{\mu^2}\right)$$

Problem: Guthrie [2] provides experimental data from a physical model: A particle of Bakelite of 1.33 mm in diameter rises at a speed of 5.4 cm/s. Estimate the rising speed and diameter of the slag particle.

Materials	ρ, g/cm^3	μ, P	Materials	ρ, g/cm^3	μ, P
Molten copper	8.25	0.033	Bromoform	1.55	0.011
Slag particle	2.35		Bakelite	1.27	

Solution: The motion of the particle is defined by two dimensionless numbers, Re number and a dimensionless weight group. If the two systems are similar, they should have similar values for those dimensionless groups. Re numbers includes both velocity and particle diameter as unknowns and the second dimensionless group only the particle diameter. Using the second group is used to define the slag particle size and then the Re number to solve for the velocity.

In order to use the second dimensionless group, the weight of the particle should be expressed in terms of the particle diameter. The weight of an immersed body in a fluid is defined as follows,

$$W = Buoyancy\ force - Weight\ immersed\ in\ a\ fluid$$

$$W = \rho_l g V_{ld} - \rho_s g V_s = \frac{4}{3}\pi R^3 (\rho_l - \rho_s)g$$

$$W = \rho_l g V_{ld} - \rho_s g V_s = \frac{4}{3}\pi \frac{d^3}{8}(\rho_l - \rho_s)g$$

Therefore, the alternate form of the second dimensionless group, π_2 is;

$$\pi_2 = \frac{W\rho}{\mu^2} = \frac{(\rho_l - \rho_s)\rho_l g d^3}{\mu^2}$$

The constants are not needed in the general functional relationship of dimensionless group. Their final value is obtained with experimental data, on the other hand, $(4/24)\pi$ vanishes when the dimensionless group are made equal for model and prototype.

$$\left\{ \frac{(\rho_l - \rho_s)\rho_l g d^3}{\mu^2} \right\}_m = \left\{ \frac{(\rho_l - \rho_s)\rho_l g d^3}{\mu^2} \right\}_p$$

In this equation all values are known, except the diameter of the particle in the prototype.

$$d_p^3 = \left\{ \frac{(\rho_l - \rho_s)\rho_l d^3}{\mu^2} \right\}_m \left\{ \frac{\mu^2}{(\rho_l - \rho_s)\rho_l} \right\}_p$$

$$d_p = \sqrt[3]{\left\{ \frac{(\rho_l - \rho_s)\rho_l d^3}{\mu^2} \right\}_m \left\{ \frac{\mu^2}{(\rho_l - \rho_s)\rho_l} \right\}_p}$$

Replacing values;

$$d_p = \sqrt[3]{\left\{ \frac{(1.55 - 1.27)(1.55)(0.013)^3}{(0.011)^2} \right\}_m \left\{ \frac{(0.033)^2}{(8.25 - 2.35)(8.25)} \right\}_p}$$

$$d_p = 5.61 \times 10^{-3}\ cm = 0.056\ mm = 56\ \mu m$$

The velocity of the slag particle is calculated from the Re number.

$$\left(\frac{\rho U d}{\mu}\right)_m = \left(\frac{\rho U d}{\mu}\right)_p$$

$$(U)_p = \text{Re}_m \left(\frac{\mu}{\rho d}\right)_p$$

Replacing values,

$$(U)_p = \left(\frac{1.55 \times 5.4 \times 0.013}{0.011}\right)_m \left(\frac{0.033}{8.25 \times 5.61 \times 10^{-3}}\right)_p$$

$$(U)_p = 9.89 \times \left(\frac{0.033}{8.25 \times 5.61 \times 10^{-3}}\right)_p$$

$$(U)_p = 7.05 \text{ cm/s}$$

Comparing this velocity with Stokes law;

$$U_t = \frac{(\rho_l - \rho_s)gd^2}{18\mu}$$

$$U_t = \frac{(8.25 - 2.35)(981)(5.61 \times 10^{-3})^2}{(18)(0.033)}$$

$$U_t = 0.31 \text{ cm/s}$$

Both calculations report high rising velocities in comparison with typical values for non-metallic inclusions in steel. Two reasons for the high velocity are; first, the large size of the slag particle and second the large density difference between liquid copper and the slag particle. About 80% of the non-metallic inclusions in liquid steel is below 5–10 μm. Another reason is that the Stokes law is valid for Re < 1. In this example the Re number is 9.89. To compare the effect of particle size on rising velocity, two additional cases are shown below [2, 3].

In this example [2], a non-metallic inclusion of 1 μm rises through stagnant liquid steel at 1600 °C. Data; density and viscosity of steel, 7000 kg/m^3 and 7 cP, respectively. Particle density, 3000 kg/m^3

$$U_t = \frac{(7000 - 3000)(9.81)(1 \times 10^{-6})^2}{(18)(0.7)}$$

$$U_t = 3.1 \times 10^{-9} \text{ m/s} = 10^{-2} \text{mm/h}$$

In this proposed example [3], hollow glass spheres of 120 μm, rising in water, are used to represent alumina inclusions. Its rising velocity was 3.2 mm/s. Estimate the rising velocity and particle size of alumina in liquid steel. Data: density and viscosity of steel, 7000 kg/m³ and 7 cP, respectively. Alumina density, 3970 kg/m³. Density and viscosity of water at room temperature are 1000 kg/m³ and 1 cP, respectively. Density of hollow glass spheres, 910 kg/m³.

$$d_p = \sqrt[3]{\left\{\frac{(\rho_l - \rho_s)\rho_l d^3}{\mu^2}\right\}_m \left\{\frac{\mu^2}{(\rho_l - \rho_s)\rho_l}\right\}_p}$$

Replacing values;

$$d_p = \sqrt[3]{\left\{\frac{(1000 - 910)(1000)(120 \times 10^{-6})^3}{(0.1)^2}\right\}_m \left\{\frac{(0.7)^2}{(7000 - 3970)(7000)}\right\}_p}$$

$$d_p = 7.1 \times 10^{-5} \text{ m} = 71 \ \mu\text{m}$$

$$U_t = \frac{(7000 - 3970)(9.81)(7.1 \times 10^{-5})^2}{(18)(0.7)}$$

$$U_t = 1.18 \times 10^{-5} \text{ m/s} = 4.2 \text{ cm/h}$$

The previous examples clearly show that under natural convection, the removal of inclusions is impossible considering the duration of a heat in a ladle furnace is less than an hour. Bottom gas injection is applied to decrease thermal stratification and improve alloy homogenization and also to assist in the removal of non-metallic inclusions. Zhang and Taniguchi [4] reviewed the optimum bubble size for inclusion flotation in steel refining and continuous casting, indicating a bubble diameter lower than 5 mm, however as the bubble size decreases the rising time decreases and the small bubbles can get trapped. To avoid re-entrainment into the bulk, the bubble velocity at the free surface should be higher than 0.1 m/s, an estimated velocity at the free surface for ladle conditions, therefore the range of bubble size should be from 1–5 mm.

10.1.3   Thermal Stresses in Continuous Casting

The thermal stresses (σ) in the center of a solidified product in continuous casting, according with Roy [5], depends on the following variables: Modulus of elasticity (E), Coefficient of linear thermal expansion (β), initial temperature difference between cooling medium and melting point of metal ($\Delta\theta$), characteristic linear dimension of solid (L), heat transfer coefficient of cooling medium at surface of solid (h), thermal conductivity of solid (k), density of solid (ρ), Specific heat capacity of solid (C_p), time from start of cooling (t), specific latent heat (ΔH).

Dimensional analysis: Let´s choose the MLTθ system of dimensions and the π-theorem.

Variable	Symbol	Units	Dimensions
Thermal stresses	σ	N/m^2	$[ML^{-1}T^{-2}]$
Modulus of elasticity	E	N/m^2	$[ML^{-1}T^{-2}]$
Coefficient of linear thermal expansion	β	m/°C	$[L\theta^{-1}]$
Temperature gradient	$\Delta\theta$	°C	$[\theta]$
Characteristic length	L	m	$[L]$
Heat transfer coefficient	h	W/m^{2}K	$[MT^{-3}\theta^{-1}]$
Thermal conductivity	k	W/mK	$[MLT^{-3}\theta^{-1}]$
Specific heat capacity	C_p	J/kgK	$[L^2T^{-2}\theta^{-1}]$
Fluid's density	ρ	kg/m^3	$[ML^{-3}]$
Time	t	s	$[T]$
Specific latent heat	ΔH	J/kg	$[L^2T^{-2}]$

There are 11 variables and 4 dimensions. According with the π-theorem this yields 7 dimensionless groups. As repeating variables we choose; L, ρ, E and $\Delta\theta$. Next, we define the values for the four dimensionless groups as follows:

$$\pi_1 = \sigma(L)^{a_1}(\rho)^{b_1}(E)^{c_1}(\Delta\theta)^{d_1}$$

$$\pi_2 = t\,(L)^{a_2}(\rho)^{b_2}(E)^{c_2}(\Delta\theta)^{d_2}$$

$$\pi_3 = \beta(L)^{a_3}(\rho)^{b_3}(E)^{c_3}(\Delta\theta)^{d_3}$$

$$\pi_4 = h(L)^{a_4}(\rho)^{b_4}(E)^{c_4}(\Delta\theta)^{d_4}$$

$$\pi_5 = k(L)^{a_5}(\rho)^{b_5}(E)^{c_5}(\Delta\theta)^{d_5}$$

$$\pi_6 = C_p(L)^{a_6}(\rho)^{b_6}(E)^{c_6}(\Delta\theta)^{d_6}$$

$$\pi_7 = \Delta H(L)^{a_7}(\rho)^{b_7}(E)^{c_7}(\Delta\theta)^{d_7}$$

First π-group:

$$\pi_1 = \sigma(L)^{a_1}(\rho)^{b_1}(E)^{c_1}(\Delta\theta)^{d_1}$$

$$\pi_1 = \left(ML^{-1}T^{-2}\right)(L)^{a_1}\left(ML^{-3}\right)^{b_1}\left(ML^{-1}T^{-2}\right)^{c_1}(\Delta\theta)^{d_1} = M^0T^0L^0$$

Balance:

$$M: \quad 1 + b_1 + c_1 = 0$$
$$L: \quad -1 + a_1 - 3b_1 - c_1 = 0$$
$$T: \quad -2 - 2c_1 = 0$$
$$\theta: \quad d_1 = 0$$

Solution:

$$d_1 = 0$$
$$c_1 = -1$$
$$b_1 = 0$$
$$a_1 = 0$$

Substituting values:

$$\pi_1 = \frac{\sigma}{E}$$

Second π-group:

$$\pi_2 = t(L)^{a_2}(\rho)^{b_2}(E)^{c_2}(\Delta\theta)^{d_2}$$

$$\pi_1 = (T)(L)^{a_2}\left(ML^{-3}\right)^{b_2}\left(ML^{-1}T^{-2}\right)^{c_2}(\theta)^{d_2} = M^0T^0L^0$$

Balance:

$$M: b_2 + c_2 = 0$$
$$L: a_2 - 3b_2 - c_2 = 0$$
$$T: 1 - 2c_2 = 0$$
$$\theta: d_2 = 0$$

Solution:

$$d_2 = 0$$
$$c_2 = 1/2$$
$$b_2 = -1/2$$
$$a_2 = -1$$

Substituting values:

$$\pi_2 = \frac{tE^{\frac{1}{2}}}{L\rho^{\frac{1}{2}}} = \frac{Et^2}{\rho L^2}$$

Third π-group:

$$\pi_3 = \beta(L)^{a_3}(\rho)^{b_3}(E)^{c_3}(\Delta\theta)^{d_3}$$

$$\pi_3 = \left(L\theta^{-1}\right)(L)^{a_3}\left(ML^{-3}\right)^{b_3}\left(ML^{-1}T^{-2}\right)^{c_3}(\theta)^{d_3} = M^0T^0L^0$$

Balance:

$$M : b_3 + c_3 = 0$$
$$L : 1 + a_3 - 3b_3 - c_3 = 0$$
$$T : -2c_3 = 0$$
$$\theta : -1 + d_3 = 0$$

Solution:

$$c_3 = 0$$
$$d_3 = 1$$
$$b_3 = 0$$
$$a_3 = -1$$

Substituting values:

$$\pi_3 = \frac{\beta\Delta\theta}{L}$$

Fourth π-group:

$$\pi_4 = h(L)^{a_4}(\rho)^{b_4}(E)^{c_4}(\Delta\theta)^{d_4}$$

$$\pi_4 = \left(MT^{-3}\theta^{-1}\right)(L)^{a_4}\left(ML^{-3}\right)^{b_4}\left(ML^{-1}T^{-2}\right)^{c_4}(\theta)^{d_4} = M^0T^0L^0$$

Balance:

$$M \ \ 1 + b_4 + c_4 = 0$$
$$L : a_4 - 3b_4 - c_4 = 0$$
$$T : -3 - 2c_4 = 0$$
$$\theta : -1 + d_4 = 0$$

Solution:

$$c_4 = -3/2$$
$$d_4 = 1$$
$$b_4 = 1/2$$
$$a_4 = 0$$

Substituting values:

$$\pi_4 = \frac{h\rho^{\frac{1}{2}}\Delta\theta}{E^{\frac{3}{2}}}$$

Fifth π-group:

$$\pi_5 = k(L)^{a_5}(\rho)^{b_5}(E)^{c_5}(\Delta\theta)^{d_5}$$

$$\pi_5 = \left(MLT^{-3}\theta^{-1}\right)(L)^{a_5}\left(ML^{-3}\right)^{b_5}\left(ML^{-1}T^{-2}\right)^{c_5}(\theta)^{d_5} = M^0T^0L^0$$

Balance:

$$M: 1 + b_5 + c_5 = 0$$
$$L: 1 + a_5 - 3b_5 - c_5 = 0$$
$$T: -3 - 2c_5 = 0$$
$$\theta: -1 + d_5 = 0$$

Solution:

$$c_5 = -3/2$$
$$d_5 = 1$$
$$b_5 = 1/2$$
$$a_5 = -1$$

Substituting values:

$$\pi_5 = \frac{k\rho^{\frac{1}{2}}\Delta\theta}{LE^{\frac{3}{2}}}$$

Sixth π-group:

$$\pi_6 = C_p(L)^{a_6}(\rho)^{b_6}(E)^{c_6}(\Delta\theta)^{d_6}$$

$$\pi_6 = \left(L^2T^{-2}\theta^{-1}\right)(L)^{a_6}\left(ML^{-3}\right)^{b_6}\left(ML^{-1}T^{-2}\right)^{c_6}(\theta)^{d_6} = M^0T^0L^0$$

Balance:

$$M: \quad b_6 + c_6 = 0$$
$$L: \quad 2 + a_6 - 3b_6 - c_6 = 0$$
$$T: \quad -2 - 2c_6 = 0$$
$$\theta: \quad -1 + d_6 = 0$$

Solution:

$$c_6 = -1$$
$$d_6 = 1$$
$$b_6 = 1$$
$$a_6 = 0$$

Substituting values:

$$\pi_6 = \frac{C_p \rho \Delta \theta}{E}$$

Seventh π-group:

$$\pi_7 = \Delta H (L)^{a_7} (\rho)^{b_7} (E)^{c_7} (\Delta \theta)^{d_7}$$

$$\pi_7 = \left(L^2 T^{-2}\right)(L)^{a_7} \left(ML^{-3}\right)^{b_7} \left(ML^{-1}T^{-2}\right)^{c_7} (\theta)^{d_7} = M^0 T^0 L^0$$

Balance:

$$M: b_7 + c_7 = 0$$
$$L: 2 + a_7 - 3b_7 - c_7 = 0$$
$$T: -2 - 2c_7 = 0$$
$$\theta: d_7 = 0$$

Solution:

$$c_7 = -1$$
$$d_7 = 0$$
$$b_7 = 1$$
$$a_7 = 0$$

Substituting values:

$$\pi_7 = \frac{\rho \Delta H}{E}$$

Then the expression in terms of dimensionless number becomes:

$$\pi_1 = f(\pi_2)(\pi_3)(\pi_4)(\pi_5)(\pi_6)(\pi_7)$$

$$\frac{\sigma}{E} = f\left(\frac{Et^2}{\rho L^2}\right)\left(\frac{\beta \Delta \theta}{L}\right)\left(\frac{h\rho^{\frac{1}{2}}\Delta\theta}{E^{\frac{3}{2}}}\right)\left(\frac{k\rho^{\frac{1}{2}}\Delta\theta}{LE^{\frac{3}{2}}}\right)\left(\frac{C_p\rho\Delta\theta}{E}\right)\left(\frac{\rho\Delta H}{E}\right)$$

The number of dimensionless groups can be reduced observing that π_3 and π_4 share similar variables

$$\left(\frac{\pi_3}{\pi_4}\right) = \left(\frac{h\rho^{\frac{1}{2}}\Delta\theta}{E^{\frac{3}{2}}}\right) \Big/ \left(\frac{k\rho^{\frac{1}{2}}\Delta\theta}{LE^{\frac{3}{2}}}\right) = \frac{hL}{k}$$

Which is the Nusselt number.

Also, π_6 and π_7 can be simplified:

$$\left(\frac{\pi_6}{\pi_7}\right) = \left(\frac{C_p\rho\Delta\theta}{E}\right) \Big/ \left(\frac{\rho\Delta H}{E}\right) = \frac{C_p\Delta\theta}{\Delta H}$$

Therefore, the final number of dimensionless is reduced to 5 groups:

$$\frac{\sigma}{E} = f\left(\frac{Et^2}{\rho L^2}\right)\left(\frac{\beta\Delta\theta}{L}\right)\left(\frac{hL}{k}\right)\left(\frac{C_p\Delta\theta}{\Delta H}\right)$$

These results are slightly different to those reported Roy [5], this is because the original source has several mistakes. This author also provides information related to similarity; given a characteristic length and time for a solidified bar, the task is to define the values for another bar with a different thermal conductivity. This can be obtained equating dimensionless groups that contain the variable L and t.

10.1.4 Mass Transfer Coefficient Under Natural Convection

The mass transfer coefficient is a very important parameter in metallurgical processes. Its value can be predicted using dimensional numbers. The mass transfer coefficient depends on the fluid's properties; density (ρ), viscosity (μ) its diffusion coefficient (D), buoyancy forces due to concentration differences ($g\Delta\rho$) and reactor's dimension or characteristic length (L).

$$k = f(D, \mu, \rho, L, g\Delta\rho)$$

Dimensional analysis: Let's choose the MLT system of dimensions and the π-theorem.

Variable	Symbol	Units	Dimensions
Mass transfer coefficient	k	m/s	$[ML^{-2}T^{-1}]$
Characteristic length	L	m	$[L]$
Fluid's viscosity	μ	kg m/s	$[ML^{-1}T^{-1}]$
Fluid's density	ρ	kg/m³	$[ML^{-3}]$
Diffusion coefficient	D	m²/s	$[L^2T^{-1}]$
Buoyancy forces	$g\Delta\rho$	kg/m²s²	$[ML^{-2}T^{-2}]$

There are 6 variables and 3 dimensions. According with the π-theorem this yields 3 dimensionless groups. As repeating variables we choose; D, L and ρ Next, we define the values for the three dimensionless groups as follows:

$$\pi_1 = k(D)^{a_1}(L)^{b_1}(\rho)^{c_1}$$

$$\pi_2 = \mu(D)^{a_2}(L)^{b_2}(\rho)^{c_2}$$

$$\pi_3 = g\Delta\rho(D)^{a_3}(L)^{b_3}(\rho)^{c_3}$$

First π-group:

$$\pi_1 = k(D)^{a_1}(L)^{b_1}(\rho)^{c_1}$$

$$\pi_1 = \left(ML^{-2}T^{-1}\right)\left(L^2T^{-1}\right)^{a_1}(L)^{b_1}\left(ML^{-3}\right)^{c_1} = M^0T^0L^0$$

Balance:

$$M: \quad 1 + c_1 = 0$$
$$L: \quad -2 + 2a_1 + b_1 - 3c_1 = 0$$
$$T: \quad -1 - a_1 = 0$$

Solution:

$$c_1 = -1$$
$$a_1 = -1$$
$$b_1 = 1$$

Substituting values:

$$\pi_1 = \frac{kL}{D\rho}$$

Second π-group:

$$\pi_2 = \mu(D)^{a_2}(L)^{b_2}(\rho)^{c_2}$$

$$\pi_2 = \left(ML^{-1}T^{-1}\right)\left(L^2T^{-1}\right)^{a_2}(L)^{b_2}\left(ML^{-3}\right)^{c_2} = M^0T^0L^0$$

Balance:

$$M: \quad 1 + c_2 = 0$$
$$L: \quad -1 + 2a_2 + b_2 - 3c_2 = 0$$
$$T: \quad -1 - a_2 = 0$$

Solution:

$$c_2 = -1$$
$$a_2 = -1$$
$$b_2 = 1 - 2a_2 + 3c_2 = 0$$

Substituting values:

$$\pi_2 = \frac{\mu}{D\rho}$$

Third π-group:

$$\pi_3 = g\Delta\rho(D)^{a_3}(L)^{b_3}(\rho)^{c_3}$$

$$\pi_3 = \left(ML^{-2}T^{-2}\right)\left(L^2T^{-1}\right)^{a_3}(L)^{b_3}\left(ML^{-3}\right)^{c_3} = M^0T^0L^0$$

Balance:

$$M: 1 + c_3 = 0$$
$$L: -2 + 2a_3 + b_3 - 3c_3 = 0$$
$$T: -2 - a_3 = 0$$

Solution:

$$c_3 = -1$$
$$a_3 = -2$$
$$b_3 = 2 - 2a_3 + 3c_3 = 3$$

Substituting values:

$$\pi_3 = \frac{g\Delta\rho L^3}{D^2\rho}$$

Then the expression in terms of dimensionless number becomes:

$$\pi_1 = f(\pi_2)(\pi_3)$$

$$\frac{kL}{D\rho} = f\left(\frac{\mu}{D\rho}\right)\left(\frac{g\Delta\rho L^3}{D^2\rho}\right)$$

π_1 and π_2 represent the Sherwood (Sh) and Schmidt (Sc) numbers, respectively;

$$(Sh) = f(Sc)\left(\frac{g\Delta\rho L^3}{D^2\rho}\right)$$

The third group can be defined as the Grashof number $\left(\frac{g\Delta\rho L^3}{\mu^2}\right)$, dividing π_3/π_2^2

$$\frac{\pi_3}{\pi_2^2} = \frac{\left(\frac{g\Delta\rho L^3}{D^2\rho}\right)}{\left(\frac{\mu}{D\rho}\right)^2} = \frac{g\Delta\rho L^3}{\mu^2}$$

Then, it is obtained the general expression relating Sherwood, Schmidt and Grashof numbers:

$$(Sh) = f(Sc)(Gr) = f(Sc, Gr)$$

10.1.5 Blast Furnace Raceway Dimensions

The blast furnace is the main ironmaking reactor to produce liquid iron from iron ore. The process includes the combustion of a fuel, typically coke with air injected through a number of tuyeres. The combustion reaction between the gas and the solid creates a cavity in front of the tuyere, called the raceway. The combustion rate of the

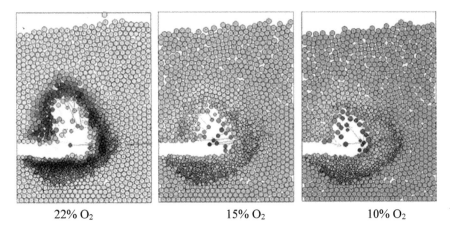

22% O_2	15% O_2	10% O_2

Fig. 10.2 Raceway geometry predictions (After Nogami et al. [8])

solid produces several gases. To replace its consumption the burden descends. The geometry of the raceway; diameter and shape, has a significant role on the burden descent. Elliot et al. [6] reported that the changes of dimensions in the raceway are not due to chemical but due to physical effects, therefore the raceway dimensions can be studied by cold models, neglecting heat and mass transfer effects. Rajneesh et al. [7] summarized the results of raceway dimensions using both hot and cold models. Nogami et al. [8] described the effect of nitrogen enrichment and temperature on the geometry of the raceway by numerical modelling, as shown in Fig. 10.2, corresponding to 800 °C and various levels of nitrogen enrichment. It is shown that the height of the raceway decreases due to nitrogen enrichment.

Gupta el al. [7, 9–11] described the formation of the raceway penetration depth (D_r) as a function of three forces; pressure exerted by the gas, bed weight and frictional forces (wall-particle and inter particle). The pressure force includes inertial and viscous forces. The variables involved derive from the previous forces. The inertial force includes the density of the gas (ρ_g), velocity of the air blast (U_b) and tuyere diameter (D_T). The viscous forces are related to the viscosity of the gas (μ_g) and the particle diameter (d_p). The weight of the bed results from the density of the solids (ρ_s), gravity force (g), height of the bed (H) and bed's porosity (ε). The frictional forces are due to wall-particle interactions (ϕ_w) and interparticle interactions (ϕ_p). The force balance can be expressed as follows:

$$\text{Pressure force} - \text{bed weight} \pm \text{frictional forces} = 0$$

Depending on the velocity of the blast, the volume of the raceway expands or contracts but the frictional forces act in a different way. If velocity decreases, wall friction acts upward resulting in a positive sign in the previous relationship.

The following relationship describes the variables that affect the raceway penetration depth in accordance with Gupta el al. The effective particle diameter (d_{eff}) results from the product $d_p S_f$, where S_f is a shape factor. The effective density of the bed is defined as $\rho_{eff} = \varepsilon\rho_g + (1-\varepsilon)\rho_s$. Wall-particle and interparticle friction depends on the internal angle of friction wall-particle and between particles, respectively. Both friction terms are dimensionless. In addition to the previous variables, Gupta et al. also considered the width of the bed (W). The following expression summarizes the raceway penetration depth as a function of the relevant variables:

$$D_r = f\left(\rho_g, U_b, D_T, \mu_g, d_{eff}, \rho_{eff}, g, H, W, \phi_w, \phi_p\right)$$

Variable	Symbol	Units	Dimensions
Raceway penetration depth	D_r	M	[L]
Tuyere diameter	D_T	M	[L]
Effective particle diameter	d_{eff}	M	[L]
Bed height	H	M	[L]
Bed width	W	M	[L]
Gas density	ρ_g	kg/m^3	$[ML^{-3}]$
Effective density of the bed	ρ_{eff}	kg/m^3	$[ML^{-3}]$
Gas viscosity	μ_g	Pa·s	$[ML^{-1}T^{-1}]$
Air blast velocity	U_b	m/s	$[LT^{-1}]$
Gravity constant	g	m/s^2	$[LT^{-2}]$
Wall-particle friction	ϕ_w	–	–
Interparticle friction	ϕ_p	–	–

This problem involves 12 variables and 3 dimensions; therefore, the number of dimensionless groups would be $n - k = 12 - 3 = 9$. As repeating variables, the effective density of the particle, blast velocity and tuyere diameter are chosen.

The general relationship to be developed has the following form:

$$\pi_1 = f(\pi_2)^{x_1} (\pi_3)^{x_2} (\pi_4)^{x_3} (\pi_5)^{x_4} (\pi_6)^{x_5} (\pi_7)^{x_6} (\pi_8)^{x_7} (\pi_9)^{x_8}$$

The last two groups are dimensionless and there is no need to include them. Also, for simplicity the density and effective viscosity of the solid are represented with the subindex p.

$$\pi_1 = D_r\left(\rho_g\right)^{a_1} (U_b)^{b_1} (D_T)^{c_1}$$

$$\pi_2 = d_p\left(\rho_g\right)^{a_2} (U_b)^{b_2} (D_T)^{c_2}$$

$$\pi_3 = \mu_g\left(\rho_g\right)^{a_3} (U_b)^{b_3} (D_T)^{c_3}$$

$$\pi_4 = \rho_s \left(\rho_g \right)^{a_4} (U_b)^{b_4} (D_T)^{c_4}$$

$$\pi_5 = g \left(\rho_g \right)^{a_5} (U_b)^{b_5} (D_T)^{c_5}$$

$$\pi_6 = H \left(\rho_g \right)^{a_6} (U_b)^{b_6} (D_T)^{c_6}$$

$$\pi_7 = W \left(\rho_g \right)^{a_7} (U_b)^{b_7} (D_T)^{c_7}$$

$$\pi_8 = \phi_w \left(\rho_g \right)^{a_8} (U_b)^{b_8} (D_T)^{c_8}$$

$$\pi_9 = \phi_p \left(\rho_g \right)^{a_9} (U_b)^{b_9} (D_T)^{c_9}$$

First π-group:

$$\pi_1 = D_r \left(\rho_g \right)^{a_1} (U_b)^{b_1} (D_T)^{c_1}$$

$$\pi_1 = (L) \left(ML^{-3} \right)^{a_1} \left(LT^{-1} \right)^{b_1} (L)^{c_1} = M^0 T^0 L^0$$

Balance:

$$
\begin{aligned}
\text{M:} \quad & a_1 = 0 \\
\text{L:} \quad & 1 - 3a_1 + b_1 + c_1 = 0 \\
\text{T:} \quad & -b_1 = 0
\end{aligned}
$$

Solution:

$$
\begin{aligned}
a_1 &= 0 \\
b_1 &= 0 \\
c_1 &= -1
\end{aligned}
$$

Substituting values:

$$\pi_1 = \frac{D_r}{D_T}$$

Second π-group:

$$\pi_2 = d_p \left(\rho_g \right)^{a_2} (U_b)^{b_2} (D_T)^{c_2}$$

$$\pi_2 = (L) \left(ML^{-3} \right)^{a_2} \left(LT^{-1} \right)^{b_2} (L)^{c_2} = M^0 T^0 L^0$$

Balance:

$$M: \quad a_2 = 0$$
$$L: \quad 1 - 3a_2 + b_2 + c_2 = 0$$
$$T: \quad -b_2 = 0$$

Solution:

$$a_2 = 0$$
$$b_2 = 0$$
$$c_2 = -1$$

Substituting values:

$$\pi_2 = \frac{d_p}{D_T}$$

Third π-group:

$$\pi_3 = \mu_g \left(\rho_g\right)^{a_3} \left(U_b\right)^{b_3} \left(D_T\right)^{c_3}$$

$$\pi_3 = \left(ML^{-1}T^{-1}\right)\left(ML^{-3}\right)^{a_3}\left(LT^{-1}\right)^{b_3}(L)^{c_3} = M^0T^0L^0$$

Balance:

$$M: \quad 1 + a_3 = 0$$
$$L: \quad -1 - 3a_3 + b_3 + c_3 = 0$$
$$T: \quad -1 - b_3 = 0$$

Solution:

$$a_3 = -1$$
$$b_3 = -1$$
$$c_3 = -1$$

Substituting values:

$$\pi_3 = \frac{\mu_g}{\rho_g U_b D_T}$$

Fourth π-group:

$$\pi_4 = \rho_s \left(\rho_g\right)^{a_4} \left(U_b\right)^{b_4} \left(D_T\right)^{c_4}$$

$$\pi_4 = \left(ML^{-3}\right)\left(ML^{-3}\right)^{a_4}\left(LT^{-1}\right)^{b_4}(L)^{c_4} = M^0T^0L^0$$

Balance:

$$M: \quad 1 + a_4 = 0$$
$$L: \quad -3 - 3a_4 + b_4 + c_4 = 0$$
$$T: \quad -b_4 = 0$$

Solution:

$$a_4 = -1$$
$$b_4 = 0$$
$$c_4 = 0$$

Substituting values:

$$\pi_4 = \frac{\rho_s}{\rho_g}$$

Fifth π-group:

$$\pi_5 = g\left(\rho_g\right)^{a_5}(U_b)^{b_5}(D_T)^{c_5}$$

$$\pi_5 = \left(LT^{-2}\right)\left(ML^{-3}\right)^{a_5}\left(LT^{-1}\right)^{b_5}(L)^{c_5} = M^0T^0L^0$$

Balance:

$$M: \quad a_5 = 0$$
$$L: \quad 1 - 3a_5 + b_5 + c_5 = 0$$
$$T: \quad -2 - b_5 = 0$$

Solution:

$$: \quad a_5 = 0$$
$$b_5 = -2$$
$$c_5 = 1$$

Substituting values:

$$\pi_5 = \frac{gD_T}{U_b^2}$$

Sixth π-group:

$$\pi_6 = H(\rho_g)^{a_6} (U_b)^{b_6} (D_T)^{c_6}$$

$$\pi_6 = (L)(ML^{-3})^{a_6} (LT^{-1})^{b_6} (L)^{c_6} = M^0 T^0 L^0$$

Balance:

$$M: \quad a_6 = 0$$
$$L: \quad 1 - 3a_6 + b_6 + c_6 = 0$$
$$T: \quad -b_6 = 0$$

Solution:

$$a_6 = 0$$
$$b_6 = 0$$
$$c_6 = -1$$

Substituting values:

$$\pi_6 = \frac{H}{D_T}$$

Seventh π-group:

$$\pi_7 = W(\rho_g)^{a_7} (U_b)^{b_7} (D_T)^{c_7}$$

$$\pi_7 = (L)(ML^{-3})^{a_7} (LT^{-1})^{b_7} (L)^{c_7} = M^0 T^0 L^0$$

Balance:

$$M: \quad a_7 = 0$$
$$L: \quad 1 - 3a_7 + b_7 + c_7 = 0$$
$$T: \quad -b_7 = 0$$

Solution:

$$a_7 = 0$$
$$b_7 = 0$$
$$c_7 = -1$$

Substituting values:

$$\pi_7 = \frac{W}{D_T}$$

Performing the same procedure to the last two π-groups results in $\pi_8 = \phi_w$ and $\pi_9 = \phi_p$.

Then the general expression in terms of dimensionless number becomes:

$$\pi_1 = f(\pi_2)^{x_1}(\pi_3)^{x_2}(\pi_4)^{x_3}(\pi_5)^{x_4}(\pi_6)^{x_5}(\pi_7)^{x_6}(\pi_8)^{x_7}(\pi_9)^{x_8}$$

$$\frac{D_r}{D_T} = C\left(\frac{d_p}{D_T}\right)^{x_1}\left(\frac{\mu_g}{\rho_g U_b D_T}\right)^{x_2}\left(\frac{\rho_s}{\rho_g}\right)^{x_3}\left(\frac{gD_T}{U_b^2}\right)^{x_4}\left(\frac{H}{D_T}\right)^{x_5}\left(\frac{W}{D_T}\right)^{x_6}(\phi_w)^{x_7}(\phi_p)^{x_8}$$

This result is similar to that reported by Gupta et al. except that in π_3 and π_5 they reported d_p instead of D_T, however both expressions are equivalent because they can be obtained by combination of existing groups: $(\pi_2)^{-1} \times (\pi_3)$ and $(\pi_2) \times (\pi_5)$

$$\frac{D_r}{D_T} = C\left(\frac{d_p}{D_T}\right)^{x_1}\left(\frac{\mu_g}{\rho_g U_b d_p}\right)^{x_2}\left(\frac{\rho_s}{\rho_g}\right)^{x_3}\left(\frac{gd_p}{U_b^2}\right)^{x_4}\left(\frac{H}{D_T}\right)^{x_5}\left(\frac{W}{D_T}\right)^{x_6}(\phi_w)^{x_7}(\phi_p)^{x_8}$$

Choosing a different set of repeating variables usually gives a different value of dimensionless groups. For example, choosing the effective density of the solid, gravity constant and tuyere diameter, a similar expression is obtained which differs only in one dimensionless group:

$$\frac{D_r}{D_T} = C\left(\frac{d_p}{D_T}\right)^{x_1}\left(\frac{\mu_g}{\rho_s g^{\frac{1}{2}} D_T^{\frac{3}{2}}}\right)^{x_2}\left(\frac{\rho_s}{\rho_g}\right)^{x_3}\left(\frac{gD_T}{U_b^2}\right)^{x_4}\left(\frac{H}{D_T}\right)^{x_5}\left(\frac{W}{D_T}\right)^{x_6}(\phi_w)^{x_7}(\phi_p)^{x_8}$$

To proof that π_3 is equivalent in both relationships, the value of π_3 in the second expression can be reduced to the first expression doing the following algebraic operation: $(\pi_3)(\pi_2)^{\frac{3}{2}}(\pi_5)^{\frac{1}{2}}$.

10.2   Engineering Problems with Experimental Data

10.2.1   Deformation of an Elastic Ball Striking a Wall

Sonin [12] described in detail the deformation of an elastic ball striking a wall using the experiments conducted by M. Bathe. Although this example is well described, is expanded in order to define a particular relationship with three dimensionless groups.

A spherical dyed elastic ball is dropped on a wall and once it strikes the wall, leaves a circular imprint of diameter (d) and rebounds. This diameter has been found to depend on the ball diameter (D), velocity at the moment of contact (U), modulus of elasticity (E), material's density (ρ) or mass (m) and Poisson's ratio. The Poisson's ratio describes the deformation in two directions in a plane. In most materials this ratio ranges from 0 to 0.5. It is assumed the fluid's properties have a small influence on the ball's motion. The velocity on free fall is defined by the product gt, so gravity is included in the value of the final velocity.

Dimensional analysis: Let's choose the MLT system of dimensions and the π-theorem.

Variable	Symbol	Units	Dimensions
Imprint diameter on the wall	d	m	$[L]$
Ball diameter	D	m	$[L]$
Ball's velocity	U	m/s	$[LT^{-1}]$
Ball's density	ρ	kg/m^3	$[ML^{-3}]$
Ball's modulus of elasticity	E	N/m^2	$[ML^{-1}T^{-2}]$
Poisson's ratio	ν	–	–

There are 6 variables and 3 dimensions. According with the π-theorem this yields 3 dimensionless groups. One of the variables is already dimensionless, say π_3, therefore only two dimensionless groups should be defined. As repeating variables we choose; D, L and ρ Next, we define the values for the three dimensionless groups as follows:

$$\pi_1 = d(D)^{a_1}(U)^{b_1}(\rho)^{c_1}$$

$$\pi_2 = E(D)^{a_2}(U)^{b_2}(\rho)^{c_2}$$

First π-group:

$$\pi_1 = d(D)^{a_1}(U)^{b_1}(\rho)^{c_1}$$

$$\pi_1 = (L)(L)^{a_1}(LT^{-1})^{b_1}(ML^{-3})^{c_1} = M^0T^0L^0$$

Balance:

$$M: -3c_1 = 0$$
$$L: 1 + a_1 + b_1 - 3c_1 = 0$$
$$T: -b_1 = 0$$

Solution:

$$b_1 = 0$$

$$c_1 = 0$$
$$a_1 = -1$$

Substituting values:

$$\pi_1 = \frac{d}{D}$$

Second π-group:

$$\pi_2 = E(D)^{a_2}(U)^{b_2}(\rho)^{c_2}$$

$$\pi_2 = \left(ML^{-1}T^{-2}\right)(L)^{a_2}\left(LT^{-1}\right)^{b_2}\left(ML^{-3}\right)^{c_2} = M^0T^0L^0$$

Balance:

$$M: \quad 1 + c_2 = 0$$
$$L: \quad -1 + a_2 + b_2 - 3c_2 = 0$$
$$T: \quad -2 - b_2 = 0$$

Solution:

$$b_2 = -2$$
$$c_2 = -1$$
$$a_2 = 0$$

Substituting values:

$$\pi_2 = \frac{E}{\rho U^2}$$

Then the expression in terms of dimensionless number becomes:

$$\pi_1 = f(\pi_2)(\pi_3)$$

$$\frac{d}{D} = f\left(\frac{E}{\rho U^2}\right)(v)$$

The general function can be expressed as follows:

$$\frac{d}{D} = C\left(\frac{E}{\rho U^2}\right)^{x_1}(v)^{x_2}$$

Experimental data (Bathe, 2001):

Material	E, Mpa	ρ, kg/m^3	U, m/s	v	π_1	π_2	π_3
Alumina	366,000	3960	43	0.22	0.15	49,986	0.22
	366,000	3960	59	0.22	0.17	26,551	0.22
	366,000	3960	77	0.22	0.19	15,589	0.22
Aluminum	69,000	2705	80	0.33	0.25	3986	0.33
	69,000	2705	126	0.33	0.3	1607	0.33
	69,000	2705	345	0.33	0.45	214	0.33
Rubber	3.93	1060	5	0.47	0.5	148	0.47
	3.93	1060	7	0.47	0.55	76	0.47
	3.93	1060	12	0.47	0.7	26	0.47

The values of the constants are defined by multiple regression analysis. The expression is linearized as follows:

$$\ln\left(\frac{d}{D}\right) = \ln C + x_1\ln\left(\frac{E}{\rho U^2}\right) + x_2\ln(v)$$

Applying the statistical tools of excel[1] the following values are obtained: $\ln C = 0.13$, $x_1 = -0.2$ and $x_2 = 0.03$. The regression coefficient (R^2) is 0.999. Replacing these values in the general expression we obtain the final solution:

$$\frac{d}{D} = 1.36\left(\frac{E}{\rho U^2}\right)^{-0.2}(v)^{0.03}$$

Figure 10.3 illustrates the high accuracy between the calculated and experimental values.

10.2.2 Pressure Drop of a Fluid in a Long Circular Pipe

Fundamental variables: The pressure drop in a long pipe depends on the forces acting on the fluid, geometry of the pipe and fluid properties. If the pressure drop is low the fluid will move slowly. First assume the fluid contains only one phase, pipe is horizontal and not affected by gravity, is Newtonian because its viscosity is not affected by the velocity of the fluid and also is a smooth pipe so that wall friction is low and can be neglected. A highly viscous fluid moves more slowly than a fluid with lower viscosity due to more friction among the layers of the fluid. The velocity of the fluid depends on the ratio between flow rate and cross-sectional area. Fluid´s

---

[1] https://www.youtube.com/watch?v=TkiB1xBnjn4.

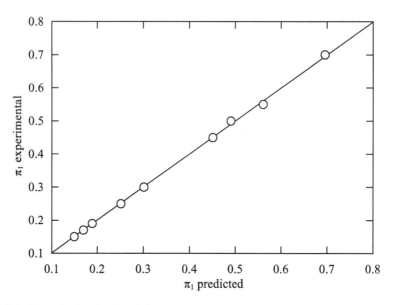

Fig. 10.3 Dimensionless imprinted diameter of a ball striking a wall. After Bathe, Ref. [6]

density measures how light or heavy is the fluid. The Reynolds number directly increases by increasing fluid´s density and therefore a fluid with higher density can become turbulent faster. Density is particularly considered when there is turbulent flow. In laminar flow the inertial forces (represented by density) are low and flow is dominated by viscous forces (represented by viscosity).

The pressure drop in a pipe can be visualized in the same form as ohm's law:

$$\Delta P = QR$$

$$V = IR$$

Considering the previous assumptions, the variables that affect pressure drop (ΔP) are: fluid´s velocity (U), fluid´s density (ρ), fluid´s viscosity (μ), pipe length (l) and pipe diameter (D). A total of 6 variables.

$$\Delta P = f(U, l, D, \rho, \mu)$$

Dimensional analysis: Let's choose the MLT system of dimensions and the π-theorem.

Variable	Symbol	Units	Dimensions
Pressure drop	ΔP	kg/(m × s^2)	$[ML^{-1}T^{-2}]$
Fluid's velocity	U	m/s	$[LT^{-1}]$

(continued)

(continued)

Variable	Symbol	Units	Dimensions
Pipe length	L	m	[L]
Pipe diameter	D	m	[L]
Fluid's viscosity	μ	kg/(m × s)	$[ML^{-1}T^{-1}]$
Fluid's density	ρ	kg/m³	$[ML^{-3}]$

The π-theorem defines three dimensionless groups because there are 6 variables— 3 dimensions = 3 π groups. As non-repeating variables we choose; ΔP, U and l. Next we define the values for the three dimensionless groups as follows:

$$\pi_1 = \Delta P(D)^{a_1}(\mu)^{b_1}(\rho)^{c_1}$$

$$\pi_2 = U(D)^{a_2}(\mu)^{b_2}(\rho)^{c_2}$$

$$\pi_3 = l(D)^{a_3}(\mu)^{b_3}(\rho)^{c_3}$$

$$\pi_1 = f(\pi_2)(\pi_3)$$

First π-group:

$$\pi_1 = \Delta P(D)^{a_1}(\mu)^{b_1}(\rho)^{c_1}$$

$$\pi_1 = \left(ML^{-1}T^{-2}\right)(L)^{a_1}\left(ML^{-1}T^{-1}\right)^{b_1}\left(ML^{-3}\right)^{c_1} = M^0T^0L^0$$

Balance:

$$
\begin{aligned}
M: & \quad 1 + b_1 + c_1 = 0 \\
L: & \quad -1 + a_1 - b_1 - 3c_1 = 0 \\
T: & \quad -2 - b_1 = 0
\end{aligned}
$$

Solution:

$$a_1 = 2, b_1 = -2, c_1 = 1$$

Substitution of these values on the π_1-group yields:

$$\pi_1 = \frac{\rho D^2 \Delta P}{\mu^2}$$

Second π-group:

$$\pi_2 = U(D)^{a_2}(\mu)^{b_2}(\rho)^{c_2}$$

$$\pi_2 = \left(LT^{-1}\right)(L)^{a_2}\left(ML^{-1}T^{-1}\right)^{b_2}\left(ML^{-3}\right)^{c_2} = M^0T^0L^0$$

Balance:

$$
\begin{aligned}
M: &\quad b_2 + c_2 = 0 \\
L: &\quad 1 + a_2 - b_2 - 3c_2 = 0 \\
T: &\quad -1 - b_2 = 0
\end{aligned}
$$

Solution:

$$a_2 = 1, b_2 = -1, c_2 = 1$$

Substitution of these values on the π_2-group yields:

$$\pi_2 = \frac{UD\rho}{\mu}$$

Third π-group:

$$\pi_3 = l(D)^{a_3}(\mu)^{b_3}(\rho)^{c_3}$$

$$\pi_3 = (L)(L)^{a_3}\left(ML^{-1}T^{-1}\right)^{b_3}\left(ML^{-3}\right)^{c_3} = M^0T^0L^0$$

Balance:

$$
\begin{aligned}
M: &\quad b_3 + c_3 = 0 \\
L: &\quad 1 + a_3 - b_3 - 3c_3 = 0 \\
T: &\quad -b_3 = 0
\end{aligned}
$$

Solution:

$$a_3 = -1, b_3 = 0, c_3 = 0$$

Substitution of these values on the π_1-group yields:

$$\pi_3 = \frac{l}{D}$$

The final equation describing the pressure drop is:

Table 10.1 Pressure Taken from Ref. [7]

D, cm	L, m	Q, m³/hr	ΔP, Pa	ρ, kg/m³	μ, kg/(m × s)	U, m/s	π_1	π_2	π_3
1.0	5.0	0.3	4,680	680[a]	$2.92 \cdot 10^{-4}$	1.06	$3.73 \cdot 10^9$	$2.47 \cdot 10^4$	500
1.0	7.0	0.6	22,300	680	$2.92 \cdot 10^{-4}$	2.12	$1.78 \cdot 10^{10}$	$4.94 \cdot 10^4$	700
1.0	9.0	1.0	70,800	680	$2.92 \cdot 10^{-4}$	3.54	$5.65 \cdot 10^{10}$	$8.24 \cdot 10^4$	900
2.0	4.0	1.0	2,080	998[b]	$1.0 \cdot 10^{-3}$	0.88	$8.30 \cdot 10^8$	$1.76 \cdot 10^4$	200
2.0	6.0	2.0	10,500	998	$1.0 \cdot 10^{-3}$	1.77	$4.19 \cdot 10^9$	$3.53 \cdot 10^4$	300
2.0	8.0	3.1	30,400	998	$1.0 \cdot 10^{-3}$	2.74	$1.21 \cdot 10^{10}$	$5.47 \cdot 10^4$	400
3.0	3.0	0.5	540	13,550[c]	$1.56 \cdot 10^{-3}$	0.20	$2.71 \cdot 10^9$	$5.21 \cdot 10^4$	100
3.0	4.0	1.0	2,480	13,550	$1.56 \cdot 10^{-3}$	0.39	$1.24 \cdot 10^{10}$	$1.02 \cdot 10^5$	133.3
3.0	5.0	1.7	9,600	13,550	$1.56 \cdot 10^{-3}$	0.67	$4.81 \cdot 10^{10}$	$1.75 \cdot 10^5$	166.7

Fluids: [a]Gasoline, [b]Water and [c]Mercury

$$\frac{\rho D^2 \Delta P}{\mu^2} = f\left(\frac{UD\rho}{\mu}\right)\left(\frac{l}{D}\right)$$

$$\pi_1 = f(\pi_2)(\pi_3)$$

The second dimensionless group correspond to the Reynolds number and the third group is the aspect ratio. The general function has the following form:

$$\pi_1 = C(\pi_2)^a (\pi_3)^b$$

where: C, a and b are constants.

This is the first step in dimensional analysis. The final form of this function can be obtained with experimental data. The experimental data reported by White [13] is shown in Table 10.1. These results correspond to experiments with 3 different pipe diameters, each case with a different fluid. In each experiment two variables were also changed, therefore the total was 3^2 experiments.

The values of the constants are defined by multiple regression analysis. The expression is linearized as follows:

$$\pi_1 = C(\pi_2)^a (\pi_3)^b$$

$$\ln \pi_1 = \ln C + a \ln(\pi_2) + b \ln(\pi_3)$$

Applying the statistical tools of excel[2] the following values are obtained: $\ln C = -25$, $a = 1.82$ and $b = 0.98$. The regression coefficient (R^2) is 0.99. Replacing this values in the general expression we obtain the final solution:

---

[2]https://www.youtube.com/watch?v=TkiB1xBnjn4.

$$\ln \pi_1 = \ln C + a \ln(\pi_2) + b \ln(\pi_3)$$

$$\pi_1 = C'(\pi_2)^a (\pi_3)^b$$

$$\pi_1 = 0.08(\pi_2)^{1.82} (\pi_3)^{0.98}$$

$$\frac{\rho D^2 \Delta P}{\mu^2} == 0.08 \left(\frac{UD\rho}{\mu} \right)^{1.82} \left(\frac{l}{D} \right)^{0.98}$$

Since the exponent in the aspect ratio is close to unity, the expression can be simplified as follows:

$$\frac{\rho D^3 \Delta P}{l \mu^2} == 0.08 \left(\frac{UD\rho}{\mu} \right)^{1.82}$$

$$\frac{\rho D^3 \Delta P}{l \mu^2} == 0.08 (Re)^{1.82}$$

The result is shown in Fig. 10.4.

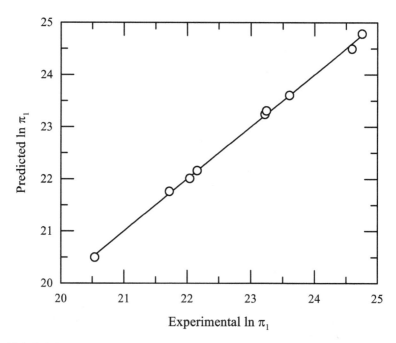

Fig. 10.4 Relationships between dimensionless groups

The book by Zirtes [14] which in addition for being an original source to understand the matrix method is also an excellent source of interesting examples from different fields, with more than 250 examples.

References

1. A. Smolianski, H. Haario, P. Luukka, Numerical study of dynamics of single bubbles and bubble swarms. Appl. Math. Model. **32**(5), 641–659 (2008)
2. R. I. L. Guthrie, Dimensional analysis and reactor design, in *Chapter 3 in Engineering in process metallurgy* (Oxford University Press, UK, 1993), pp. 151–178.
3. D. Mazumdar, J. Evans, Elements of physical modeling, in *Ch. 3 in Modelling of steelmaking processes* (CRC Press, Boca Raton FL, USA, 2010), pp. 99–138.
4. L. Zhang, S. Taniguchi, Fundamentals of inclusion removal from liquid steel by bubble flotation. Int. Matls. Reviews. **45**(2), 59–82 (2000)
5. G.G. Roy, *Transport Processes, Dimensional Analysis and Physical Simulation in Steel Making, in Steelmaking* (PHI learning, New Delhi India, 2012), pp. 175–197
6. J. Elliot, R. Buchanan, J. Wagstaff, Physical conditions in the combustion and smelting zones of a blast furnace. Trans. AIME **194**, 709–717 (1952)
7. S. Rajneesh, S. Sarkar, G.S. Gupta, Prediction of raceway size in blast furnace from two dimensional experimental correlations. ISIJ Int. **44**, 1298–1307 (2004)
8. H. Nogami, H. Yamaoka, K. Takatani, Raceway design for the innovative blast furnace. ISIJ Int. **44**, 2150–2158 (2004)
9. G.S. Gupta, V. Rudolph, Comparison of blast furnace raceway size with theory. ISIJ Int. **46**, 195–201 (2006)
10. G.S. Gupta, Prediction of cavity size in the packed bed systems using new correlations and mathematical model. US Patent US7209871B2, April 24, 2007.
11. S. Sarkar, G.S. Gupta, J.D. Litster, V. Rudolph, E.T. White, S.K. Choudhary, A cold model study of raceway hysteresis. Metall. Mater. Trans. B Process Metall. Mater. Process. Sci. **34**, 183–191 (2003)
12. A.A. Sonin, *The Physical Basis of Dimensional Analysis*, 2nd edn (MIT, Cambridge MA, USA, 2001)
13. F.M. White, *Fluid Dynamics*, 7th edn. (McGraw-Hill, USA, 2011), pp. 329–331
14. T. Zirtes, *Applied Dimensional Analysis and Modeling*, 2nd edn. (UK, Elsevier-BH, 2007), p. 633

Chapter 11
Applications of Dimensional Analysis in Metallurgy

11.1 Slag Foaming

11.1.1 Introduction to the Problem

Slag foaming is a very important phenomenon in steelmaking, either to avoid it or to promote it. Slag foaming consists in an increase of its original height due to the escape of a gas phase. Foaming stability can be increased by:

– Slag chemistry: increasing P_2O_5 (decreases surface tension), decreasing basicity (silica increases viscosity and decreases surface tension), decreasing FeO (increases viscosity). A higher viscosity prevents fast draining of liquid slag from the bubbles.
– Temperature: Decreasing temperature increases the viscosity. Surface tension increases by decreasing temperature and increasing basicity.

Foaming index: The foaming index (Σ) is a measure of the time it takes for the gas phase to go through the slag and escape. Its experimental value can be computed from the ratio between slag height and the superficial gas velocity:

$$\Sigma = \frac{L}{U_g^s} = \frac{\alpha H_f}{\alpha U_g} = \frac{H_f}{U_g}$$

where: α is the gas void fraction, L is the thickness of the slag, H_f is the height of the foaming slag in cm, U_g^s is the superficial velocity of the gas crossing the foaming slag in cm/s and U_g is the actual gas velocity. The gas void fraction ranges from 0.8 to 0.9 [1].

The slag foaming capacity depends on its thermophysical properties; density, viscosity and surface tension. If the crucible diameter is above 3 cm it has no effect on the foaming index [1, 2], also, if the bubble size is small (about 10 mm) its effect on the foaming index can be neglected [3].

© The Author(s), under exclusive license to Springer Nature Singapore Pte Ltd. 2021
A. N. Conejo, *Fundamentals of Dimensional Analysis*,
https://doi.org/10.1007/978-981-16-1602-0_11

$$\Sigma = f(\rho, \mu, \sigma, g)$$

where: Σ is the foaming index (s), ρ is the slag density (kg/m^3), μ is the slag viscosity (kg/m $\times$ s), σ is the slag surface tension (kg/s^2) and g the gravitational constant (m/s^2). The problem involves 5 variables and 3 fundamental dimensions.

Using the method of repeating variables: We choose μ, σ, g as the repeating variables. The number of dimensionless groups is two (n − k = 5 − 3 = 2).

$$\pi_1 = \Sigma(g)^a(\mu)^b(\sigma)^c$$
$$\pi_1 = (T)(LT^{-2})^a(ML^{-1}T^{-1})^b(MT^{-2})^c = M^0L^0T^0$$

The solution to the system of equations gives: $a = 1, b = 1, c = -1$.
Replacing these values in the variables of the first dimensionless group:

$$\pi_1 = \frac{\Sigma g \mu}{\sigma}$$

In a similar way for the second dimensionless group:

$$\pi_2 = \rho(g)^a(\mu)^b(\sigma)^c$$
$$\pi_2 = (ML^{-3})(LT^{-2})^a(ML^{-1}T^{-1})^b(MT^{-2})^c = M^0L^0T^0$$

The solution to the system of equations gives: $a = -1, b = -4, c = 3$.
Replacing these values in the variables of the first dimensionless group:

$$\pi_2 = \frac{\rho \sigma^3}{\mu^4 g}$$
$$\pi_1 = f(\pi_2)$$
$$\frac{\Sigma g \mu}{\sigma} = f\left(\frac{\rho \sigma^3}{\mu^4 g}\right)$$

The previous expression is the **general solution**. The dimensionless group on the right side is the inverse of the Morton number which represents a balance between gravitational, viscous and surface tension forces. The name is given after the American mathematician and hydrodynamics researcher Rose Morton.

A particular solution is obtained with experimental data. The functional relationship involves experimental data on foaming index and the slag thermophysical properties. This information is reported below in two parts; (i) experimental work on foaming index and (ii) slag thermophysical properties.

11.1.2   Experimental Data on Foaming Index from Fruehan et al.: [1–9]

Equation is the basis to measure the foaming index. In this equation, the actual gas velocity is computed from the following relationship:

$$U_g = \frac{Q_g}{A}$$

where: Q_g is the gas flow rate and A is the cross-sectional area of the crucible.

Fruehan and co-workers [1–9] carried out extensive research work to obtain the foaming index for different types of slags using the experimental set-up shown in Fig. 11.1. The height of the foaming slag was measured with an electrical probe method consisting of a stainless-steel rod. These authors defined the foaming slag height as the difference between the top foam surface position and the original slag surface without gas injection. Previous to the experiments the crucible wall was coated with slag. The slag was contained in a crucible of alumina (i.d. = 32– 50 mm and h = 200 mm). Argon was injected through a stainless-steel pipe of 2.1 mm in diameter. A pressure transducer was employed to measure bubble formation frequency. The mean bubble size is obtained from the gas flow rate and the bubble formation frequency. The bubble size was fairly constant, about 10 mm. The amount of slag was about 150 g equivalent to a height of 40 mm in a crucible of 45 mm. The foaming index was taken as the slope of the ratio between foaming height and gas velocity. After the experiments, the pick-up of alumna ranged from 3 to 5%.

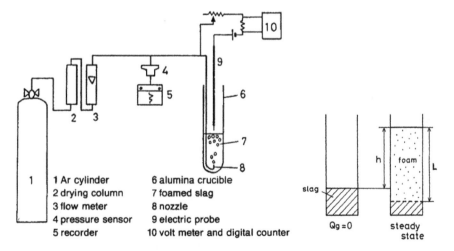

1 Ar cylinder	6 alumina crucible
2 drying column	7 foamed slag
3 flow meter	8 nozzle
4 pressure sensor	9 electric probe
5 recorder	10 volt meter and digital counter

Fig. 11.1 Experimental set-up to obtain foaming height as a function of gas velocity [1]

Slag thermophysical properties:

These properties can be computed as a function of the slag chemical composition using different models, as will be described below.

Density: The molar volume is estimated using the following equation:

$$V = x_1 \overline{V}_1 + x_2 \overline{V}_2 + x_3 \overline{V}_3 + \ldots\ldots$$

where: x is the mole fraction and $\overline{V}$ the partial molar volume of pure components, taken from Shu and Chou [10], listed in Table 11.1. The partial molar volumes at 1500 °C are taken from Mills and Keene [11].

Surface tension: Is computed using the following equation:

$$\sigma = x_1 \overline{\sigma}_1 + x_2 \overline{\sigma}_2 + x_3 \overline{\sigma}_3 + \ldots\ldots$$

where: x is the mole fraction and $\overline{\sigma}$ the surface tension of pure components, taken from Mills and Keene and shown in Table 11.2.

Table 11.1 Density of pure components, $\overline{V} = V_0 + \alpha T$

Component	V_0, cm^3 mol^{-1}	α, cm^3 mol^{-1} K^{-1}	$\overline{V}$, 1500 °C, cm^3 mol^{-1}
CaO	16.49	0.0027	20.7
SiO$_2$	22.54	0.00175	$19.55 + 7.966\ X_{Al_2O_3}$
FeO	13.54	0.00125	15.8
Fe$_2$O$_3$			38.4
MgO	10.27	0.00122	16.1
MnO	12.83	0.00156	15.6
Al$^{(VI)}{}_2$O$_3$	6.87	0.0118	$28.31 + 32\ X_{Al_2O_3} - 31 X^2_{Al_2O_3}$
Al$^{(IV)}{}_2$O$_3$	15.8	0.00875	
CaF$_2$			31.3
P$_2$O$_5$			65.7
Na$_2$O	27.15	0.0033	33.0
K$_2$O	42.62	0.00518	51.8
Li$_2$O	15.82	0.0019	
TiO$_2$			24

Table 11.2 Partial molar surface tension at 1500 °C. From Ref. [11]

Oxide	Al$_2$O$_3$	CaO	FeO	MgO	MnO	SiO$_2$	TiO$_2$
$\overline{\sigma}$, mN m^{-1}	655	625	645	635	645	260	360

Viscosity: Is calculated using Urbain's model. Urbain reported the first version in 1981 and a second version in 1987 [12]. Urbain classified the slag components in three groups: glass or network formers (G), modifier cations (M) and amphoteric (A), as follows:

$$X_G = X_{SiO_2} + X_{P_2O_5}$$
$$X_M = X_{CaO} + X_{MgO} + X_{FeO} + X_{MnO} + X_{CrO} + X_{NiO}$$
$$+ X_{Na_2O} + X_{K_2O} + X_{Li_2O} + 2X_{TiO_2} + 2X_{ZrO_2} + 3X_{CaF_2}$$
$$X_A = X_{Al_2O_3} + X_{Fe_2O_3} + X_{B_2O_3} + X_{Cr_2O_3}$$

This gives the mole fraction of a hypothetical ternary system: TO_2-A_2O_3-MO. Two normalized parameters, are defined as follows:

$$X_G^* = \frac{X_G}{X_G + X_M + X_A}$$
$$\alpha = \frac{X_M}{X_M + X_A}$$

The calculation of B depends on X_G^* and α, as follows:

$$B = B_0 + B_1 \cdot X_G^* + B_2 \cdot \left(X_G^*\right)^2 + B_3 \cdot \left(X_G^*\right)^3$$
$$B_i = a(i) + b(i)\alpha + c(i) \cdot (\alpha)^2$$

The final B value is obtained from the expression:

$$B_{global} = \frac{X_{CaO}B_{CaO} + X_{MgO}B_{MgO} + X_{MnO}B_{MnO}}{X_{CaO} + X_{MgO} + X_{MnO}}$$

Table 11.3 summarizes the values of the constants a, b and c. Term A is calculated from the following relationship:

$$-\ln A = 0.29\, B_{global} + 11.57$$

Table 11.3 Parameters a, b and c. Ternary system SiO_2-Al_2O_3-MO, where M represents Ca, Mg or Mn. From Ref. [12]

i	a(i)	b(i)			c (i)		
	All	Mg	Ca	Mn	Mg	Ca	Mn
0	13.2	15.9	41.5	20.0	−18.6	−45	−25.6
1	30.5	−54.1	−117.2	26	33	130	−56
2	−40.4	138	232.1	−110.3	−112	−298.6	186.2
3	60.8	−99.8	−156.4	64.3	97.6	213.6	−104.6

Finally, viscosity is computed as follows:

$$\mu = AT \exp \left(\frac{1000 \, B_{global}}{T} \right)$$

where: A, B are viscosity parameters, T is the absolute temperature (K) and μ is the dynamic viscosity (Poise).

The predicted thermophysical properties and the experimental data reported by Fruehan and co-workers [1–9]. is summarized in Tables 11.4 and 11.5.

The general relationship between the two dimensionless groups is as follows:

$$\pi_1 = b(\pi_2)^m$$

Table 11.4 Slag chemical composition

	CaO	SiO$_2$	FeO	Al$_2$O$_3$	MgO	CaF$_2$	PbO
1	45.5	45.5	5	4			
2	44.3	44.3	7.5	4			
3	43.0	43.0	10	4			
4	41.8	41.8	12.5	4			
5	40.5	40.5	15	4			
6	53.3	42.7	0	4			
7	51.7	41.3	3	4			
8	50.6	40.4	5	4			
9	49.2	39.3	7.5	4			
10	47.8	38.2	10	4			
11	45.0	36.0	15	4			
12	45	30	0	15	10		
13	44.6	29.7	1	14.9	9.9		
14	43.7	29.1	3	14.6	9.7		
15	42.3	28.2	6	14.1	9.4		
16	41.0	27.3	9	13.7	9.1		
17	45	5		50		0	
18	37.5	4.2		41.7		16.7	
19	35.6	4.0		39.5		21	
20	33.8	3.8		37.5		25	
21		55					45
22		55					45
23		55					45
24		55					45
25	38.2	31.8	30				

Table 11.5 Foaming index and calculated π-groups

	Σ	σ	ρ	μ	π_1	π_2	$\log \pi_1$	$\log \pi_2$
	–	mN/m	g/cm^3	Ns/m^2	–	–	–	–
1	1.4	461.8	2.753	0.380	11.301	1325.5	1.053	3.1
2	1.2	465.9	2.797	0.364	9.197	1642.5	0.964	3.2
3	0.9	470.1	2.841	0.35	6.573	2004.9	0.818	3.3
4	0.8	474.4	2.886	0.336	5.558	2464.4	0.745	3.4
5	0.75	478.6	2.930	0.325	4.996	2934.8	0.699	3.5
6	0.6	473.3	2.693	0.304	3.781	3407.9	0.578	3.5
7	1.3	477.6	2.743	0.293	7.824	4133.1	0.893	3.6
8	0.9	480.6	2.777	0.287	5.272	4631.6	0.722	3.7
9	0.8	484.3	2.82	0.279	4.521	5389.0	0.655	3.7
10	0.8	488.1	2.862	0.272	4.373	6198.0	0.641	3.8
11	0.7	495.7	2.949	0.259	3.588	8137.0	0.555	3.9
12	2.9	502.3	2.761	0.528	29.905	458.9	1.476	2.7
13	2	503.6	2.78	0.521	20.298	491.2	1.307	2.7
14	1.6	506.5	2.817	0.509	15.773	555.9	1.198	2.7
15	1.3	510.5	2.872	0.491	12.266	670.2	1.089	2.8
16	1.2	514.7	2.928	0.475	10.864	799.4	1.036	2.9
17	1	613.6	2.816	0.707	11.303	265.4	1.053	
18	0.8	613.4	2.816	0.707	9.046	265.2	0.956	
19	0.5	613.3	2.816	0.707	5.654	265.0	0.752	
20	0.72	260	2.184					
21	1.3	260	2.184					
22	3.37	260	2.184					
23	6.8	260	2.184					
24	1.1	511.4	3.182	0.155	3.271	75160.0	0.515	4.9

In order to obtain a linear relationship between π_1 and π_2 to define the values of the constant b and the exponent m, the equation is expressed in logarithmic form:

$$\log \pi_1 = m \log(\pi_2) + \log b$$

Figure 11.2 is the relationship between $\log(\pi_1)$ and $\log(\pi_2)$. From this figure the value of $\log b = 4.8907$ and $m = -1.7193$. These values are different than those reported by Fruehan et al. [3] because we have considered a smaller number of slags due to the lack of information and also because the values on viscosity applying Urbain's model are slightly different.

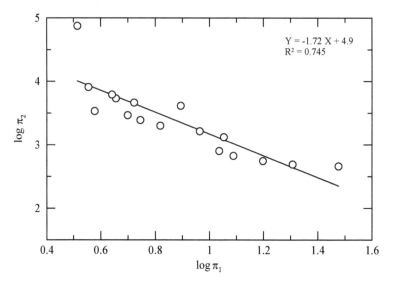

Fig. 11.2 Relationship between dimensionless groups

$$\log \pi_2 = -1.72 \log(\pi_1) + \log b$$
$$\log b = 4.8907 \quad \therefore b = 10^{4.8907} = 77,750$$
$$\pi_2 = b(\pi_1)^m$$

Then, the **particular solution** is:

$$\frac{\rho \sigma^3}{\mu^4 g} = 77,750 \left(\frac{\Sigma g \mu}{\sigma} \right)^{-1.72}$$

Re-arranging the previous expression to define the dependent variable, foaming index in terms of slag properties.

$$\Sigma = 268 \frac{\mu^{1.6}}{\rho^{0.58} \sigma^{0.75}}$$

This result indicates that slag viscosity has a stronger effect on the foaming index as compared with density and surface tension.

The previous equation predicts slightly lower values for the foaming index when compared with the expression reported by Jiang and Fruehan [3]:

$$\Sigma = 115 \frac{\mu}{\sqrt{\rho \sigma}}$$

On a practical basis, it is more useful to define foaming in terms of foaming height:

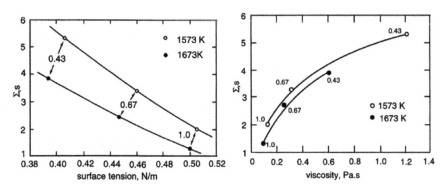

Fig. 11.3 Effect of surface tension, viscosity and temperature on the foaming index. From Ref. [13]

$$H = 268 \frac{\mu^{1.6}}{\rho^{0.58}\sigma^{0.75}} U_g$$

It is important to point out that this result was obtained from estimated values of slag properties.

Gaskell et al. [13, 14] repeated the same dimensional analysis using experimental measurements of densities, viscosities and surface tension in CaO-SiO_2-30%FeO slags. This analysis is limited because only 7 slags were used in obtaining the following relationship.

$$\Sigma = 100 \frac{\mu^{0.54}}{\rho^{0.39}\sigma^{0.15}}$$

As the temperature increases, foaming index, surface tension and viscosity, all decrease, as shown in Fig. 11.3. It is observed that a decrease in the surface tension from 0.5 to 0.4 N/m causes a 2.5-fold increase in the foaming index at 1573 K.

11.1.3 Experimental Data Including the Effect of Bubble Diameter from Fruehan et al. [4, 5]

Initially the effect of bubble diameter as an additional variable was ruled out by Ito and Fruehan [1, 2] because they used large bubbles, in the order of 10 mm. The experimental work by Zhang and Fruehan [4] with bubbles from 1 to 5 mm indicated that bubble diameter has an important effect on foaming.

Dimensional analysis including bubble diameter:

$$\Sigma = f(\rho, \mu, \sigma, g, D_b)$$

The number of dimensionless groups is three ($n - k = 6 - 3 = 3$). Using the π-method of repeating variables, we choose g, μ and σ as the repeating variables.

$$\pi_1 = \Sigma(g)^a(\mu)^b(\sigma)^c$$
$$\pi_2 = \rho(g)^a(\mu)^b(\sigma)^c$$
$$\pi_3 = D_b(g)^a(\mu)^b(\sigma)^c$$

The first two dimensionless groups are identical to the previous analysis [3]. The form of these dimensionless numbers is:

$$\pi_1 = \frac{\Sigma g \mu}{\sigma}$$

$$\pi_2 = \left(\frac{\rho\sigma^3}{\mu^4 g}\right)$$

For the third dimensionless group:

$$\pi_3 = (L)\left(LT^{-2}\right)^a\left(ML^{-1}T^{-1}\right)^b\left(MT^{-2}\right)^c = M^0 L^0 T^0$$

The solution to the system of equations gives: $a = 1$, $b = 2$, $c = -2$. Replacing these values in the third dimensionless group:

$$\pi_3 = \left(\frac{D_b g \mu^2}{\sigma^2}\right)$$

The general solution is:

$$\pi_1 = C\,(\pi_2)^{\alpha_1}(\pi_3)^\beta$$

$$\frac{\Sigma g \mu}{\sigma} = C\left(\frac{\rho\sigma^3}{\mu^4 g}\right)^{\alpha_1}\left(\frac{D_b g \mu^2}{\sigma^2}\right)^\beta$$

Equivalent to:

$$\frac{\Sigma g \mu}{\sigma} = C\left(\frac{\mu^4 g}{\rho\sigma^3}\right)^{\alpha}\left(\frac{D_b g \mu^2}{\sigma^2}\right)^\beta$$

The π_2-group is the Morton number (Mo). Zhang and Fruehan [4] reported a variation of the π_3-group, in the form of the Archimedes number. The equation involves a relationship between the foaming index and the bubble diameter. Zhang and Fruehan did some measurements on the bubble diameter and measured the foaming index and then compared the experimental value on bubble diameter with the Ruff's model [15].

Due to satisfactory agreement in most of their experimental data they used that model to quantify the bubble diameter, arriving to the following equation:

$$\Sigma = 115 \frac{\mu^{1.2}}{D^{0.9} \rho \sigma^{0.2}}$$

The previous equation indicates that foaming is inversely proportional to the bubble diameter.

$$H = 115 \frac{\mu^{1.2}}{D^{0.9} \rho \sigma^{0.2}} U_g$$

11.1.4 Experimental Data Including the Effect of Bubble Diameter from Ghag et al. [16, 17]

Ghag et al. [16, 17] reported only two dimensionless groups by regrouping the variables ρg and additionally, instead of the absolute value on surface tension, they used the decrease in surface tension (surface tension depression) due to the presence of surfactants ($\Delta \sigma$) on the basis that pure liquids do not produce foams.

$$\Sigma = f(\rho g, \mu, \Delta \sigma, D_b)$$

Dimensional analysis: Let's choose the MLT system of dimensions and the π-theorem.

Variable	Symbol	Units	Dimensions
Bubble diameter	D	m	$[L]$
Product	ρg	kg/(m² × s²)	$[ML^{-2}T^{-2}]$
Viscosity	μ	kg/m × s	$[ML^{-1}T^{-1}]$
Surface tension	$\Delta \sigma$	N/m	$[MT^{-2}]$
Foaming index	Σ	s	$[T]$

The π-theorem defines two dimensionless groups because there are 5 variables − 3 dimensions = 2π groups. As non-repeating variables we choose; $\Delta \sigma$, μ and D_b. Next, we define the values for the two dimensionless groups as follows:

$$\pi_1 = \Sigma (\Delta \sigma)^{a_1} (\mu)^{b_1} (D_b)^{c_1}$$
$$\pi_2 = \rho g (\Delta \sigma)^{a_2} (\mu)^{b_2} (D_b)^{c_2}$$
$$\pi_1 = f(\pi_2)$$

First π-group:

$$\pi_1 = \Sigma(\Delta\sigma)^{a_1}(\mu)^{b_1}(D_b)^{c_1}$$
$$\pi_1 = (T)\left(MT^{-2}\right)^{a_1}\left(ML^{-1}T^{-1}\right)^{b_1}(L)^{c_1} = M^0T^0L^0$$

Balance:
  M: $a_1 + b_1 = 0$
  L: $-b_1 + c_1 = 0$
  T: $1 - 2a_1 - b_1 = 0$

Solution:
  $a_1 = -b_1$
  $b_1 = c_1 \therefore a_1 = -b_1 = -c_1$
  $1 - 2a_1 + a_1 = 1 - a_1 = 0 \therefore a_1 = 1$
  $a_1 = 1, b_1 = -1, c_1 = -1$

Substitution of these values on the π_1-group yields:

$$\pi_1 = \frac{\Sigma\Delta\sigma}{\mu D_b}$$

Second π-group:

$$\pi_2 = \rho g(\Delta\sigma)^{a_2}(\mu)^{b_2}(D_b)^{c_2}$$
$$\pi_2 = \left(ML^{-2}T^{-2}\right)\left(MT^{-2}\right)^{a_2}\left(ML^{-1}T^{-1}\right)^{b_2}(L)^{c_2} = M^0T^0L^0$$

Balance:
  M: $1 + a_2 + b_2 = 0$
  L : $-2 - b_2 + c_2 = 0$
  T : $-2 - 2a_2 - b_2 = 0$

Solution:
  $L_1 : 2a_2 + 2b_2 = -2$
  $L_3 : -2a_2 - b_2 = 2 \therefore b_2 = 0$
  $L_1 : a_2 = -1 - b_2 = -1$
  $L_2 : c_2 = 2 + b_2 \therefore c_2 = 2$
  $a_2 = -1, b_2 = 0, c_2 = 2$

Substitution of these values on the π_2-group yields:

$$\pi_2 = \frac{(\rho g)D_b^2}{\Delta\sigma} = \frac{\Delta\sigma}{(\rho g)D_b^2}$$

$$\frac{\Sigma\Delta\sigma}{\mu D_b} = f\left(\frac{\Delta\sigma}{(\rho g)D_b^2}\right)$$

The final relationship to define the foaming index has the following form:

$$\Sigma = f\left(\frac{\mu D_b}{\Delta\sigma}\right)^x\left(\frac{\Delta\sigma}{(\rho g)D_b^2}\right)^y = f\left(\frac{\mu\Delta\sigma}{(\rho g)D_b}\right)^z$$

Experimental data:

In order to define the particular solution to the general expression it is necessary to measure the foaming index, liquid properties (density, surface tension depression, viscosity) and bubble diameter. Experiments were made with water-glycerol solutions [16]. A surfactant was employed to modify the surface tension depression. The measurement of $\Delta\sigma$ was made using the Wilhemy plate method and applying the Langmuir and Frumkin isotherms to process the data. The bubble size was measured using a video camera, taking an average of 200 bubbles for each experiment. Foaming was generated injecting air. The foam height was also measured with the video camera. The foaming index was obtained from the ratio height/superficial velocity. The results are shown in Table 11.6 and the data processed showing the dimensionless groups shown in Table 11.7.

Figure 11.4 shows the linear relationship between π_1 and π_2 in log scale.

The linear relationship between π_1 and π_2 is defined by the following equation:

$$\log\pi_1 = 2.32 \log(\pi_2) + \log b$$
$$\log b = 6.3014 \therefore b = 10^{6.3014} = 2 \times 10^6$$

$$\pi_1 = b(\pi_2)^m$$
$$\pi_1 = 2 \times 10^6(\pi_2)^{2.32}$$

$$\frac{\Sigma\Delta\sigma}{\mu D_b} = 2 \times 10^6\left(\frac{\Delta\sigma}{(\rho g)D_b^2}\right)^{2.32}$$

$$\Sigma = 2 \times 10^6\mu\frac{\Delta\sigma^{1.32}}{(\rho g)^{2.32}D_b^{3.64}}$$

This result indicates that bubble diameter has the strongest effect. It should be noticed that the bubble size was small, in the range from 1.5 to 2.2 mm. As the bubble size increases, its effect becomes smaller. The previous equation was developed for a water-glycerol system. It was applied to predict the foaming index of metallurgical slags. The predicted values on foaming index are shown in Table 11.8.

As can be shown from the previous table, the predicted values are extremely large; three to four orders of magnitude higher than observed values [17]. The authors attributed this behavior to the non-equilibrium nature of the foaming process and concluded that the model cannot be applied to steelmaking slags. Comparing the previous equation with that reported by Zhang and Fruehan [4] it is observed two major differences in the value of the exponents for $\Delta\sigma$ and particularly on the bubble diameter. Foaming is a non-equilibrium phenomenon because three process occur

Table 11.6 Experimental data on foaming and index and surface tension depression

Glycerol wt%	$\Delta\sigma$ (mN/m)	ρ (kg/m^3)	μ (mPa s)	Bubble size (mm)	$\sum$ (s)
78%	0	1204	46	1.56	1.04
	1	1204	46	1.5	26.8
	2	1204	46	1.47	149
	3.5	1204	46	1.54	452
	5	1204	46	1.51	702
85%	0	1223	102	1.78	0.64
	1	1223	102	1.71	20.4
	2	1223	102	1.58	423
	3.5	1223	102	1.63	627
	5	1223	102	1.67	749
	10	1223	102	1.56	1132
90%	0	1238	225	2.1	0.3
	1	1238	225	2.06	100.9
	2	1238	225	1.96	391
	3.5	1238	225	1.89	1064
	5	1238	225	1.93	1408
	10	1238	225	1.94	1435
95%	0	1251	522	2.23	0.54
	0.5	1251	522	2.14	46.13
	1	1251	522	2.21	438
	2	1251	522	2.03	906
	3.5	1251	522	2.11	1899
	5	1251	522	2.14	4077
	10	1251	522	2.04	4882

simultaneously; generation of bubbles, coalescence and rupture. A modified version of the previous model replaces $\Delta\sigma$ for the Marangoni dilatational modulus (E_M). E_M represents the maximum elasticity of the film and is related to the surface tension depression, as follows:

$$E_M = \frac{d\Delta\sigma}{d \ln \Gamma_i}$$

where: Γ_i is the Gibbs relative surface excess of the surface-active species.

The surface tension depression can be defined as a function of Γ_i applying either the Langmuir or Frumkin adsorption isotherms. From these relationships, Γ_i can be computed. The final expressions using Langmuir or Frumkin isotherms are:

Table 11.7 Dimensionless groups

Glycerol wt%	$\Delta\sigma$ (mN/m)	π_1	$\pi_2(\times 10^{-5})$	$\log \pi_1$	$\log \pi_2$
78%	0	0.39	3.77	−0.41	−4.42
	1	4.41	7.84	0.64	−4.11
	2	22.33	12.5	1.35	−3.90
	3.5	50.53	18.6	1.70	−3.73
	5	0.12	2.85	−0.93	−4.54
85%	0	5.25	6.68	0.72	−4.17
	1	13.20	11.0	1.12	−3.96
	2	21.99	15.0	1.34	−3.83
	3.5	71.14	34.3	1.85	−3.46
	5	0.22	1.94	−0.66	−4.71
	10	1.77	4.29	0.25	−4.37
90%	0	8.76	8.08	0.94	−4.09
	1	16.21	11.1	1.21	−3.96
	2	32.88	21.9	1.52	−3.66
	3.5	0.02	0.891	−1.69	−5.05
	5	0.38	1.67	−0.42	−4.78
	10	1.71	3.96	0.23	−4.40
95%	0	6.03	6.41	0.78	−4.19
	0.5	18.25	8.91	1.26	−4.05
	1	45.85	19.6	1.66	−3.71
	2	0.39	3.77	−0.41	−4.42
	3.5	4.41	7.84	0.64	−4.11
	5	22.33	12.5	1.35	−3.90
	10	50.53	18.6	1.70	−3.73

$$E_M = \frac{RT\,\Gamma_{i,\infty}\Gamma_i}{(\Gamma_{i,\infty} - \Gamma_i)}$$

$$E_M = \frac{RT\,\Gamma_{i,\infty}\Gamma_i}{(\Gamma_{i,\infty} - \Gamma_i)} - 2a'\left(\frac{\Gamma_i}{\Gamma_{i,\infty}}\right)^2$$

where: $\Gamma_{i,\infty}$ is the Gibbs relative surface excess at saturation of the surface and a' is related to the enthalpy of adsorption (zero for the Langmuir isotherm).

The values for $\Gamma_{i,\alpha}$ and a' were measured experimentally with water-glycerol solutions. The results are shown in Table 11.9. It can be observed that in the range from 78 to 85% water-glycerol the solution obeys the Langmuir adsorption isotherm and from 90 to 95%, obeys the Frumkin isotherm.

Since dimensions are the same, replacement of $\Delta\sigma$ by E_M leads to the following dimensionless groups:

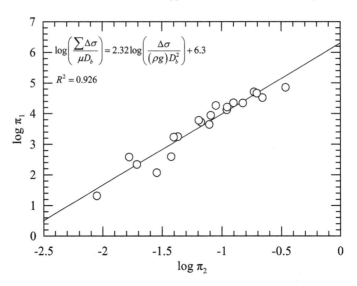

Fig. 11.4 Relationship between π_1 and π_2 in the study of foaming. Adapted from [16]

Table 11.8 Calculated properties and gas residence times of CaO-SiO$_2$ slags at 1873 K (K. C. Mills, IMR, 1987. Taken from Ref. [16])

Basicity	Density (kg/m^3)	Viscosity (Pa s)	$\Delta\sigma$ (mN/m)	D_b (mm)	Σ (s)
0.67	2545	0.86	219	5	4798
0.8	2576	0.55	203	5	2645
1	2613	0.35	183	5	1397
1.35	2655	0.23	155	5	698

Table 11.9 Fitted parameters for Langmuir and Frumkin isotherms for water-glycerol-SDBS solutions at 20 °C, (from Ref. [16])

Glycerol Wt%	$\Gamma_{i,\alpha}$ ($\times 10^{-6}$ mol/m^2)	$C_{i,L}$ ($\times 10^{-4}$ M)	a' (mN/m)
78	4.21	2.69	0
85	4.38	3.30	0
90	4.90	4.05	11.94
95	5.26	4.84	21.95

$$\frac{\Sigma \cdot E_M}{\mu D_b} = f\left(\frac{E_M}{(\rho g)D_b^2}\right)$$

The experimental data reported by Ghag et al. is shown in Table 11.10.
Figure 11.5 shows the linear relationship between π_1 and π_2 in log scale.
The linear relationship between π_1 and π_2 is defined by the following equation:

Table 11.10 Experimental data

Glycerol wt%	$\Delta\sigma$ (N/m)	ρ (kg/m^3)	μ (Pa s)	D_b (m)	$\sum$ (s)	E_M ($\times 10^{-3}$) (N/m)	Γ_i ($\times 10^{-6}$)	Γ_∞ ($\times 10^{-6}$)
78%	0	1204	0.046	0.00156	1.04	0	0	4.21
	0.001	1204	0.046	0.0015	26.8	1.05	0.391	4.21
	0.002	1204	0.046	0.00147	149	2.21	0.746	4.21
	0.0035	1204	0.046	0.00154	452	4.17	1.22	4.21
	0.005	1204	0.046	0.00151	702	6.44	1.62	4.21
85%	0	1223	0.102	0.00178	0.64	0	0	4.38
	0.001	1223	0.102	0.00171	20.4	1.05	0.392	4.38
	0.002	1223	0.102	0.00158	423	2.20	0.749	4.38
	0.0035	1223	0.102	0.00163	627	4.14	1.22	4.38
	0.005	1223	0.102	0.00167	749	6.38	1.64	4.38
	0.01	1223	0.102	0.00156	1132	1.66	2.66	4.38
90%	0	1238	0.225	0.0021	0.3	0	0	4.90
	0.001	1238	0.225	0.00206	100.9	0.889	0.394	4.90
	0.002	1238	0.225	0.00196	391	1.61	0.756	4.90
	0.0035	1238	0.225	0.00189	1064	2.52	1.25	4.90
	0.005	1238	0.225	0.00193	1408	3.41	1.68	4.90
	0.01	1238	0.225	0.00194	1435	7.97	2.78	4.90
95%	0	1251	0.522	0.00223	0.54	0	0	5.26
	0.0005	1251	0.522	0.00214	46.13	0.446	0.201	5.26
	0.001	1251	0.522	0.00221	438	0.793	0.395	5.26
	0.002	1251	0.522	0.00203	906	1.25	0.760	5.26
	0.0035	1251	0.522	0.00211	1899	1.52	1.26	5.26
	0.005	1251	0.522	0.00214	4077	1.53	1.70	5.26
	0.01	1251	0.522	0.00204	4882	2.26	2.85	5.26

$$\log \pi_1 = 1.91 \log(\pi_2) + \log b$$
$$\log b = 4.567 \therefore b = 10^{4.567} = 8.194 \times 10^5$$
$$\pi_1 = b(\pi_2)^m$$
$$\pi_1 = 8.194 \times 10^5 (\pi_2)^{1.91}$$
$$\frac{\sum \Delta E_M}{\mu D_b} = 8.194 \times 10^5 \left(\frac{E_M}{(\rho g) D_b^2} \right)^{1.91}$$
$$\Sigma = 8.194 \times 10^5 \mu \frac{E_M^{0.91}}{(\rho g)^{1.91} D_b^{2.82}}$$

The regression coefficient R^2 = 0.81, clearly lower than that for the first model, therefore the modified model is not an improvement. A third model by the same

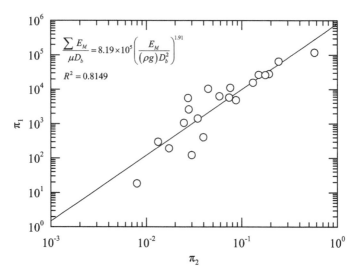

Fig. 11.5 Relationship between π_1 and π_2 in the study of foaming. Adapted from [16]

authors [17] introduces the adsorption time in the analysis. The results shown cannot confirm a reliable application to metallurgical slags.

11.1.5 Experimental Database from Pilon et al. [18–22]

Pilon et al. [18–22] carried out dimensional analysis to describe foam thickness. They followed two approaches. First, they developed the governing differential equation of the foaming process and making it dimensionless defined the dimensionless numbers. In a second approach they started defining the relevant variables and followed a conventional method of dimensional analysis to define the relationship. Both methods are summarized below and is shown that they arrive at similar results. They applied their equations to a large database, including previous reported experimental data from slags and aqueous systems. It is so far, the most complete analysis to unify all experimental work on foaming phenomena.

Dimensional analysis from differential equation:

The governing differential equation defines the rate of change of foam thickness as a function of liquid drainage and gas leaving the foam. An important element in the analysis was to describe the foaming process not in terms of the superficial gas velocity but a reduced gas velocity. The reduced gas velocity subtracts the minimum gas velocity because below that value foaming doesn't occur. The reported differential equation is the following:

$$\frac{dH}{dt} = \frac{d(z_2 - z_2)}{dt} = \frac{U}{X_g^{z_{2,t}}} - h_g^{z_{2,t}} - \frac{h_g^{z_{1,t}} X_g^{z_{1,t}}}{1 - X_g^{z_{1,t}}}$$

Thee first term on the right side represents the increase of the foam thickness due to incoming gas at the inlet position z_2 and the last two terms correspond to foaming suppression due to liquid drainage and gas released at position z_1, as shown schematically in Fig. 11.6. An expression to compute h_g (z, t) was also developed which depends on the process variables; fluid's density, fluid's viscosity, fluid's surface tension, average bubble radius, and gravity force.

The differential equation was transformed in dimensionless form using reference values, for example; H_∞ is the steady-state foam thickness, r_0 is the bubble average radius and j_m is the superficial gas velocity at the onset of foaming. These results in the two following dimensionless groups:

$$\frac{dH}{dt} = \frac{d(z_2 - z_1)}{dt} = \frac{j}{X_g^{z_{2,t}}} - h_g^{z_{2,t}} - \frac{h_g^{z_{1,t}} X_g^{z_{1,t}}}{1 - X_g^{z_{1,t}}}$$

$$\pi_1 = \frac{\mu H_\infty (j - j_m)}{\sigma r_0}$$

$$\pi_2 = \frac{\rho g r_0^2}{\mu (j - j_m)}$$

Fig. 11.6 Foaming model developed by Pilon et al. (Taken from Ref. [18])

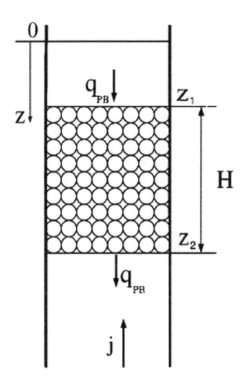

The authors interpreted the first group as a ratio of gravitational to viscous forces on an average bubble of radius r_0 with a velocity $(j - j_m)$. The second group was interpreted as a product of two ratios; one is the ratio of viscous to surface tension forces and the second, a ratio for the steady state foam characteristic height to the bubble average radius.

$$\pi_1 = \frac{\mu(j - j_m)}{\sigma} \times \frac{H_\infty}{r_0} = Ca\frac{H_\infty}{r_0}$$

$$\pi_2 = \frac{\rho g r_0^3}{\mu(j - j_m)r_0} = \frac{\frac{\rho(j-j_m)r_0}{\mu}}{\frac{(j-j_m)^2}{gr_0}} = \frac{Re}{Fr}$$

where: Ca is the capillary number representing a ratio between viscous drag forces and surface tension forces across an interface between a liquid and a gas. Re is the Reynolds number and Fr is the Froude number.

The general dimensionless form of the functional relationship is as follows:

$$\frac{\mu H_\infty (j - j_m)}{\sigma r_0} = C\left(\frac{\rho g r_0^2}{\mu(j - j_m)}\right)^n$$

$$Ca\frac{H_\infty}{r_0} = C\left(\frac{Re}{Fr}\right)^n$$

where: C and n are constants to be defined by regression analysis.

Experimental data:

Pilon et al. [18–20] carried out an extensive review of available experimental data. The slags included in the database correspond to steelmaking slags (system CaO-SiO$_2$-FeO) with FeO in the range from 0 to 30% and ladle furnace slags (system CaO-SiO$_2$-Al$_2$O$_3$) with alumina in the range from 10 to 15%. The variables required are: superficial velocity (j), minimum superficial gas velocity at the onset of foaming (j$_m$), steady-state foam thickness (H$_\infty$), bubble radius (r$_0$), viscosity (μ), density (ρ) and surface tension (σ). The superficial velocity is calculated from the ratio between gas flow rate and the cross-sectional area of the reactor.

A phenomenological equation [19] was developed to compute the minimum superficial gas velocity at the onset of foaming (j$_m$). It is based on the ratio of drainage time (t$_d$) to collision duration time (t$_c$). If t$_c$ > t$_d$ bubble coalescence occurs, otherwise there is bubble bouncing.

The general expression is defined as follows:

$$j_m = U_\infty f(r^*)\alpha_m(1 - \alpha_m)^{n-1}$$

where: U_∞ is the terminal velocity in m^2/s, α_m is the gas fraction in the foam, $f(r^*)$ is a function of the dimensionless radius r^*, both $f(r^*)$ and the parameter n depend on the flow regime. For viscous flow in the range of application of the Stokes law the $f(r^*)$ is the unity, and n has a value of 3.

$$j_m = U_\infty \alpha_m (1 - \alpha_m)^2$$

$$U_\infty = \frac{2}{9} \frac{\Delta \rho g r^2}{\mu_l}$$

In order to account for the gas fraction in the foam, there is a distinction between foam height (h) and steady-state foaming thickness (H_∞). Ito and Fruehan [1] reported that the value of the gas fraction in a foaming slag is almost constant, from 0.7 to 0.9. Pilon et al. [18–20] assumed a value of 0.8 indicating that any value in the range from 0.7 to 0.9 has a small effect on the regression analysis of experimental data.

$$h = H_\infty \alpha_m$$

Prof. Laurent Pilon [23] from UCLA, kindly provided a large data base comprising the whole experimental work from Fruehan's lab and other researchers. Table 11.11 shows only one group of data from Ito and Fruehan [1] corresponding to the original data in Fig. 5 of this reference, for CaO-SiO_2-30%FeO slags with a constant binary basicity ratio of 0.67. For this system, the calculated slag physical properties are: dynamic viscosity = 1.6 Pa s, density = 3055 kg/m^3, and surface tension = 0.477 N/m. The bubble radius was 12 mm.

$$\pi_1 = f(\pi_2)$$

$$\frac{\pi(j - j_m) H_\infty}{\sigma r_0} = C \left(\frac{\rho g r_0^2}{\mu (j - j_m)} \right)^x$$

Figure 11.7 includes the database with more than 120 experimental data points. Regression analysis of the experimental data reports C = 2905 and n = −1.8, with a regression coefficient of 0.95.

$$Ca \frac{H_\infty}{r_0} = 2905 \left(\frac{Fr}{Re} \right)^{1.8}$$

Defining the foam thickness in terms of the original variables:

$$H_\infty = 2905 \frac{\sigma}{r_0^{2.6}} \frac{[\mu(j - j_m)]^{1.8}}{(\rho g)^{1.8}}$$

This equation is valid for high viscosity fluids, like metallurgical slags. The range of validity in properties is: dynamic viscosity from 46 to 12,100 mPa s, density from

Table 11.11 Database on slag foaming. Taken from Ref. [23]

h (mm)	j (mm/s)	H_∞ (mm)	π_1	π_2
6.1	1.76	7.66	0.001	5001.28
3.5	2.05	4.44	0.001	3243.40
11.1	3.20	13.90	0.008	1358.72
6.9	4.66	8.62	0.008	781.36
13.1	5.20	16.40	0.018	675.97
16.2	5.65	20.28	0.025	606.63
9.4	6.37	11.80	0.017	522.37
25.1	9.28	31.39	0.071	333.60
23.5	10.29	29.34	0.074	296.41
32.8	13.49	40.96	0.141	219.08
39.4	13.58	49.26	0.170	217.59
37.1	15.39	46.34	0.184	189.80
41.8	16.32	52.25	0.221	178.05
45.2	17.41	56.47	0.256	166.02
50.4	18.51	63.05	0.305	155.50
53.7	19.42	67.10	0.342	147.75
66.1	22.91	82.61	0.502	123.94
78.3	24.31	97.82	0.632	116.46
81.3	26.87	101.67	0.730	104.84

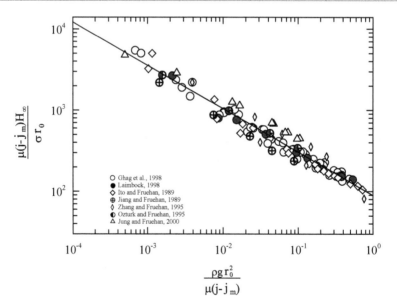

Fig. 11.7 Dimensionless relationship describing foam height. After Pilon [23]

1200 to 3000 kg/m^3, surface tension from 69.5 to 478 mN/m, bubble radius from 0.7 to 20 mm and superficial gas velocities from 0 to 40 mm/s.

Most previous models on slag foaming have reported the foaming thickness to decrease as surface tension increases [1–9]. Pilon et al. [18–22] explained this discrepancy on the following reasoning: Experimental evidence indicates that if the bubble radius decreases, the bubble surface tension also decreases, in a linear way, therefore the net effect of surface tension should be evaluated based on the ratio $\sigma/r_0^{2.6}$, which indicates that if surface tension decreases, the net effect of this ratio is to increase.

Pilon et al. [18, 19] found the model is highly sensitive to the bubble radius, suggesting further improvements to include a bubble size distribution and also include a third dimensionless group to account for the thermophysical properties of the gas (solubility in the liquid phase, diffusion coefficient, etc.)

Dimensional analysis applying conventional methods: [20]

The previous analysis indicates foam thickness (H) depends on six variables: Bubble radius (r_0), viscosity (μ), density (ρ) surface tension (σ), gravity force (g) and the reduced superficial gas velocity ($j - j_m$).

Let's choose the MLT system of dimensions and the π-theorem.

Variable	Symbol	Units	Dimensions
Foam thickness	H	m	[L]
Bubble radius	r_0	m	[L]
Dynamic viscosity	μ	kg/(m × s)	$[ML^{-1}T^{-1}]$
Density	ρ	kg/m^3	$[ML^{-3}]$
Surface tension	σ	N/m	$[MT^{-2}]$
Gravitational acceleration	g	m/s^2	$[LT^{-2}]$
Reduced Sup. gas velocity	U_r	m/s	$[LT^{-1}]$

The π-theorem defines four dimensionless groups because there area a total of 7 variables $-$ 3 dimensions $= 4\pi$ groups. As non-repeating variables we can choose; r_0, μ, and U_r. Next, we define the values for the four dimensionless groups as follows:

$$\pi_1 = H\,(r_0)^{a_1}(\mu)^{b_1}(U_r)^{c_1}$$

$$\pi_2 = \sigma(r_0)^{a_2}(\mu)^{b_2}(U_r)^{c_2}$$

$$\pi_3 = \rho(r_0)^{a_3}(\mu)^{b_3}(U_r)^{c_3}$$

$$\pi_4 = g\,(r_0)^{a_4}(\mu)^{b_4}(U_r)^{c_4}$$

$$\pi_1 = f\,(\pi_2)(\pi_3)(\pi_4)(\pi_5)$$

First π-group:

$$\pi_1 = H\,(r_0)^{a_1}\,(\mu)^{b_1}\,(U_r)^{c_1}$$
$$\pi_1 = (L)(L)^{a_1}\left(ML^{-1}T^{-1}\right)^{b_1}\left(LT^{-1}\right)^{c_1} = M^0T^0L^0$$

Balance:
   M $: b_1 = 0$
   L $: 1 + a_1 - b_1 + c_1 = 0$
   T $: -b_1 - c_1 = 0 \therefore c_1 = 0$

Solution:
   $b_1 = 0$
   $c_1 = 0$
   $a_1 = -1$

Substitution of these values on the π_1-group yields:

$$\pi_1 = \frac{H}{r_0}$$

Second π-group:

Balance:
   M $: 1 + b_2 = 0$
   L $: a_2 - b_2 + c_2 = 0$
   T $: -2 - b_2 - c_2 = 0$

Solution:
   $b_2 = -1$
   $c_2 = -2 - b_2 \therefore c_2 = -1$
   $a_2 = b_2 - c_2 \therefore a_2 = 0$

Substitution of these values on the π_2-group yields:

$$\pi_2 = \frac{\sigma}{\mu U_r}$$

Third π-group:

$$\pi_3 = \rho(r_0)^{a_3}\,(\mu)^{b_3}\,(U_r)^{c_3}$$
$$\pi_3 = \left(ML^{-3}\right)(L)^{a_3}\left(ML^{-1}T^{-1}\right)^{b_3}\left(LT^{-1}\right)^{c_3} = M^0T^0L^0$$

Balance:
   M $: 1 + b_3 = 0$
   L $: -3 + a_3 - b_3 + c_3 = 0$
   T $: -b_3 - c_3 = 0$

Solution:

$$b_3 = -1.$$
$$c_3 = -b_3 \therefore c_3 = 1$$
$$a_3 = b_3 - c_3 + 3 \therefore a_3 = 1.$$

Substitution of these values on the π_2-group yields:

$$\pi_3 = \frac{\rho U_r r_0}{\mu}$$

Fourth π-group:

$$\pi_4 = g \, (r_0)^{a_4} (\mu)^{b_4} (U_r)^{c_4}$$
$$\pi_4 = (LT^{-2})(L)^{a_4} (ML^{-1}T^{-1})^{b_4} (LT^{-1})^{c_4} = M^0 T^0 L^0$$

Balance:

M: $b_4 = 0$

L: $1 + a_4 - b_4 + c_4 = 0$

T: $-2 - b_4 - c_4 = 0$

Solution:

$$b_4 = 0$$
$$c_4 = -2 - b_4 \therefore c_4 = -2$$
$$a_4 = b_4 - c_4 - 1 \therefore a_4 = 1$$
$$\pi_4 = \frac{g r_0}{U_r^2}$$

Finally,

$$\pi_1 = f \, (\pi_2)(\pi_3)(\pi_4)$$
$$\frac{H}{r_0} = C_1 \left(\frac{\sigma}{\mu U_r} \right)^{x_1} \left(\frac{\rho U_r r_0}{\mu} \right)^{y_1} \left(\frac{g r_0}{U_r^2} \right)^{z}$$

π_2 is the capillary number, π_3 is the inverse of the Froude number and π_4 is the Reynolds number.

$$\frac{H}{r_0} = C_1 \left(\frac{1}{Ca} \right)^{x_1} \left(\frac{1}{Fr} \right)^{y_1} (Re)^{z}$$

Regression analysis is more convenient, to eliminate the inverse functions in the following form:

$$\frac{H}{r_0} = C(Ca)^x (Fr)^y (Re)^z$$

Using the previous database, Pilon et al. [20] defined the following values by regression analysis: $C = 2617 \pm 1$, $x = -1.01 \pm 0.02$, $y = 1.77 \pm 0.04$ and $z = -1.74 \pm 0.02$, with a regression coefficient $R^2 = 0.98$.

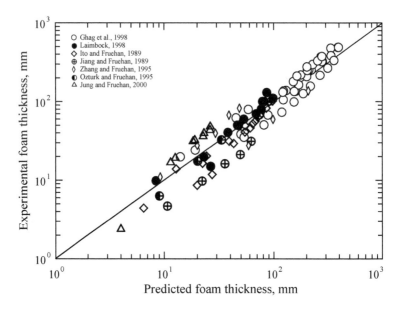

Fig. 11.8 Comparison between predicted and experimental values on foam thickness. After Pilon [23]

$$\frac{H_\infty}{r_0} = 2905(\text{Ca})^{-1.01}(\text{Fr})^{1.77}(\text{Re})^{-1.74}$$

Once again, defining the foam thickness in terms of the original variables:

$$H_\infty = 2617 \frac{\sigma^{1.01}}{r_0^{2.51}} \frac{\mu^{0.73}\left(j - j_m\right)^{0.79}}{\rho^{1.74}g^{1.77}}$$

This expression is quite similar to the previous one [18]. Figure 11.8 compares predicted and experimental data using this expression.

11.2 Slag Open Eye Area

11.2.1 Introduction to the Problem

Slag open eye area is an important phenomenon in the ladle furnace. It is the result of gas injection from the bottom of the reactor. The gas is injected from the bottom and the bubbles formed escape at the top, pushing aside the top slag layer. Gas injection is an important method to improve the mixing conditions of liquid steel, both from a chemical and thermal viewpoint. While gas injection is needed to stir the fluids in the vessel, the ladle eye is a negative effect, however it cannot be eliminated. The problems presented in this section are related to the development of relationships to predict ladle eye and to the definition of conditions to decrease its magnitude.

11.2.2 Experimental Data from Yonezawa and Schwerdtfeger [24]

During bottom gas injection in ladles to homogenize liquid steel the gas bubbles rise to the top and after achieving a critical gas flow rate the top slag layer is opened, exposing liquid steel to the ambient atmosphere. The ladle eye area increases as the gas flow rate increases.

The variables that influence the ladle eye area (A) are: gas flow rate (Q), gravity constant (g), height of liquid steel in the ladle (H), slag thickness (h), nozzle diameter (d), density of metal or water (ρ_m), density of slag or oil (ρ_s), density of gas (ρ_g), viscosity of metal or water (μ_m), viscosity of slag or oil (μ_s), viscosity of gas (μ_g), interfacial tension slag-metal (σ_{sm}), interfacial tension slag-gas (σ_{sg}), interfacial tension metal-gas (σ_{mg}), ambient pressure (p_a) and pressure at nozzle exit (p_b).

In order to compare with the experimental data from Yonezawa and Schwerdtfeger [24] who made experiments with one system (mercury/silicon oil/nitrogen) at atmospheric pressure, then the variables to be considered are:

$$A = f(Q, H, h, g, d)$$

Dimensional analysis: Let's choose the MLT system of dimensions and the π-theorem.

Variable	Symbol	Units	Dimensions
Ladle eye area	A	m^2	$[L^2]$
Gas flow rate	Q	m^3/h	$[L^3T^{-1}]$
Metal height	H	m	$[L]$
Slag height	h	m	$[L]$
Gravitational acceleration	g	m/s^2	$[LT^{-2}]$
Nozzle diameter	d	m	$[L]$

The π-theorem defines four dimensionless groups because there are 6 variables − 2 dimensions = 4π groups. As non-repeating variables we can choose; g and d. Next, we define the values for the four dimensionless groups as follows:

$$\pi_1 = A(g)^{a_1}(h)^{b_1}$$
$$\pi_2 = Q(g)^{a_2}(h)^{b_2}$$
$$\pi_3 = H(g)^{a_3}(h)^{b_3}$$

$$\pi_4 = d(g)^{a_3}(h)^{b_3}$$

$$\pi_1 = f(\pi_2)(\pi_3)(\pi_4)$$

First π-group:

$$\pi_1 = A(g)^{a_1}(h)^{b_1}$$
$$\pi_1 = (L^2)(LT^{-2})^{a_1}(L)^{b_1} = M^0T^0L^0$$

Balance:
  L: $2 + a_1 + b_1 = 0$
  T: $-a_1 = 0$

Solution:
  $a_1 = 0$
  $\therefore b_1 = -2 - a_1 = -2.$

Substitution of these values on the π_1-group yields:

$$\pi_1 = \frac{A}{h^2}$$

Second π-group:

$$\pi_2 = Q(g)^{a_2}(h)^{b_2}$$
$$\pi_2 = (L^3T^{-1})(LT^{-2})^{a_2}(L)^{b_2} = M^0T^0L^0$$

Balance:
  L: $3 + a_2 + b_2 = 0$
  T: $-1 - 2a_2 = 0$

Solution:
  $2a_2 = -1 \therefore a_2 = 1/2$
  $b_2 = -3 - a_2 \therefore b_2 = -5/2$

Substitution of these values on the π_2-group yields:

$$\pi_2 = \frac{Q}{g^{\frac{1}{2}}d^{\frac{5}{2}}} = \frac{Q^2}{gh^5}$$

Notice that depending on the repeating variables chosen this group can have other similar expressions. Using the nozzle diameter instead of the height of the liquid (H) or the slag thickness (h) is possible, however considering that the nozzle diameter (d) has a smaller effect compared to either H or h, and in order to get the same expression reported by Yonezawa and Schwerdtfeger [24], h was selected.

Third π-group:

$$\pi_3 = H(g)^{a_3}(h)^{b_3}$$
$$\pi_2 = (L)(LT^{-2})^{a_2}(L)^{b_2} = M^0T^0L^0$$

Balance:
L: $1 + a_3 + b_3 = 0$
T: $-a_3 = 0$

Solution:
$a_3 = 0$.
$b_3 = -1 - a_3 \therefore b_3 = -1$.

Substitution of these values on the π_2-group yields:

$$\pi_3 = \frac{H}{h}$$

Fourth π-group:

$$\pi_4 = d(g)^{a_3}(h)^{b_3}$$

Similar to π_3

$$\pi_4 = \frac{d}{h}$$

Finally,

$$\pi_1 = f(\pi_2)(\pi_3)(\pi_4)$$
$$\frac{A}{h^2} = C\left(\frac{Q^2}{gh^5}\right)^x \left(\frac{H}{h}\right)^y \left(\frac{d}{h}\right)^z$$

Yonezawa and Schwerdtfeger [24] made experimental measurements of ladle eye as a function of Q, H, h and d. A water model was made of transparent acrylic plastic with an internal diameter of 290 mm. The ladle eye was measured using a video-camera. Silicon oil was used to represent the slag with a thickness from 3 to 30 mm (equivalent to 1.3–13% of height of liquid mercury). Mercury was used to represent liquid steel, with a height of 225 mm. Nitrogen was used as stirring gas using one nozzle located in the center. The fluid's properties are shown in the Table below.

Steel plant measurements: 350-ton ladle, ladle diameter 4.4 m, height of liquid steel 3.5 m, slag thickness approximately 50 mm (equivalent to 1.4% of height of liquid steel), argon flow rate from 6 to 30 Nm^3/h, porous plug diameter of 90 mm.

In both types of experiments, it is important to notice the gas flow rate reported and the gas flow rate included in the analysis, because are different. The gas flow rate in the water model is reported at an ambient pressure of 707 mm Hg (Q_a). The gas flow rate for the industrial ladle is reported at standard conditions (Q_0). The gas flow rate in the analysis correspond to the nozzle exit, at the bottom of the vessels (Q_b). This implies to make adjustments for pressure and temperature. The pressure at the bottom is given by the following relationship (Table 11.12):

$$P_b = P_A + \rho_s g h + \rho_l g H$$

The flowrate for the water model needs to be adjusted for pressure only:

$$Q_b = Q_a \frac{P_a}{P_b}$$

The flowrate for the industrial case needs adjustments for both pressure and temperature:

$$Q_b = Q_0 \frac{P_{std}}{P_b} \frac{1873}{T_{std}}$$

	Mercury	Silicon oil	Steel	Slag	Nitrogen	Argon
Density, kg/m^3	13,600	960	7000	3000	1.429	0.876
Pa, mm Hg	707		760			
Height, m	0.225		3.5			

The following analysis is based on the previous relationship:

$$\ln\left(\frac{A}{h^2}\right) = \ln C + x \ln\left(\frac{Q_b^2}{gh^5}\right) + y \ln\left(\frac{H}{h}\right) + z \ln\left(\frac{d}{h}\right)$$

Applying the statistical tools of excel, by regression analysis, the following values are obtained: $\log C = -0.89$, $x_1 = 0.644$, $x_2 = 1.22$, $x_3 = 0.0498$. The regression coefficient (R^2) is 0.96 and the standard error 0.28. Replacing these values in the general expression which describes ladle eye in both a water model and an industrial plant, it is obtained the following particular equation:

$$\log\left(\frac{A}{h^2}\right) = -0.89 + 0.644 \log\left(\frac{Q_b^2}{gh^5}\right) + 1.22 \log\left(\frac{H}{h}\right) + 0.0498 \log\left(\frac{d}{h}\right)$$

or the alternate form:

Table 11.12 Experimental data from Yonezawa and Schwerdtfeger on ladle eye formation Ref. [24]

Slag height, h, m	Height of metal, H, m	Nozzle diameter, d, m	Q_b, Nm³/h	Ladle eye area, A, m²	π_1	π_2	π_3	π_4
0.003	0.225	0.0005	0.0379	0.00231	256.44	46.5	75.0	0.167
0.003	0.225	0.0005	0.0758	0.00338	375.56	186.1	75.0	0.167
0.003	0.225	0.0005	0.1517	0.00883	981.33	744.5	75.0	0.167
0.003	0.225	0.0005	0.2275	0.01364	1515.44	1675.0	75.0	0.167
0.003	0.225	0.0005	0.3033	0.01686	1873.44	2977.8	75.0	0.167
0.003	0.225	0.0005	0.3791	0.01876	2084.22	4652.9	75.0	0.167
0.003	0.225	0.0005	0.4550	0.01955	2172.11	6700.2	75.0	0.167
0.003	0.225	0.0005	0.5308	0.02045	2271.89	9119.6	75.0	0.167
0.005	0.225	0.0005	0.0379	0.00103	41.08	3.6	45.0	0.100
0.005	0.225	0.0005	0.0758	0.00201	80.20	14.5	45.0	0.100
0.005	0.225	0.0005	0.1516	0.00374	149.52	57.9	45.0	0.100
0.005	0.225	0.0005	0.2275	0.00530	211.80	130.2	45.0	0.100
0.005	0.225	0.0005	0.3033	0.00695	278.08	231.5	45.0	0.100
0.005	0.225	0.0005	0.3791	0.00931	372.40	361.7	45.0	0.100
0.005	0.225	0.0005	0.4549	0.01031	412.24	520.8	45.0	0.100
0.005	0.225	0.0005	0.5307	0.01143	457.24	708.9	45.0	0.100
0.01	0.225	0.0005	0.0379	0.00007	0.70	0.1	22.5	0.050
0.01	0.225	0.0005	0.0758	0.00024	2.38	0.5	22.5	0.050
0.01	0.225	0.0005	0.1516	0.00091	9.11	1.8	22.5	0.050
0.01	0.225	0.0005	0.2274	0.00141	14.07	4.1	22.5	0.050
0.01	0.225	0.0005	0.3032	0.00211	21.11	7.2	22.5	0.050
0.01	0.225	0.0005	0.3789	0.00247	24.67	11.3	22.5	0.050
0.01	0.225	0.0005	0.4547	0.00282	28.20	16.3	22.5	0.050
0.01	0.225	0.0005	0.5305	0.00334	33.35	22.1	22.5	0.050
0.015	0.225	0.0005	0.0379	0.00001	0.02	0.0	15.0	0.033
0.015	0.225	0.0005	0.0758	0.00002	0.10	0.1	15.0	0.033
0.015	0.225	0.0005	0.1515	0.00024	1.05	0.2	15.0	0.033
0.015	0.225	0.0005	0.2273	0.00035	1.53	0.5	15.0	0.033
0.015	0.225	0.0005	0.3030	0.00061	2.70	1.0	15.0	0.033
0.015	0.225	0.0005	0.3788	0.00120	5.35	1.5	15.0	0.033
0.015	0.225	0.0005	0.4546	0.00128	5.67	2.1	15.0	0.033
0.015	0.225	0.0005	0.5303	0.00132	5.87	2.9	15.0	0.033
0.03	0.225	0.0005	0.3027	0.00001	0.02	0.0	7.5	0.017
0.03	0.225	0.0005	0.4540	0.00004	0.05	0.1	7.5	0.017

(continued)

Table 11.12 (continued)

Slag height, h, m	Height of metal, H, m	Nozzle diameter, d, m	Q_b, Nm^3/h	Ladle eye area, A, m^2	π_1	π_2	π_3	π_4
0.004	0.225	0.0001	0.1516	0.00853	533.13	176.6	56.3	0.025
0.004	0.225	0.0001	0.2275	0.012574	785.88	397.4	56.3	0.025
0.004	0.225	0.0001	0.3033	0.01409	880.63	706.5	56.3	0.025
0.004	0.225	0.0001	0.3791	0.016026	1001.63	1104.0	56.3	0.025
0.004	0.225	0.0001	0.4549	0.017524	1095.25	1589.7	56.3	0.025
0.006	0.225	0.0001	0.0758	0.001062	29.50	5.8	37.5	0.017
0.006	0.225	0.0001	0.1516	0.004936	137.11	23.3	37.5	0.017
0.006	0.225	0.0001	0.2274	0.006811	189.19	52.3	37.5	0.017
0.006	0.225	0.0001	0.3032	0.008846	245.72	93.0	37.5	0.017
0.006	0.225	0.0001	0.3791	0.009636	267.67	145.3	37.5	0.017
0.006	0.225	0.0001	0.4549	0.010242	284.50	209.3	37.5	0.017
0.01	0.225	0.0001	0.1516	0.001113	11.13	1.8	22.5	0.010
0.01	0.225	0.0001	0.3032	0.00303	30.30	7.2	22.5	0.010
0.01	0.225	0.0001	0.4547	0.004414	44.14	16.3	22.5	0.010
0.01	0.225	0.0001	0.5305	0.005487	54.87	22.1	22.5	0.010
0.015	0.225	0.0001	0.1515	0.000414	1.84	0.2	15.0	0.007
0.015	0.225	0.0001	0.3030	0.001393	6.19	1.0	15.0	0.007
0.015	0.225	0.0001	0.4546	0.002393	10.64	2.1	15.0	0.007
0.015	0.225	0.0001	0.5303	0.002639	11.73	2.9	15.0	0.007
0.02	0.225	0.0001	0.1515	0.000147	0.37	0.1	11.3	0.005
0.02	0.225	0.0001	0.3029	0.000484	1.21	0.2	11.3	0.005
0.02	0.225	0.0001	0.4544	0.001044	2.61	0.5	11.3	0.005
0.02	0.225	0.0001	0.5301	0.001354	3.39	0.7	11.3	0.005
0.03	0.225	0.0001	0.3027	0.000057	0.06	0.0	7.5	0.003
0.03	0.225	0.0001	0.4540	0.000155	0.17	0.1	7.5	0.003
0.005	0.225	0.0015	0.3033	0.010003	400.12	231.5	45.0	0.300
0.005	0.225	0.0015	0.4549	0.011866	474.64	520.8	45.0	0.300
0.005	0.225	0.0015	0.5307	0.012506	500.24	708.9	45.0	0.300
0.01	0.225	0.0015	0.3032	0.002855	28.55	7.2	22.5	0.150
0.01	0.225	0.0015	0.4547	0.003806	38.06	16.3	22.5	0.150
0.01	0.225	0.0015	0.5305	0.004882	48.82	22.1	22.5	0.150
0.015	0.225	0.0015	0.3030	0.001087	4.83	1.0	15.0	0.100
0.015	0.225	0.0015	0.4546	0.002223	9.88	2.1	15.0	0.100
0.015	0.225	0.0015	0.5303	0.002634	11.71	2.9	15.0	0.100

(continued)

Table 11.12 (continued)

Slag height, h, m	Height of metal, H, m	Nozzle diameter, d, m	Q_b, Nm³/h	Ladle eye area, A, m²	π_1	π_2	π_3	π_4
0.02	0.225	0.0015	0.3029	0.000383	0.96	0.2	11.3	0.075
0.02	0.225	0.0015	0.4544	0.001101	2.75	0.5	11.3	0.075
0.02	0.225	0.0015	0.5301	0.001422	3.56	0.7	11.3	0.075
0.03	0.225	0.0015	0.3027	0.000042	0.05	0.0	7.5	0.050
0.03	0.225	0.0015	0.4540	0.000156	0.17	0.1	7.5	0.050
0.03	0.225	0.0015	0.5297	0.000291	0.32	0.1	7.5	0.050
0.05	3.5	0.09	54.7	1.91	764.00	75.3	70.0	1.800
0.05	3.5	0.09	72.9	2.06	824.00	133.9	70.0	1.800
0.05	3.5	0.09	76.0	2.06	824.00	145.3	70.0	1.800
0.05	3.5	0.09	76.0	2.01	804.00	145.3	70.0	1.800
0.05	3.5	0.09	91.2	2.06	824.00	209.2	70.0	1.800
0.05	3.5	0.09	42.5	1.58	632.00	45.6	70.0	1.800
0.05	3.5	0.09	72.9	1.77	708.00	133.9	70.0	1.800
0.05	3.5	0.09	97.2	2.43	972.00	238.0	70.0	1.800
0.05	3.5	0.09	54.7	2.01	804.00	75.3	70.0	1.800
0.05	3.5	0.09	72.9	2.16	864.00	133.9	70.0	1.800
0.05	3.5	0.09	91.2	2.22	888.00	209.2	70.0	1.800

$$\frac{A}{h^2} = -0.128 \left(\frac{Q_b^2}{gh^5} \right)^{0.644} \left(\frac{H}{h} \right)^{1.22} \left(\frac{d}{h} \right)^{0.0498}$$

Comparison between the predicted values using the previous expression and the experimental results is shown in Fig. 11.9. The agreement is satisfactory except for low and high values of the dimensionless ladle slag eye area.

Yonezawa and Schwerdtfeger [24] analyzed a modified version of the previous dimensionless relationship. Redefining the first-group as follows:

$$\pi_1' = \frac{\pi_1}{\pi_3} = \frac{A}{hH}$$

Then, the modified expression becomes:

$$\frac{A}{hH} = C \left(\frac{Q^2}{gh^5} \right)^x \left(\frac{d}{h} \right)^y$$

Analysis of the experimental data, according with the modified version is illustrated in Fig. 11.10. As can be observed both expressions yield similar results.

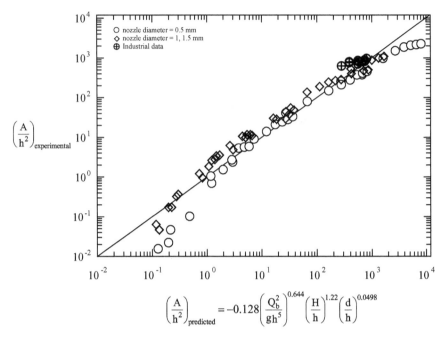

$$\left(\frac{A}{h^2}\right)_{predicted} = -0.128\left(\frac{Q_b^2}{gh^5}\right)^{0.644}\left(\frac{H}{h}\right)^{1.22}\left(\frac{d}{h}\right)^{0.0498}$$

Fig. 11.9 Comparison between predicted and experimental values using four dimensionless groups. Database from Yonezawa and Schwerdtfeger [24]

The low value of the exponent in the fourth dimensionless group indicate the ratio d/h plays a minor influence on the dimensionless slag eye area. This result suggest that ladle eye can be analyzed only as a function of the Froude number, as follows:

$$\frac{A}{hH} = C\left(\frac{Q^2}{gh^5}\right)^x$$

Yonezawa and Schwerdtfeger [24] did the analysis using the dimensionless groups in the previous equation. Observing that they followed a polynomial function, they reported two expressions, as shown in Fig. 11.11.

The results are the following:

For d = 0.5 mm y = −0.683 + 0.905 X − 0.160 X^2 + 0.019 X^3.

For d = 1 and 1.5 mm y = −0.427 + 0.785 X − 0.146 X^2 + 0.023 X^3.

where: y = $\log\left(\frac{A}{hH}\right)$ and x = $\log\left(\frac{Q^2}{gh^5}\right)$

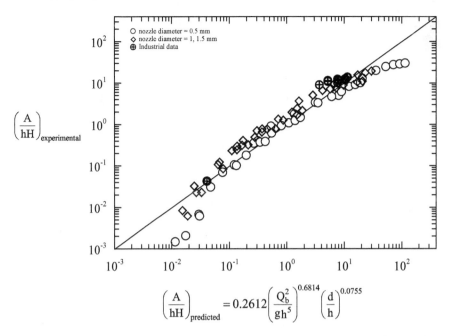

$$\left(\frac{A}{hH}\right)_{predicted} = 0.2612\left(\frac{Q_b^2}{gh^5}\right)^{0.6814}\left(\frac{d}{h}\right)^{0.0755}$$

Fig. 11.10 Comparison between predicted and experimental values using three dimensionless groups. Database from Yonezawa and Schwerdtfeger [24]

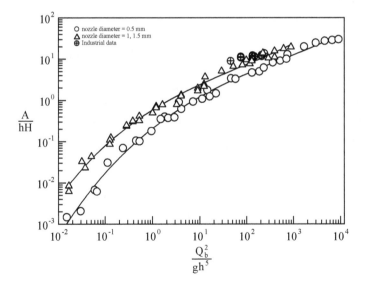

Fig. 11.11 Comparison between predicted and experimental values using two dimensionless groups. Database from Yonezawa and Schwerdtfeger [24]

11.2.3   Experimental Data from Peranandhanthan and Mazumdar [25]

Peranandhanthan and Mazumdar [25] involved the following variables: ladle eye (A), gas flow rate (Q) or plume velocity (U_p), gravity constant (g), height of liquid steel in the ladle (H), slag thickness (h), metal density (ρ_m), density difference between liquid metal and upper phase ($\Delta\rho$) and kinematic viscosity of the upper phase (ν_s).

This system contains 8 variables and 3 dimensions.

$$A = f(Q, H, h, g, \mu_s, \rho_m, \Delta\rho, \nu_s)$$

Dimensional analysis: Let's choose the MLT system of dimensions.

Variable	Symbol	Units	Dimensions
Ladle eye area	A	m^2	$[L^2]$
Gas flow rate	Q	m/min	$[LT^{-1}]$
Metal height	h	m	$[L]$
Slag height	H	m	$[L]$
Gravitational acceleration	g	m/s^2	$[LT^{-2}]$
Metal density	ρ_m	kg/m^3	$[ML^{-3}]$
Density difference oil-water	$\Delta\rho$	kg/m^3	$[ML^{-3}]$
Slag kinematic viscosity	ν_s	m^2/min	$[L^2T^{-1}]$

This system, according with the π-theorem yields five π-dimensionless groups: 8 variables − 3 dimensions. Rayleigh's method would define a system with more unknowns than equations. By choosing the π-method we need to choose as repeating variables a set that includes the three dimensions (MLT) otherwise the system of equations would also give more unknowns than equations, for example, if we choose as repeating variables h, H and U, this system only includes the dimensions LT, or when choosing h, H and ρ_m would include only the dimensions ML. Therefore, once convenient set of repeating variables can be: $\Delta\rho$, H, Q.

There are several relationships that define the velocity of the plume in a ladle in terms of the gas flow rate, therefore Q and U_p can be used indistinctly. For example, Castello-Branco and Schwerdtfeger reported the following relationship [26]:

$$U_p = 17.4Q^{0.244} h^{-0.08} \left(\rho_g/\rho_l\right)^{0.0218} d^{-0.0288}$$

where: U_P is the velocity of the plume in cm/s, Q is the gas flow rate in cm/s, h is the height of the primary liquid in cm and d is the nozzle diameter in cm. Since the effect of d is very small, it can be neglected. If air is injected in the water model, the term involving the density ratio air/water is 0.866, then the equation reduces to

$$U_p = 15.1Q^{0.244} h^{-0.08}$$

Which is almost identical to the equation proposed by Ebneth and Pluschkell [27]. Caution is suggested on two things. First; commonly, capital H is used for the height of the primary liquid, however, to respect the original analysis from Peranandhanthan and Mazumdar [25] we use capital H for the height of the slag thickness. Second; Dimensional analysis involves meter as a unit of length but the equation for the plume velocity is in cm.

$$A = f\left(U_p, H, h, g, \mu_s, \rho_m, \Delta\rho, v_s\right)$$

$$\pi_1 = A(\Delta\rho)^{a_1} (H)^{b_1} \left(U_p\right)^{c_1}$$
$$\pi_2 = g(\Delta\rho)^{a_2} (H)^{b_2} \left(U_p\right)^{c_2}$$
$$\pi_3 = \rho_m(\Delta\rho)^{a_3} (H)^{b_3} \left(U_p\right)^{c_3}$$
$$\pi_4 = h(\Delta\rho)^{a_4} (H)^{b_4} \left(U_p\right)^{c_4}$$
$$\pi_5 = v_s(\Delta\rho)^{a_5} (H)^{b_5} \left(U_p\right)^{c_5}$$

$$\pi_1 = f\left(\pi_2\right)(\pi_3)(\pi_4)(\pi_5)$$

First π-group:

$$\pi_1 = A(\Delta\rho)^{a_1} (H)^{b_1} \left(U_p\right)^{c_1}$$
$$\pi_1 = \left(L^2\right)\left(ML^{-3}\right)^{a_1} (L)^{b_1} \left(LT^{-1}\right)^{c_1} = M^0T^0L^0$$

Balance:
M: $a_1 = 0$
L: $2 - 3a_1 + b_1 + c_1 = 0$
T: $-c_1 = 0$

Solution:
$a_1 = 0, c_1 = 0$
$b_1 = -2 + 3a_1 - c_1 \therefore b_1 = -2$.

Substitution of these values on the π_1-group yields:

$$\pi_1 = \frac{A}{H^2}$$

Second π-group:

$$\pi_2 = g(\Delta\rho)^{a_2} (H)^{b_2} \left(U_p\right)^{c_2}$$
$$\pi_2 = \left(LT^{-2}\right)\left(ML^{-3}\right)^{a_2} (L)^{b_2} \left(LT^{-1}\right)^{c_2} = M^0T^0L^0$$

Balance:

M: $a_2 = 0$

L: $1 - 3a_2 + b_2 + c_2 = 0$

T: $-2 - c_2 = 0 \therefore c_2 = -2$

Solution:

$a_2 = 0, c_2 = -1.$

$1 - 3a_2 + b_2 + c_2 = 0. \therefore b_2 = -c_2 - 1 = 2 - 1 = 1$

$\therefore b_2 = 1$

Substituting values:

$$\pi_2 = \frac{gH}{U^2}$$

Notice that π_2 is the Froude number.

Third π-group:

$$\pi_3 = \rho_m (\Delta\rho)^{a_3} (H)^{b_3} (U_p)^{c_3}$$
$$\pi_3 = (ML^{-3})(ML^{-3})^{a_3} (L)^{b_3} (LT^{-1})^{c_3} = M^0 T^0 L^0$$

Balance:

M: $1 + a_3 = 0 \therefore a_3 = -1$

L: $-3 - 3a_3 + b_3 + c_3 = 0$

T: $-c_3 = 0$

Solution:

$a_3 = -1, c_3 = 0$

$-3 - 3a_3 + b_3 + c_3 = 0$ then, $b_3 = 3 + 3a_3 \therefore b_3 = 0$

Substituting values:

$$\pi_3 = \frac{\rho_m}{\Delta\rho}$$

Fourth π-group:

$$\pi_4 = h(\Delta\rho)^{a_4} (H)^{b_4} (U_p)^{c_4}$$
$$\pi_4 = (L)(ML^{-3})^{a_4} (L)^{b_4} (LT^{-1})^{c_4} = M^0 T^0 L^0$$

Balance:

M: $a_4 = 0$

L: $1 - 3a_4 + b_4 + c_4 = 0$

T: $-c_4 = 0$

Solution:

$a_4 = 0, c_4 = 0$

$1 - 3a_4 + b_4 + c_4 = 0$ then, $b_4 = -1$

Substituting values:

$$\pi_4 = \frac{h}{H}$$

Fifth π-group:

$$\pi_5 = v_s(\Delta\rho)^{a_5}(H)^{b_5}\left(U_p\right)^{c_5}$$
$$\pi_5 = \left(L^2T^{-1}\right)\left(ML^{-3}\right)^{a_5}(L)^{b_5}\left(LT^{-1}\right)^{c_5} = M^0T^0L^0$$

Balance:

M: $a_5 = 0$

L: $2 - 3a_5 + b_5 + c_5 = 0$

T: $-1 - c_5 = 0 \therefore c_5 = -1$

Solution:

$a_5 = 0, c_5 = -1$

$2 - 3a_5 + b_5 + c_5 = 0$, then: $b_5 = 3a_5 - 2 - c_5 = -2 + 1 \therefore b_5 = -1$

Substituting values:

$$\pi_5 = \frac{v_s}{HU_p}$$

Substituting into the general expression:

$$\pi_1 = f(\pi_2)(\pi_3)(\pi_4)(\pi_5)$$
$$\frac{A}{H^2} = f\left(\frac{gH}{U^2}\right)\left(\frac{\rho_m}{\Delta\rho}\right)\left(\frac{h}{H}\right)\left(\frac{v_s}{HU_p}\right)$$

Equivalent to:

$$\frac{A}{H^2} = C\left(\frac{U^2}{gH}\right)^{x_1}\left(\frac{\rho_m}{\Delta\rho}\right)^{x_2}\left(\frac{h}{H}\right)^{x_3}\left(\frac{v_s}{HU_p}\right)^{x_4}$$

This result includes 5 dimensionless groups, different to the one reported in the original paper by Peranandhanthan and Mazumdar [25] with 4 dimensionless groups. This is possible if either h and H remain constant or it is assumed the exponent x_3 is approximately the unity, then it can be reduced to:

$$\frac{A}{H^2}\left(\frac{H}{h}\right) = \frac{A}{Hh} = C\left(\frac{U^2}{gH}\right)^{x_1}\left(\frac{\rho_m}{\Delta\rho}\right)^{x_2}\left(\frac{v_s}{HU_p}\right)^{x_4}$$

Table 11.13 Physical properties of oils. Ref. [25]

	Fluid	μ_s, Pa s	ρ_s, kg/m^3	v_s, m^2/s
1	Water	0.9125×10^{-3}	1000	0.9125×10^{-6}
2	Petroleum ether	0.38×10^{-3}	640	0.5937×10^{-6}
3	Mustard oil	70×10^{-3}	895	78.2122×10^{-6}
4	Soya bean oil	69.3×10^{-3}	910	76.1538×10^{-6}
5	Tetrachloro ethelene (C_2Cl_4)	0.84×10^{-3}	1622	0.5206×10^{-6}
6	Silicon oil I	50×10^{-3}	960	52.083×10^{-6}
7	Silicon oil II	100×10^{-3}	960	104.16×10^{-6}
8	Perfumed coconut oil	21×10^{-3}	843	24.9110×10^{-6}
9	Steel @ 1873K, RT	6×10^{-3}	7014	0.8554×10^{-6}
10	Slag @ 1873K, RT	85×10^{-3}	3150	26.9841×10^{-6}

The following analysis explores the two possibilities.

Experimental work:

A 1:10 scale water model of a 150 ton steel ladle was employed to measure the slag eye area. Air gas injected through one nozzle placed in the center, at the bottom of the model. After the flow reached steady state in a matter of seconds the slag eye area was measured using a video recording method. About 15 images were captured during 5 min. Image-J software was used to process the areas. An average value is reported at each gas flow rate. Dimensional analysis indicates 5 dimensionless groups and 8 variables $\left(U_p, H, h, g, \mu_s, \rho_m, \Delta\rho, v_s\right)$. One of the current problems of water models to reproduce the same behavior as in the real reactors is the lack of similarity in the density ratio oil/water. In the water models this ratio is close to 1 but in the real case is less than half. The results in this condition can mislead in some cases, therefore in order to compensate and decrease the role of oil density a large number of oils in this work were employed.

Table 11.13 shows the physical properties of the oils employed. In this case the slag has been assigned a large value but depending on its chemical composition, it can reach very low values, in the order of 2000 kg/m^3.

Table 11.14 shows the dimensions of the water model and prototype as well as the range of some of the variables in the experimental work. Most of the experiments were carried out with this model except a few experiments with a smaller ladle using C_2Cl_4 as the primary phase.

In order to define values for the five dimensionless groups ladle eye is measured as a function of Q, H, h, g, μ_s, ρ_m, ρ_s and v_s. Table 11.15 summarize the general data base.

Results:

Analysis of results:

Table 11.16 summarize the calculated dimensionless groups.

Table 11.14 Dimensions of water model and prototype. Ref. [25]

	Parameter	1:10	Variables	150 ton
1	Base diameter, m	0.30		3.0
2	Height of liquid, m	0.30	0.35, 0.3	3.0
3	Nozzle diameter, m	0.008		–
4	Gas flow rate, m^3/s	0.66×10^{-5}	$(1.6–6.6) \times 10^{-5}$	2.08×10^{-3}
5	Slag layer thickness, mm	15		150
6	Slag layer thickness, %		0.6–5	5
7	Aspect ratio (H/D)	1	0.75–1.0	1

The final expression is linearized as follows:

$$\ln \frac{A}{H^2} = \ln C + x_1 \ln \left(\frac{U^2}{gH}\right) + x_2 \ln \left(\frac{\rho_m}{\Delta\rho}\right) + x_3 \ln \left(\frac{v_s}{HU_p}\right) + x_4 \ln \left(\frac{h}{H}\right)$$

$$\ln \pi_1 = \ln C + x_1 \ln (\pi_2) + x_2 \ln (\pi_3) + x_3 \ln (\pi_4) + x_4 \ln (\pi_5)$$

Applying the statistical tools of excel the following values are obtained: $\ln C = 1.127$, $x_1 = 1.09$, $x_2 = 1.184$, $x_3 = -0.167$, $x_4 = 1.424$. The regression coefficient (R^2) is 0.992. Replacing these values in the general expression we obtain the following solution:

$$\frac{A}{H^2} = 0.12 \left(\frac{U^2}{gH}\right)^{1.09} \left(\frac{\rho_m}{\Delta\rho}\right)^{1.184} \left(\frac{v_s}{HU_p}\right)^{-0.167} \left(\frac{h}{H}\right)^{1.424}$$

The assumption of x_4 to be the unity is not accurate. In the original work, Peranandhanthan and Mazumdar [25], using 4 dimensionless groups, the regression coefficient was smaller, 0.977, for the following relationship.

$$\frac{A}{hH} = 3.25 \left(\frac{U^2}{gH}\right)^{1.28} \left(\frac{\rho_m}{\Delta\rho}\right)^{0.55} \left(\frac{v_s}{HU_p}\right)^{-0.05}$$

Figure 11.12 compares the agreement between π_1 experimental and the calculated values. It can be seen the good relationship using the relationship with 5 dimensionless groups.

In order to compare the result using 4 dimensionless groups, the following relationship was obtained:

$$\frac{A}{hH} = 5.352 \left(\frac{U^2}{gH}\right)^{1.303} \left(\frac{\rho_m}{\Delta\rho}\right)^{0.428} \left(\frac{v_s}{HU_p}\right)^{0.019}$$

Table 11.15 Experimental database from Ref. [25]

No	Bath liquid	Top liquid	Q, m³/s	ρ_s, kg/m³	ν_s, m²/s	h, m	H, m	A, m²
1	C2Cl4	Water	0.0000167	1000	9.125E−07	0.1377	0.0041	0.007
2	C2Cl4	Water	0.0000167	1000	9.125E−07	0.162	0.0049	0.0072
3	C2Cl4	Water	0.0000167	1000	9.125E−07	0.1377	0.0069	0.0057
4	C2Cl4	Water	0.0000167	1000	9.125E−07	0.162	0.0081	0.0064
5	C2Cl4	Water	0.0000333	1000	9.125E−07	0.1377	0.0041	0.0091
6	C2Cl4	Water	0.0000333	1000	9.125E−07	0.162	0.0049	0.0097
7	C2Cl4	Water	0.0000333	1000	9.125E−07	0.1377	0.0069	0.0078
8	C2Cl4	Water	0.0000333	1000	9.125E−07	0.162	0.0081	0.0086
9	Water	Pet. Ether	0.0000333	640	5.9375E−07	0.255	0.0026	0.0235
11	Water	Pet. Ether	0.0000333	640	5.9375E−07	0.255	0.0077	0.0123
17	Water	Pet. Ether	0.00005	640	5.9375E−07	0.255	0.0026	0.0281
19	Water	Pet. Ether	0.00005	640	5.9375E−07	0.255	0.0077	0.0159
25	Water	Pet. Ether	0.0000667	640	5.9375E−07	0.255	0.0026	0.0373
27	Water	Pet. Ether	0.0000667	640	5.9375E−07	0.255	0.0077	0.0194
10	Water	Pet. Ether	0.0000333	640	5.9375E−07	0.3	0.003	0.0241
12	Water	Pet. Ether	0.0000333	640	5.9375E−07	0.3	0.009	0.0132
18	Water	Pet. Ether	0.00005	640	5.9375E−07	0.3	0.003	0.0334
20	Water	Pet. Ether	0.00005	640	5.9375E−07	0.3	0.009	0.0166
26	Water	Pet. Ether	0.0000667	640	5.9375E−07	0.3	0.003	0.0395
28	Water	Pet. Ether	0.0000667	640	5.9375E−07	0.3	0.009	0.0201
13	Water	Mustard oil	0.0000333	895	7.8212E−05	0.255	0.0026	0.0381
14	Water	Mustard oil	0.0000333	895	7.8212E−05	0.3	0.003	0.0405
15	Water	Mustard oil	0.0000333	895	7.8212E−05	0.255	0.0077	0.035
16	Water	Mustard oil	0.0000333	895	7.8212E−05	0.3	0.009	0.035
21	Water	Mustard oil	0.00005	895	7.8212E−05	0.255	0.0026	0.0466
22	Water	Mustard oil	0.00005	895	7.8212E−05	0.3	0.003	0.0479
23	Water	Mustard oil	0.00005	895	7.8212E−05	0.255	0.0077	0.0405
24	Water	Mustard oil	0.00005	895	7.8212E−05	0.3	0.009	0.0409
29	Water	Mustard oil	0.0000667	895	7.8212E−05	0.255	0.0026	0.0501

(continued)

Table 11.15 (continued)

No	Bath liquid	Top liquid	Q, m³/s	ρ_s, kg/m³	ν_s, m²/s	h, m	H, m	A, m²
30	Water	Mustard oil	0.0000667	895	7.8212E−05	0.3	0.003	0.0507
31	Water	Mustard oil	0.0000667	895	7.8212E−05	0.255	0.0077	0.045
32	Water	Mustard oil	0.0000667	895	7.8212E−05	0.3	0.009	0.0454
33	Water	Coconut oil	0.0000333	845	2.4852E−05	0.255	0.0026	0.0302
34	Water	Coconut oil	0.0000333	845	2.4852E−05	0.3	0.003	0.0322
35	Water	Coconut oil	0.0000333	845	2.4852E−05	0.255	0.0077	0.0209
36	Water	Coconut oil	0.0000333	845	2.4852E−05	0.3	0.009	0.0214
37	Water	Coconut oil	0.00005	845	2.4852E−05	0.255	0.0026	0.0341
38	Water	Coconut oil	0.00005	845	2.4852E−05	0.3	0.003	0.0356
39	Water	Coconut oil	0.00005	845	2.4852E−05	0.255	0.0077	0.024
40	Water	Coconut oil	0.00005	845	2.4852E−05	0.3	0.009	0.0261
41	Water	Coconut oil	0.0000667	845	2.4852E−05	0.255	0.0026	0.0414
42	Water	Coconut oil	0.0000667	845	2.4852E−05	0.3	0.003	0.0455
43	Water	Coconut oil	0.0000667	845	2.4852E−05	0.255	0.0077	0.0274
44	Water	Coconut oil	0.0000667	845	2.4852E−05	0.3	0.009	0.0306

Both equations with 4 dimensionless groups give similar results in spite of the differences in the constants.

Taking into consideration that the experiments with C_2Cl_4 were carried out in a different ladle size and the primary phase was not water, another relationship was developed excluding those experiments. The resulting equation was the following:

$$\frac{A}{hH} = 2.471 \left(\frac{U^2}{gH}\right)^{1.369} \left(\frac{\rho_m}{\Delta\rho}\right)^{0.678} \left(\frac{\nu_s}{HU_p}\right)^{-0.042}$$

Table 11.16 Calculated dimensionless groups. From Ref. [25]

	U_P, m/s	π_1	π_2	π_3	π_4	π_5
1	0.2400	416.4188	1.4337	2.6077	0.0009	33.5854
2	0.2369	299.8751	1.1688	2.6077	0.0008	33.0612
3	0.2400	119.7227	0.8519	2.6077	0.0006	19.9565
4	0.2369	97.5461	0.7071	2.6077	0.0005	20.0000
5	0.2840	541.3444	2.0078	2.6077	0.0008	33.5854
6	0.2804	403.9983	1.6369	2.6077	0.0007	33.0612
7	0.2840	163.8311	1.1930	2.6077	0.0005	19.9565
8	0.2804	131.0776	0.9902	2.6077	0.0004	20.0000
9	0.2732	3476.3314	2.9300	2.7778	0.0008	98.0769
11	0.2732	2677.7778	2.4742	2.7778	0.0007	100.0000
17	0.3017	207.4549	0.9894	2.7778	0.0003	33.1169
19	0.3017	162.9630	0.8247	2.7778	0.0002	33.3333
25	0.3237	5636.0947	2.9300	9.5238	0.1101	98.0769
27	0.3237	4500.0000	2.4742	9.5238	0.0967	100.0000
10	0.2697	590.3188	0.9894	9.5238	0.0372	33.1169
12	0.2697	432.0988	0.8247	9.5238	0.0322	33.3333
18	0.2978	4156.8047	3.5729	2.7778	0.0008	98.0769
20	0.2978	3711.1111	3.0170	2.7778	0.0007	100.0000
26	0.3195	268.1734	1.2064	2.7778	0.0003	33.1169
28	0.3195	204.9383	1.0057	2.7778	0.0002	33.3333
13	0.2732	6893.4911	3.5729	9.5238	0.0997	98.0769
14	0.2697	5322.2222	3.0170	9.5238	0.0875	100.0000
15	0.2732	683.0832	1.2064	9.5238	0.0337	33.1169
16	0.2697	504.9383	1.0057	9.5238	0.0292	33.3333
21	0.3017	5517.7515	4.1124	2.7778	0.0007	98.0769
22	0.2978	4388.8889	3.4726	2.7778	0.0006	100.0000
23	0.3017	327.2053	1.3886	2.7778	0.0002	33.1169
24	0.2978	248.1481	1.1575	2.7778	0.0002	33.3333
29	0.3237	7411.2426	4.1124	9.5238	0.0929	98.0769
30	0.3195	5633.3333	3.4726	9.5238	0.0816	100.0000
31	0.3237	758.9813	1.3886	9.5238	0.0314	33.1169
32	0.3195	560.4938	1.1575	9.5238	0.0272	33.3333
33	0.2732	4467.4556	2.9300	6.4516	0.0350	98.0769
34	0.2697	3577.7778	2.4742	6.4516	0.0307	100.0000
35	0.2732	352.5046	0.9894	6.4516	0.0118	33.1169
36	0.2697	264.1975	0.8247	6.4516	0.0102	33.3333

(continued)

Table 11.16 (continued)

	U_P, m/s	π_1	π_2	π_3	π_4	π_5
37	0.3017	5044.3787	3.5729	6.4516	0.0317	98.0769
38	0.2978	3955.5556	3.0170	6.4516	0.0278	100.0000
39	0.3017	404.7900	1.2064	6.4516	0.0107	33.1169
40	0.2978	322.2222	1.0057	6.4516	0.0093	33.3333
41	0.3237	6124.2604	4.1124	6.4516	0.0295	98.0769
42	0.3195	5055.5556	3.4726	6.4516	0.0259	100.0000
43	0.3237	462.1353	1.3886	6.4516	0.0100	33.1169
44	0.3195	377.7778	1.1575	6.4516	0.0086	33.3333

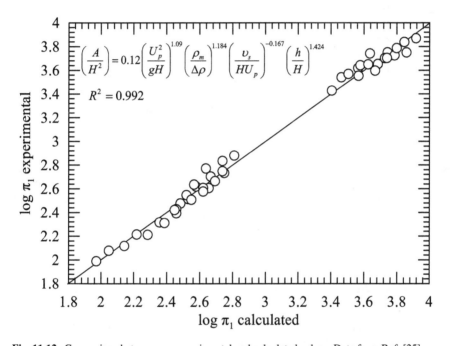

$$\left(\frac{A}{H^2}\right) = 0.12\left(\frac{U_p^2}{gH}\right)^{1.09}\left(\frac{\rho_m}{\Delta\rho}\right)^{1.184}\left(\frac{v_s}{HU_p}\right)^{-0.167}\left(\frac{h}{H}\right)^{1.424}$$

$$R^2 = 0.992$$

Fig. 11.12 Comparison between π_1 experimental and calculated values. Data from Ref. [25]

The low value of the exponent in group (π_5) is an indication that the slag's kinematic viscosity has a minor effect on ladle eye.

An additional step in the analysis is the application of the correlation developed to estimate the experimental values on ladle eye. First the size of the ladle eye area is compared in Fig. 11.13.

The absolute value of the ladle eye area, shown in the previous plot, is specific for the dimensions of the vessels employed. It is not possible to visualize if the ladle eye is small or large. In order to define a scale for comparison the total area is chosen as the reference point. In this work two ladles were employed one with a diameter

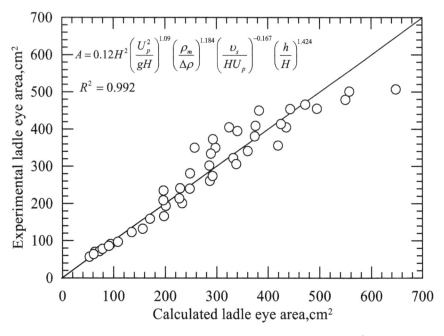

Fig. 11.13 Agreement between experimental and calculated ladle eye areas, cm^2

of 30 cm and a second one with a diameter of 16.2 cm. In Fig. 11.14 the results are reported as a percentage that results from dividing the ladle eye area by the total area of the vessels. In this form it can be observed the large range in the results, from 15 to 80%.

11.2.4 Ladle Eye as a Function of Nozzle Radial Position

Mantripragada and Sarkar [28, 29] were the first to introduce the effect of nozzle radial position on ladle eye area by dimensional analysis. They compared predictions from cold model experiments with industrial data and found large discrepancies that arise from the lack of similarity between the oils employed and actual steelmaking slags. They then developed a mathematical model using the Coupled Level Set Volume of Fluid (CLSVOF) on the basis that the conventional VOF method does not provide a sharp and continuous interface, especially when there is interfacial merging or pinching. The model studied the effect of gas flow rate, metal height, slag thickness, porous plug position and gravity force on ladle eye formation. They assumed the metal and slag physical properties as well as porous plug diameter do not have a significant role on slag eye. Separate dimensionless equations were developed for one and two porous plugs.

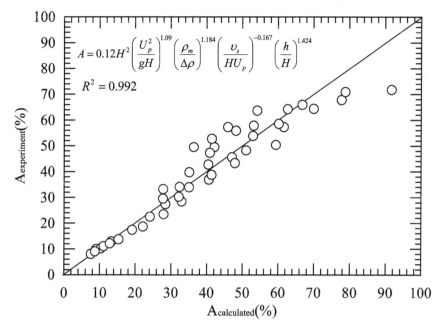

Fig. 11.14 Agreement between experimental and calculated ladle eye areas, %

Dimensional analysis for one porous plug: This system, according with the π-theorem yields 4 π-dimensionless groups: 6 variables − 2 dimensions. In order to obtain a similar set of dimensionless groups, observing that in the author's reported dimensionless groups a repeating variable is H, it is chosen H and g, as repeating variables

$$A = f(Q, H, h, R_1 \ g)$$

$$\pi_1 = A(H)^{a_1}(g)^{b_1}$$
$$\pi_2 = Q(H)^{a_2}(g)^{b_2}$$
$$\pi_3 = R_1(H)^{a_3}(g)^{b_3}$$
$$\pi_4 = h(H)^{a_4}(g)^{b_4}$$
$$\pi_1 = f(\pi_2)(\pi_3)(\pi_4)$$

<u>First π-group:</u>

$$\pi_1 = A(H)^{a_1}(g)^{b_1}$$
$$\pi_1 = (L^2)(L)^{a_1}(LT^{-2})^{b_1} = M^0 T^0 L^0$$

Balance:

$$L: 2 + a_1 + b_1 = 0$$
$$T: -2b_1 = 0$$

Solution:

$$b_1 = 0$$
$$a_1 = -b_1 - 2$$
$$a_1 = -2.$$

Substitution of these values on the π_1-roup yields:

$$\pi_1 = \frac{A}{H^2}$$

Second π-group:

$$\pi_2 = Q(H)^{a_2}(g)^{b_2}$$
$$\pi_2 = (L^3 T^{-1})(L)^{a_2}(LT^{-2})^{b_2} = M^0 T^0 L^0$$

Balance:

$$L: 3 + a_2 + b_2 = 0$$
$$T: -1 - 2b_2 = 0 \therefore b_2 = -\frac{1}{2}$$

Solution:

$$b_2 = -\frac{1}{2}$$
$$a_2 = +\frac{1}{2} - 3 \therefore a_2 = -5/2$$

Substituting values:

$$\pi_2 = \frac{Q^2}{gH^5}$$

Third π-group:

$$\pi_3 + R_1(H)^{a_3}(g)^{b_3}$$
$$\pi_3 = (L)(L)^{a_3}(LT^{-2})^{b_3} = M^0 T^0 L^0$$

Balance:

$$L: 1 + a_3 + b_3 = 0$$
$$T: -2b_3 = 0$$

Solution:

$$b_3 = 0$$
$$a_3 = -b_3 - 1 \therefore a_3 = -1$$

Substituting values:

$$\pi_3 = \frac{R_1}{H}$$

Fourth π-group:

$$\pi_4 = h(H)^{a_4}(g)^{b_4}$$
$$\pi_4 = (L)(L)^{a_4}(LT^{-2})^{b_4} = M^0 T^0 L^0$$

Balance:
  L: $1 + a_4 + b_4 = 0$
  T: $-2b_4 = 0$

Solution:
  $b_4 = 0$
  $a_4 = -b_4 - 1 \therefore a_4 = -1$

Substituting values:

$$\pi_4 = \frac{h}{H}$$

Therefore,

$$\pi_1 = f(\pi_2)(\pi_3)(\pi_4)$$
$$\frac{A}{H^2} = f\left(\frac{Q^2}{gH^5}\right)\left(\frac{R_1}{H}\right)\left(\frac{h}{H}\right)$$

Equivalent to:

$$\frac{A}{H^2} = C\left(\frac{Q^2}{gH^5}\right)^{x_1}\left(\frac{R_1}{H}\right)^{x_2}\left(\frac{h}{H}\right)^{x_3}$$

Numerical simulations: In this system, ladle eye is defined in terms of four independent variables (Q, H, h and R_1). The dimensions of the industrial ladle were taken from Tripathi et al. [30]. These correspond to an industrial size ladle of 160 ton. with a bottom diameter of 3.41 m. The range of variables using one and two porous plugs is shown in Table 11.17. In this example the position of the porous plug includes central gas injection. This would be a problem to process the data in log space. To avoid this, a value close to zero is chosen; 0.001R, solving the indefinite value for log (0).

The design of experiments, response surface method was employed to study the functional relationship among those variables. The values on slag eye area used for the present dimensionless analysis were kindly provided by Prof Sarkar from IITM,

Table 11.17 Variables in this experimental work

Gas flow rate ($\times 10^{-3}$)	Q	15, 30, 45, 60, 82	m^3/s
Metal height	H	3.0, 3.2, 3.4	m
Slag thickness	h	0.1, 0.2, 0.3	m
Single inlet	R_1	0.001R, 0.25R, 0.5R, 0.75R	m
Dual inlet	R_1	0.001R, 0.25R, 0.5R, 0.75R	m
	R_2	0.25R, 0.5R, 0.75R	m
	θ	45, 90, 135, 180	Degrees

a portion of that data was reported by Mantipragada and Sarkar in ref. [29]. The numerical model results on slag eye area and calculated dimensionless groups are shown in Tables 11.18a and 11.18b for one and two nozzles, respectively.

The final expression is linearized as follows:

$$\frac{A}{H^2} = C\left(\frac{Q^2}{gH^5}\right)^{x_1}\left(\frac{R_1}{H}\right)^{x_2}\left(\frac{h}{H}\right)^{x_3}$$

$$\log \frac{A}{H^2} = \log C + x_1 \log\left(\frac{Q^2}{gH^5}\right) + x_2 \log\left(\frac{R_1}{H}\right) + x_3 \log\left(\frac{h}{H}\right)$$

$$\log \pi_1 = \log C + x_1 \log(\pi_2) + x_2 \log(\pi_3) + x_3 \log(\pi_4)$$

Applying the statistical tools of excel the following values are obtained: $\log C = 288$, $x_1 = 0.358$, $x_2 = 0.038$, $x_3 = -0.886$. The regression coefficient (R^2) is 0.936. Replacing these values in the general expression we obtain the following particular solution:

$$\frac{A}{H^2} = 2.46\left(\frac{Q^2}{gH^5}\right)^{0.358}\left(\frac{r_1}{H}\right)^{0.038}\left(\frac{h}{H}\right)^{-0.886}$$

Figure 11.15 compares the agreement between π_1 predicted by the previous equations and those predicted by the numerical model.

Figure 11.16 compares the total slag eye area predicted by dimensionless equation and those predicted by the numerical model.

Dimensional analysis for two porous plugs: With two porous plugs, the analysis should include two additional variables; radial position of the second porous plug (r_2) and the separation angle between the two porous plugs (θ). Mantripragada and Sarkar [28] replaced the separation angle by an additional distance (r_3) that connects the two porous plugs in order to maintain only two dimensions. r_3 is defined as follows:

$$r_3 = \sqrt{r_1^2 + r_2^2 - 2r_1r_2 \cos\theta}$$

Table 11.18a Experimental and calculated dimensionless groups, N=1

Q, Nm³/s	H, m	h, m	r_1, m	A_e, m²	π_1	π_2	π_3	π_4
0.03	3.2	0.3	0.32625	0.689	0.067285	2.73E-07	0.101953	0.09375
0.015	3	0.1	0.001305	1.3061	0.145122	9.44E-08	0.000435	0.033333
0.045	3.4	0.1	0.97875	2.8448	0.24609	4.54E-07	0.287868	0.029412
0.015	3.4	0.1	0.001305	1.3273	0.114818	5.05E-08	0.000384	0.029412
0.06	3	0.2	0.32625	1.9487	0.216522	1.51E-06	0.10875	0.066667
0.06	3.2	0.1	0.32625	3.2366	0.316074	1.09E-06	0.101953	0.03125
0.082	3	0.3	0.001305	1.3312	0.147911	2.82E-06	0.000435	0.1
0.082	3	0.1	0.97875	3.948	0.438667	2.82E-06	0.32625	0.033333
0.082	3.2	0.3	0.97875	1.8263	0.17835	2.04E-06	0.305859	0.09375
0.045	3.2	0.2	0.001305	1.4746	0.144004	6.15E-07	0.000408	0.0625
0.015	3	0.2	0.97875	0.6096	0.067733	9.44E-08	0.32625	0.066667
0.015	3.2	0.1	0.6525	1.3406	0.130918	6.84E-08	0.203906	0.03125
0.015	3.2	0.2	0.6525	0.8538	0.083379	6.84E-08	0.203906	0.0625
0.045	3.2	0.2	0.6525	2.0285	0.198096	6.15E-07	0.203906	0.0625
0.045	3	0.3	0.6525	1.313	0.145889	8.49E-07	0.2175	0.1
0.045	3.2	0.2	0.97875	1.8011	0.175889	6.15E-07	0.305859	0.0625
0.015	3.4	0.3	0.97875	0.6447	0.05577	5.05E-08	0.287868	0.088235
0.015	3	0.3	0.001305	0.2282	0.025356	9.44E-08	0.000435	0.1
0.082	3.4	0.2	0.6525	2.5621	0.221635	1.51E-06	0.191912	0.058824
0.082	3.2	0.2	0.6525	2.728	0.266406	2.04E-06	0.203906	0.0625
0.03	3	0.1	0.6525	2.0801	0.231122	3.78E-07	0.2175	0.033333
0.082	3.4	0.1	0.001305	3.8098	0.329567	1.51E-06	0.000384	0.029412
0.06	3.2	0.1	0.97875	3.2366	0.316074	1.09E-06	0.305859	0.03125
0.06	3	0.2	0.6525	1.9561	0.217344	1.51E-06	0.2175	0.066667
0.082	3.4	0.3	0.001305	1.4832	0.128304	1.51E-06	0.000384	0.088235

The second system, according with the π-theorem yields 6 π-dimensionless groups: 8 variables − 2 dimensions. The chosen repeating variables are H and g.

$$A = f(Q, H, h, r_1, r_2, r_3, g)$$

$$\pi_1 = A(H)^{a_1}(g)^{b_1}$$
$$\pi_2 = Q(H)^{a_2}(g)^{b_2}$$
$$\pi_3 = r_1(H)^{a_3}(g)^{b_3}$$
$$\pi_4 = r_2(H)^{a_4}(g)^{b_4}$$

Table 11.18b Experimental and calculated dimensionless groups, N=2

Q, Nm³/s	H, m	H, m	r_1, m	r_2, m	θ	r_3, m	A_e, m²	π_1	π_2	π_3	π_4	π_5	π_6
0.015	3.00	0.30	0.326	0.979	45	0.783	0.158	0.018	9.44E-08	0.109	0.326	0.261	0.100
0.015	3.40	0.10	0.326	0.979	45	0.783	1.333	0.115	5.05E-08	0.096	0.288	0.230	0.029
0.082	3.40	0.30	0.326	0.979	45	0.783	1.280	0.111	1.51E-06	0.096	0.288	0.230	0.088
0.015	3.40	0.10	0.979	0.979	180	1.958	0.828	0.072	5.05E-08	0.288	0.288	0.576	0.029
0.082	3.40	0.30	0.979	0.979	180	1.958	0.963	0.083	1.51E-06	0.288	0.288	0.576	0.088
0.015	3.00	0.30	0.979	0.979	180	1.958	0.181	0.020	9.44E-08	0.326	0.326	0.653	0.100
0.082	3.13	0.10	0.653	0.979	135	1.512	2.322	0.237	2.29E-06	0.209	0.313	0.483	0.032
0.045	3.40	0.21	0.653	0.979	135	1.512	0.796	0.069	4.54E-07	0.192	0.288	0.445	0.062
0.082	3.16	0.10	0.326	0.326	45	0.250	2.515	0.253	2.19E-06	0.103	0.103	0.079	0.032
0.015	3.00	0.10	0.326	0.326	45	0.250	0.973	0.108	9.44E-08	0.109	0.109	0.083	0.033
0.015	3.40	0.30	0.326	0.326	45	0.250	0.232	0.020	5.05E-08	0.096	0.096	0.073	0.088
0.082	3.00	0.10	0.653	0.653	45	0.499	2.438	0.271	2.82E-06	0.218	0.218	0.166	0.033
0.015	3.20	0.20	0.653	0.653	45	0.499	0.579	0.057	6.84E-08	0.204	0.204	0.156	0.064
0.082	3.18	0.30	0.979	0.326	180	1.305	0.808	0.080	2.11E-06	0.308	0.103	0.410	0.094
0.082	3.00	0.10	0.979	0.326	180	1.305	2.222	0.247	2.82E-06	0.326	0.109	0.435	0.033
0.015	3.40	0.30	0.979	0.326	180	1.305	0.186	0.016	5.05E-08	0.288	0.096	0.384	0.088
0.082	3.00	0.30	0.330	0.980	45	0.782	1.074	0.119	2.82E-06	0.110	0.327	0.261	0.100
0.045	3.40	0.10	0.979	0.326	45	0.783	2.400	0.208	4.54E-07	0.288	0.096	0.230	0.029
0.015	3.19	0.20	0.979	0.326	45	0.783	0.660	0.065	6.97E-08	0.307	0.102	0.246	0.062
0.015	3.40	0.10	0.326	0.326	180	0.653	1.144	0.099	5.05E-08	0.096	0.096	0.192	0.029
0.082	3.40	0.30	0.326	0.326	180	0.653	1.092	0.094	1.51E-06	0.096	0.096	0.192	0.088
0.03	3.00	0.30	0.326	0.326	180	0.653	0.279	0.031	3.78E-07	0.109	0.109	0.218	0.100
0.06	3.00	0.10	0.326	0.979	180	1.305	1.968	0.219	1.51E-06	0.109	0.326	0.435	0.033
0.015	3.40	0.30	0.326	0.979	180	1.305	0.160	0.014	5.05E-08	0.096	0.288	0.384	0.088
0.082	3.00	0.19	0.326	0.653	135	0.913	1.236	0.137	2.82E-06	0.109	0.218	0.304	0.064
0.045	3.23	0.30	0.326	0.653	135	0.913	0.521	0.050	5.85E-07	0.101	0.202	0.282	0.093
0.015	3.00	0.10	0.979	0.326	90	1.032	0.961	0.107	9.44E-08	0.326	0.109	0.344	0.033
0.082	3.40	0.20	0.979	0.326	90	1.032	1.856	0.161	1.51E-06	0.288	0.096	0.303	0.058
0.082	3.40	0.10	0.979	0.979	45	0.749	3.108	0.269	1.51E-06	0.288	0.288	0.220	0.029
0.015	3.40	0.30	0.979	0.979	45	0.749	0.200	0.017	5.05E-08	0.288	0.288	0.220	0.088
0.06	3.00	0.17	0.979	0.979	45	0.749	1.357	0.151	1.51E-06	0.326	0.326	0.250	0.056
0.082	3.40	0.10	0.326	0.979	180	1.305	2.632	0.228	1.51E-06	0.096	0.288	0.384	0.029
0.082	3.00	0.30	0.326	0.979	180	1.305	0.813	0.090	2.82E-06	0.109	0.326	0.435	0.100
0.015	3.00	0.10	0.326	0.979	180	1.305	0.789	0.088	9.44E-08	0.109	0.326	0.435	0.033

$$\pi_5 = r_3(H)^{a_5}(g)^{b_5}$$
$$\pi_6 = h(H)^{a_6}(g)^{b_6}$$
$$\pi_1 = f(\pi_2)(\pi_3)(\pi_4)(\pi_5)(\pi_6)$$

First π-group:

$$\pi_1 = A(H)^{a_1}(g)^{b_1}$$
$$\pi_1 = (L^2)(L)^{a_1}(LT^{-2})^{b_1} = M^0 T^0 L^0$$

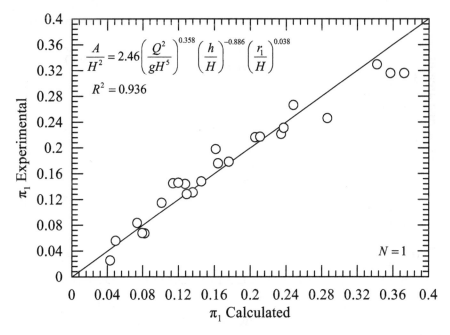

Fig. 11.15 Comparison between π_1 values predicted by dimensionless equation and those predicted by the numerical model

Balance:

    L: $2 + a_1 + b_1 = 0$

    T: $-2b_1 = 0$

Solution:

    $b_1 = 0$

    $a_1 = -b_1 - 2$

    $a_1 = -2.$

Substitution of these values on the π_1-group yields:

$$\pi_1 = \frac{A}{H^2}$$

Second π-group:

$$\pi_2 = Q(H)^{a_2}(g)^{b_2}$$
$$\pi_2 = \left(L^3 T^{-1}\right)(L)^{a_2}\left(LT^{-2}\right)^{b_2} = M^0 T^0 L^0$$

Balance:

    L: $3 + a_2 + b_2 = 0$

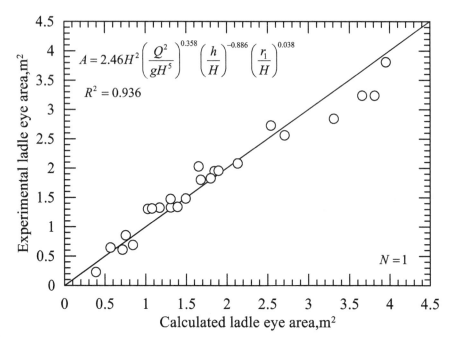

Fig. 11.16 Comparison between π_1 values predicted by dimensionless equation and those predicted by the numerical model

T: $-1 - 2b_2 = 0 \therefore b_2 = -\frac{1}{2}$

Solution:
$b_2 = -\frac{1}{2}$
$a_2 = +\frac{1}{2} - 3 \therefore a_2 = -5/2$

Substituting values:

$$\pi_2 = \frac{Q^2}{gH^5}$$

Third π-group:

$$\pi_3 = r_1 (H)^{a_3} (g)^{b_3}$$
$$\pi_3 = (L)(L)^{a_3} (LT^{-2})^{b_3} = M^0 T^0 L^0$$

Balance:
L: $1 + a_3 + b_3 = 0$
T: $-2b_3 = 0$

Solution:

$b_3 = 0$
$a_3 = -b_3 - 1 \therefore a_3 = -1$

Substituting values:

$$\pi_3 = \frac{r_1}{H}$$

Fourth π-group:

$$\pi_4 = r_2(H)^{a_4}(g)^{b_4}$$
$$\pi_4 = (L)(L)^{a_4}\left(LT^{-2}\right)^{b_4} = M^0 T^0 L^0$$

Balance:
  L: $1 + a_4 + b_4 = 0$
  T: $-2b_4 = 0$

Solution:
  $b_4 = 0$
  $a_4 = -b_4 - 1 \therefore a_4 = -1$

Substituting values:

$$\pi_4 = \frac{r_2}{H}$$

Fifth π-group:

$$\pi_5 = r_3(H)^{a_5}(g)^{b_5}$$
$$\pi_3 = (L)(L)^{a_5}\left(LT^{-2}\right)^{b_5} = M^0 T^0 L^0$$

Balance:
  L: $1 + a_5 + b_5 = 0$
  T: $-2b_5 = 0$

Solution:
  $b_5 = 0$
  $a_5 = -b_5 - 1 \therefore a_5 = -1$

Substituting values:

$$\pi_5 = \frac{r_3}{H}$$

Sixth π-group:

$$\pi_6 = h(H)^{a_6}(g)^{b_6}$$
$$\pi_4 = (L)(L)^{a_6}(LT^{-2})^{b_6} = M^0T^0L^0$$

Balance:

  L: $1 + a_6 + b_6 = 0$
  T: $-2b_6 = 0$

Solution:

  $b_6 = 0$
  $a_6 = -b_6 - 1 \therefore a_6 = -1$

Substituting values:

$$\pi_6 = \frac{h}{H}$$

Therefore,

$$\pi_1 = f(\pi_2)(\pi_3)(\pi_4)(\pi_5)(\pi_6)$$
$$\frac{A}{H^2} = f\left(\frac{Q^2}{gH^5}\right)\left(\frac{r_1}{H}\right)\left(\frac{r_2}{H}\right)\left(\frac{r_3}{H}\right)\left(\frac{h}{H}\right)$$

Equivalent to:

$$\frac{A}{H^2} = C\left(\frac{Q^2}{gH^5}\right)^{x_1}\left(\frac{r_1}{H}\right)^{x_2}\left(\frac{r_2}{H}\right)^{x_3}\left(\frac{r_3}{H}\right)^{x_4}\left(\frac{h}{H}\right)^{x_5}$$

The final expression is linearized as follows:

$$\log\frac{A}{H^2} = \log C + x_1\log\left(\frac{Q^2}{gH^5}\right) + x_2\log\left(\frac{r_1}{H}\right) + x_3\log\left(\frac{r_2}{H}\right) + x_4\log\left(\frac{r_3}{H}\right)$$
$$+ x_5\log\left(\frac{h}{H}\right)$$

$$\log\pi_1 = \log C + x_1\log(\pi_2) + x_2\log(\pi_3) + x_3\log(\pi_4) + x_4\log(\pi_5)$$
$$+ x_5\log(\pi_6)$$

Applying the statistical tools of excel, the following values are obtained: $\log C = 3.953$, $x_1 = 0.375$, $x_2 = 0.052$, $x_3 = -0.017$, $x_4 = -0.128$, $x_5 = -1.193$. The regression coefficient (R^2) is 0.933. Replacing these values in the general expression we obtain the following solution:

$$\frac{A}{H^2} = 0.597\left(\frac{Q^2}{gH^5}\right)^{0.375}\left(\frac{r_1}{H}\right)^{0.052}\left(\frac{r_2}{H}\right)^{-0.017}\left(\frac{r_3}{H}\right)^{-0.128}\left(\frac{h}{H}\right)^{-1.193}$$

Figure 11.17 compares the agreement between π_1 predicted by the previous equations and those predicted by the numerical model.

Figure 11.18 compares predicted and experimental data on ladle eye using the previous relationship, derived by dimensional analysis.

Experimental data from Liu et al.: Liu et al. [31] also carried out experimental measurements on slag eye area as a function of nozzle radial position, using a one-third scale water model. Bean oil was used to replace the slag and the injected gas was nitrogen, using one and two nozzles. The dimensions of the model were; bottom diameter 617 mm, conicity 2.44°, liquid height 700 mm, porous plug diameter 43.4 mm. The range of variables used in the experimental work is shown in Table 11.19.

The slag eye area was measured using a video-camera, then the software ImageJ was used to extract the images and taking an average area for each gas flow rate. Three groups of results were reported corresponding to: (i) One nozzle, three porous plug radial positions. One constant oil thickness, all gas flow rates, (ii) One nozzle, three oil thicknesses, one constant plug radial position, all gas flow rates, (iii) Two nozzles, all separation angles, one constant oil thickness, all gas flow rates. They found a small effect of the porous plug positions because the range was narrow, from 0.56R to 0.67R, however it still can be observed that in this range when the plugs are closer to the walls, the slag eye area increases.

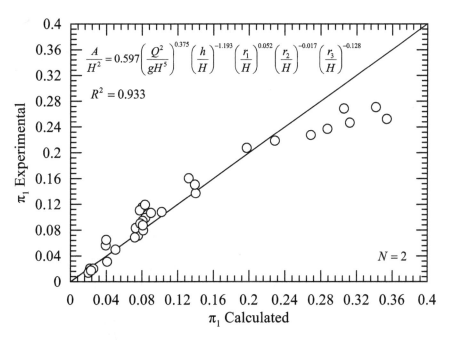

Fig. 11.17 Compares the total slag eye area predicted with the dimensionless equation and those predicted by the numerical model. The regression coefficient is 0.933

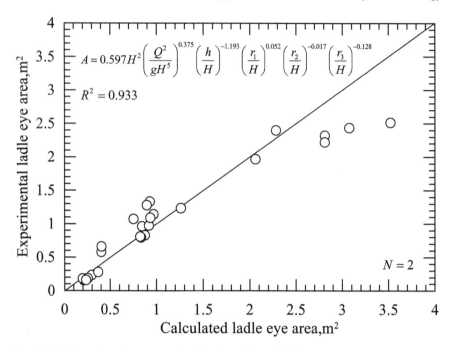

Fig. 11.18 Comparison between predicted and experimental ladle eye

Table 11.19 Variables studied in the experimental work

Gas flow rate	Q	40–170	lt/h
Metal height	H	700	mm
Slag thickness	h	20, 40, 50	mm
Single inlet	R_1	0 R, 0.5R, 0.56 R, 0.62R, 0.67 R, 0.73R	–
Dual inlet	R	0 R, 0.5R, 0.56 R, 0.62R, 0.67 R, 0.73R	–
	θ	60, 90, 120, 150, 180	Degrees

Applying the previous analysis for one porous plug:

$$\frac{A}{H^2} = C\left(\frac{Q^2}{gH^5}\right)^{x_1}\left(\frac{r_1}{H}\right)^{x_2}\left(\frac{h}{H}\right)^{x_3}$$

$$\log \frac{A}{H^2} = \log C + x_1 \log\left(\frac{Q^2}{gH^5}\right) + x_2 \log\left(\frac{r_1}{H}\right) + x_3 \log\left(\frac{h}{H}\right)$$

$$\log \pi_1 = \log C + x_1 \log(\pi_2) + x_2 \log(\pi_3) + x_3 \log(\pi_4) \log \pi_1$$

The following results are obtained by regression analysis: $\log C = 3.17$, $x_1 = 0.564$, $x_2 = 1.027$, $x_3 = -1.009$. The regression coefficient (R^2) is 0.977. Replacing these values in the general expression we obtain the following particular solution:

$$\frac{A}{H^2} = 10^{3.17}\left(\frac{Q^2}{gH^5}\right)^{0.564}\left(\frac{r_1}{H}\right)^{1.027}\left(\frac{h}{H}\right)^{-1.009}$$

Figure 11.19 compares the agreement between π_1 predicted by the previous equations and those predicted by the numerical model.

Figure 11.20 compares the total ladle eye between the calculated and experimental values using the equation defined by dimensional analysis.

Finally, to complete this example, the results from Mantripragada and Sarkar [28] and those from Liu et al. [31] are collected to define a dimensionless expression that covers both results from a water model and industrial data on slag eye areas. Applying multilinear regression analysis, the following values are obtained: $\log C = 1.927$, $x_1 = 0.204$, $x_2 = 0.06$, $x_3 = -0.905$. The regression coefficient (R^2) is 0.862. Replacing these values in the general expression we obtain the following particular solution:

$$\frac{A}{H^2} = 0.285\left(\frac{Q^2}{gH^5}\right)^{0.204}\left(\frac{r_1}{H}\right)^{0.06}\left(\frac{h}{H}\right)^{-0.905}$$

Figure 11.21 compares the agreement between π_1 predicted by the previous equation and the values reported by both Mantripragada and Sarkar [28] and Liu et al. [31].

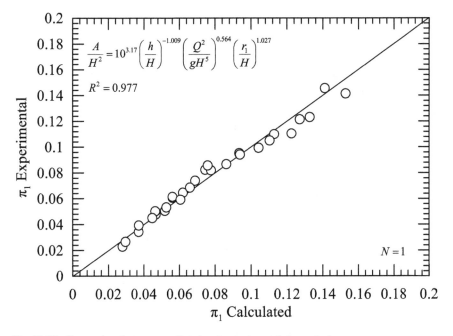

Fig. 11.19 Comparison between predicted and experimental dimensionless groups

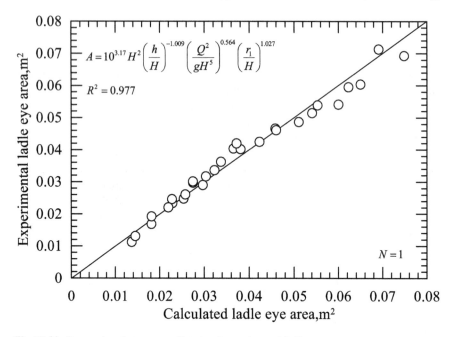

Fig. 11.20 Comparison between predicted and experimental ladle eye area

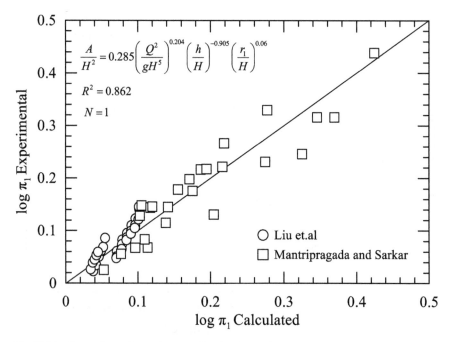

Fig. 11.21 Comparison of π_1 values, predicted and experimental results using experimental data from [28] and [31]

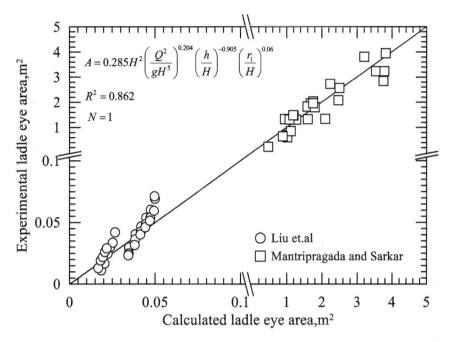

Fig. 11.22 Comparison of predicted and experimental results on ladle eye area using experimental data from [28] and [31]

Figure 11.22 compares the total slag eye area predicted with the dimensionless equation with the values reported by both Mantripragada and Sarkar [28] and Liu et al. [31]. The regression coefficient is 0.945.

11.3 Mixing Time in the Ladle Furnace

Mixing time is a measure of mixing efficiency. Mixing is a critical parameter that has a direct influence on the rate of chemical reactions, removal of impurities, homogenization not just chemically but also thermally and mechanically. One key reason to improve mixing in a ladle is to decrease the processing time and therefore improve productivity.

11.3.1 Experimental Data from Pan et al. [32]

Introduction to the problem

Conventional bottom gas injection has the problem of slag eye or ladle eye by which liquid steel is exposed to the atmosphere and becomes re-oxidized. To solve this

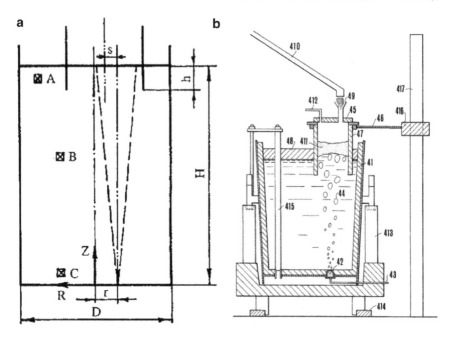

Fig. 11.23 CAS process. **a** Dimensions, **b** overall view

problem a new reactor was developed by Nippon Steel Corporation in 1974, called
CAS process which stands for "Composition Adjustment by Sealed argon bubbling".
The CAS process uses a refractory lined cylinder that is submerged, sealing the ladle
eye from the atmosphere. The alloy additions are made in this zone. The radial
position of a porous plug can be changed as well as the position of the snorkel.
Usually, the center of the snorkel coincides with the central axis of gas injection,
corresponding with a value of "s" in Fig. 11.23 of zero.

A variation of this reactor is the CAS-OB process, where OB stands for Oxygen
Blowing. The injection of oxygen is required to heat the liquid steel by exothermic
reactions, typically the controlled oxidation of aluminum.

Dimensional analysis:

Pan et al. [32] proposed the following variables to describe mixing time in the
absence an upper slag layer.

$$\tau = f(Q, D, h, r, \mu, \sigma, g)$$

where: τ is the mixing time, Q is the bottom gas flow rate (STP), D is inner diameter
of the ladle, h is the submerged depth of slag baffle into the liquid bath, r is the radial
distance between the center point of ladle bottom and that of the porous plug, μ is
the viscosity, σ is the surface tension and g the gravitational constant.

Variable	Symbol	Units	Dimensions
Mixing time	τ	s	$[T]$
Ladle inner diameter	D	m	$[L]$
Gas flow rate (STP)	Q	m^3/hr	$[L^3 T^{-1}]$
Baffle depth	h	m	$[L]$
Radial position	r	m	$[L]$
Gravitational acceleration	g	m/s^2	$[LT^{-2}]$
Surface tension	σ	N/m	$[MT^{-2}]$
Dynamic viscosity	μ	$kg/(m \times s)$	$[ML^{-1}T^{-1}]$

The problem involves 8 variables and 3 fundamental dimensions. The number of dimensionless groups is five ($n - k = 8 - 3 = 5$). Using the method of repeating variables, we choose D, g and σ_s as the repeating variables.

$$\tau = f(Q, D, h, r, \mu, \sigma, g)$$

$$\pi_1 = \tau(D)^{a_1} (g)^{b_1} (\sigma)^{c_1}$$
$$\pi_2 = Q(D)^{a_2} (g)^{b_2} (\sigma)^{c_2}$$
$$\pi_3 = \mu(D)^{a_3} (g)^{b_3} (\sigma)^{c_3}$$
$$\pi_4 = h(D)^{a_4} (g)^{b_4} (\sigma)^{c_4}$$
$$\pi_5 = r(D)^{a_5} (g)^{b_5} (\sigma)^{c_5}$$

$$\pi_1 = f(\pi_2)(\pi_3)(\pi_4)(\pi_5)$$

First π-group:

$$\pi_1 = \tau(D)^{a_1} (g)^{b_1} (\sigma)^{c_1}$$
$$\pi_1 = (T)(L)^{a_1} (LT^{-2})^{b_1} (MT^{-2})^{c_1} = M^0 T^0 L^0$$

Balance:
  M: $c_1 = 0$
  L: $a_1 + b_1 = 0$
  T: $1 - 2b_1 - 2c_1 = 0$

Solution:
  $c_1 = 0$
  $-2b_1 = 2c_1 - 1 = -1 \therefore b_1 = \frac{1}{2}$
  $a_1 = -b_1 = -\frac{1}{2}$

Substitution of these values on the π_1-group yields:

$$\pi_1 = \frac{\tau g^{1/2}}{D^{1/2}} = \frac{\tau^2 g}{D}$$

Second π-group:

$$\pi_2 = Q(D)^{a_2}(g)^{b_2}(\sigma)^{c_2}$$
$$\pi_2 = \left(L^3 T^{-1}\right)(L)^{a_2}\left(LT^{-2}\right)^{b_2}\left(MT^{-2}\right)^{c_2} = M^0 T^0 L^0$$

Balance:
  M: $c_2 = 0$
  L: $3 + a_2 + b_2 = 0$
  T: $-1 - 2b_2 - 2c_2 = 0$

Solution: $c_2 = 0$.
   $-1 - 2b_2 - 2c_2 = 0 \therefore b_2 = -\frac{1}{2}$
   $3 + a_2 + b_2 = 0 \therefore a_2 = -b_2 - 3 = \frac{1}{2} - 6/2 = -5/2$

Substituting values:

$$\pi_2 = \frac{Q}{g^{1/2}D^{\frac{5}{2}}} = \frac{Q^2}{gD^5}$$

Third π-group:

$$\pi_3 = \mu(D)^{a_3}(g)^{b_3}(\sigma)^{c_3}$$
$$\pi_3 = \left(ML^{-1}T^{-1}\right)(L)^{a_3}\left(LT^{-2}\right)^{b_3}\left(MT^{-2}\right)^{c_3} = M^0 T^0 L^0$$

Balance:
  M: $1 + c_3 = 0 \therefore c_3 = -1$
  L: $-1 + a_3 + b_3 = 0$
  T: $-1 - 2b_3 - 2c_3 = 0$

Solution:
   $c_3 = -1$
   $-1 - 2b_3 - 2c_3 = 0$ then, $-2b_3 = 1 + 2c_3 \therefore b_3 = \frac{1}{2}$
   $-1 + a_3 + b_3 = 0$ then, $a_3 = -b_3 + 1 = -\frac{1}{2} + 1 = \frac{1}{2}$

Substituting values:

$$\pi_3 = \frac{\mu D^{1/2} g^{1/2}}{\sigma_s} = \frac{Dg\mu^2}{\sigma_s^2}$$

Fourth π-group:

$$\pi_4 = h(D)^{a_4}(g)^{b_4}(\sigma)^{c_4}$$
$$\pi_4 = (L)(L)^{a_4}\left(LT^{-2}\right)^{b_4}\left(MT^{-2}\right)^{c_4} = M^0T^0L^0$$

Balance:
   M: $c_4 = 0$
   L: $1 + a_4 + b_4 = 0$
   T: $-2b_4 - 2c_4 = 0$

Solution:
   $c_4 = 0$
   $-2b_4 - 2c_4 = 0$ then, $b_4 = 0$
   $1 + a_4 + b_4 = 0$ then, $a_4 = -b_4 - 1 = 1$

Substituting values:

$$\pi_4 = \frac{h}{D}$$

Fifth π-group:

$$\pi_5 = r(D)^{a_5}(g)^{b_5}(\sigma)^{c_5}$$
$$\pi_5 = (L)(L)^{a_5}\left(LT^{-2}\right)^{b_5}\left(MT^{-2}\right)^{c_5} = M^0T^0L^0$$

Balance:
   M: $c_5 = 0$
   L: $1 + a_5 + b_5 = 0$
   T: $-2b_5 - 2c_5 = 0$

Solution:
   $c_5 = 0$
   $-2b_5 - 2c_5 = 0 \therefore b_5 = 0$
   $1 + a_5 + b_5 = 0$, then: $a_5 = -1$

Substituting values:

$$\pi_5 = \frac{r}{D}$$

Substituting into the general expression:

$$\pi_1 = f(\pi_2)(\pi_3)(\pi_4)(\pi_5)$$
$$\frac{\tau^2 g}{D} = f\left(\frac{Q^2}{gD^5}\right)\left(\frac{Dg\mu^2}{\sigma^2}\right)\left(\frac{h}{D}\right)\left(\frac{r}{D}\right)$$

Pan et al. [32] reported only 4 dimensionless groups. This is possible because one dimensionless group $(Dg\mu^2/\sigma^2)$ is a constant under the specific experimental conditions they employed, using only one liquid. In these conditions, the total number of dimensionless groups can be reduced to three.

$$\left(\frac{Q^2}{gD^5}\right)\left(\frac{Dg\mu^2}{\sigma^2}\right) = \left(\frac{Q^2\mu^2}{D^4\sigma^2}\right) = \left(\frac{Q\mu}{D^2\sigma}\right)^2$$

Then the final equation becomes:

$$\frac{\tau^2 g}{D} = C\left(\frac{Q\mu}{D^2\sigma}\right)^{x_1}\left(\frac{h}{D}\right)^{x_2}\left(\frac{r}{D}\right)^{x_3}$$
$$\pi_1 = C\,(\pi_2)^{x_1}\,(\pi_3)^{x_2}\,(\pi_3)^{x_3}$$

Dynamic similarity

The gas flow rate in the water model is defined based on dynamic similarity. This criterion states that the forces in the water model and the prototype should be similar. It is commonly accepted that this criterion is based on the Froude number and if the $(Fr)_m = (Fr)_p$ then the resulting equation is used to define the gas flow rate in the water model. In this work the authors have arrived at a different result and have obtained a different equation. The discussion of this result is in the next problem where Mandal and Mazumdar criticizes Pan's work.

The dimensional formula reported by Pan et al. suggests that the π_2 group can be used to define the gas flow rate in the water model. Therefore, its value should be the same in the water model and the prototype:

$$\left(\frac{Q\mu}{D^2\sigma}\right)_m = \left(\frac{Q\mu}{D^2\sigma}\right)_p$$

where: Q is bottom blowing gas flow rate (m^3/h), μ is viscosity (Pa s), D is inner diameter of ladle (m), σ is surface tension (N/m). The subscript m denotes the model and p the prototype. By rearranging the previous equation and considering $\lambda = D_m/D_p$, we get

$$Q_m = \frac{\mu_p\sigma_m}{\mu_m\sigma_p} \cdot \left(\frac{D_m}{D_p}\right)^2 \cdot Q_p = \frac{\mu_p\sigma_m}{\mu_m\sigma_p} \cdot \lambda^2 \cdot Q_p$$

Substituting the following values, for liquid steel at 1600 °C and water at 20 °C:

$$\mu_p = 0.0025\,\text{Pa s}$$
$$\sigma_p = 1.28\,\text{N/m}$$

$$\mu_m = 0.001 \, \text{Pa s}$$
$$\sigma_m = 0.073 \, \text{N/m}$$

$$\frac{\mu_p \sigma_m}{\mu_m \sigma_p} = \frac{0.0025 \times 0.073}{0.001 \times 1.28} = 0.143$$

Replacing this result in the previous equation it is obtained the equation that defines the criterion for dynamic similarity between model and prototype:

$$Q_m = \frac{\mu_p \sigma_m}{\mu_m \sigma_p} \cdot \lambda^2 \cdot Q_p = 0.143 \, \lambda^2 Q_p$$

The gas flow rate for the water model is defined for STP conditions and those of liquid steel at high temperature (1600 °C), therefore, to defines Q_p for STP, correction for T and P are defined as follows:

$$Q_p^{STP} = \frac{1873}{293} \cdot \frac{P_{atm}}{P_{atm} + \rho_s g H_s} Q_p^{TP}$$

where: P_{atm} is atmospheric pressure, ρ_s is density of liquid steel, H_s is height of liquid steel bath, Q_{CAS} is bottom blowing gas flow rate (at 0 °C and 101325 Pa).

$$P_{atm} = 101325 \, \text{Pa},$$
$$\rho_s = 7000 \, \text{kg/m}^3,$$
$$g = 9.81 \, \text{m/s}^2$$
$$H_s = 3.8 \, \text{m}$$

So,

$$\frac{1873}{293} \cdot \frac{P_{atm}}{P_{atm} + \rho_s g H_s} = \frac{1873}{293} \cdot \frac{101325}{101325 + 7000 \times 9.81 \times 3.8} = 1.788$$

Then, the gas flow rate for the prototype for STP conditions is finally defined as follows:

$$Q_p^{STP} = 1.788 Q_p^{TP}$$

The gas flow rate for the water model in terms of the gas flow rate of the prototype at T and P conditions:

$$Q_m = 0.143 \, \lambda^2 Q_p = 0.143 \, \lambda^2 \cdot 1.788 Q_p^{TP} = 0.2257 \, \lambda^2 Q_p^{TP}$$

The final result is:

$$Q_m = 0.2257 \, \lambda^2 Q_p^{TP}$$

Table 11.20 Dimensions of model and prototype

Items	Prototype	Model
D, Inner diameter of ladle (mm)	2800[a]	280
H, Height of liquid bath (mm)	3800–4000	392
Liquid	Liquid steel	Water
Bottom blow gas	Argon	Nitrogen
Q, Bottom blow gas ($\times 10^{-2}$ Nm^3/h)		1.14–6
Inner diameter of slag baffle (mm)	1400	140
h, Submerged depth of slag baffle (mm)	200–400[b]	20–60
r/D, radial position	0.1263	0, 0.07, 0.14, 0.21

[a] Mean value (the prototype has a conicity of 4°
[b] Excludes the slag layer thickness

This equation is the design equation to define the gas flow rate in the water model, given the gas flow rate in the prototype. It indicates that is proportional to the geometric scale factor (λ) to the second power.

Experimental work:

Table 11.20 shows the dimensions of model and prototype. The geometric scale ratio was 1:10. The aspect ratio (H/D) was kept constant at 1.4. Mixing time was measured as a function of changes in r (radial position), h (submerged depth of slag baffle) and Q (gas flow rate). Nitrogen was injected at the bottom. A tracer was added to change the water electric conductivity. Measurement of this conductivity as a function of time define the mixing time. NaCl was the tracer. Table 11.21 shows the physical properties of water employed in the calculations.

In this process the gas flow rate should be controlled below a critical gas flow rate. Above the critical value the plume overflows the slag baffle. The authors evaluated the critical value as a function of the nozzle radial position and the submerged depth (h) of the slag baffle and found that increasing h, the critical gas flow rate could

Table 11.21 Physical properties of water and liquid steel

	Water (15–20 °C)	Steel (1500–1600 °C)
Density ρ (kg/m^3)	1000	7000
Viscosity μ (Pa s)	0.001	0.0025
Surface tension σ (N/m)	0.073	1.28

Jiaxiang Chen: Handbook of diagrams and data for steelmaking, Metallurgical Industry, Beijing, China (1984), 409

be increased, on the contrary, by increasing the radial position (r/D) that value was decreased. The maximum value was reported for a ratio r/D = 0.07. A correlation was developed to define the critical gas flow rate as a function of the ratio h/D.

Results

Table 11.22 shows the results on mixing time.

Table 11.22 Mixing time results from Pan et al. [32]

D (m)	μ (Pa s)	σ (N/m)	Q (m³/s)	h/D	r/D	τ (s)
0.28	0.001	0.073	3.17E−06	0.072	0	250
0.28	0.001	0.073	3.17E−06	0.143	0	430
0.28	0.001	0.073	3.17E−06	0.143	0.072	275
0.28	0.001	0.073	3.17E−06	0.072	0.143	210
0.28	0.001	0.073	3.17E−06	0.143	0.143	270
0.28	0.001	0.073	3.17E−06	0.214	0.143	300
0.28	0.001	0.073	3.17E−06	0.143	0.214	280
0.28	0.001	0.073	6.33E−06	0.072	0	140
0.28	0.001	0.073	6.33E−06	0.143	0	230
0.28	0.001	0.073	6.33E−06	0.143	0.072	130
0.28	0.001	0.073	6.33E−06	0.143	0.143	125
0.28	0.001	0.073	6.33E−06	0.143	0.214	135
0.28	0.001	0.073	8.33E−06	0.072	0	75
0.28	0.001	0.073	8.33E−06	0.143	0	140
0.28	0.001	0.073	8.33E−06	0.214	0	260
0.28	0.001	0.073	8.33E−06	0.143	0.072	105
0.28	0.001	0.073	8.33E−06	0.072	0.143	95
0.28	0.001	0.073	8.33E−06	0.143	0.143	100
0.28	0.001	0.073	8.33E−06	0.214	0.143	120
0.28	0.001	0.073	8.33E−06	0.143	0.214	110
0.28	0.001	0.073	1.33E−05	0.143	0	100
0.28	0.001	0.073	1.33E−05	0.214	0	130
0.28	0.001	0.073	1.33E−05	0.143	0.072	85
0.28	0.001	0.073	1.33E−05	0.072	0.143	75
0.28	0.001	0.073	1.33E−05	0.143	0.143	80
0.28	0.001	0.073	1.33E−05	0.214	0.143	100
0.28	0.001	0.073	1.33E−05	0.143	0.214	90
0.28	0.001	0.073	1.67E−05	0.143	0	50
0.28	0.001	0.073	1.67E−05	0.214	0	110

Analysis of results

This is an interesting problem in terms of the values of zero in two dimensionless groups, both r/D and h/D. Central gas injection is defined by r/D = 0. Similarly, when the snorkel is not submerged and stays on top of the liquid is defined by h/D = 0. Both are dimensionless groups and in accordance with the method, to linearize the equation it is transformed into log space, as follows:

$$\log\left(\frac{\tau^2 g}{D}\right) = \log C + x_1\log\left(\frac{Q\mu}{D^2\sigma}\right) + x_2\log\left(\frac{h}{D}\right) + x_3\log\left(\frac{r}{D}\right)$$
$$\log \pi_1 = \log C + x_1\log\pi_2 + x_2\log\pi_3 + x_3\log\pi_4$$

Since the log (0) is not defined, this would bring a problem when processing the data. From the two cases, values where h/D = 0 are few (non-realistic operationally in any case) and therefore can be omitted without any practical effects, however central gas injection is a common practice and a solution is needed. Pan et al. analyzed the effect of the ratio r/D (in other words π_4) on mixing time τ (π_1), keeping constant π_2 and π_3. However, considering that the fluid's properties are constant, analyzing the effect of the r/D ratio on mixing time at constant Q and h/D values yields similar results. Figure 11.24 shows this relationship.

It is observed central gas injection is not the best position. It is observed minimum values on mixing time for a ratio r/D = 0.14, indicating that off-center gas injection is better. From this graph a functional relationship τ-r/D can thus be defined. The lines in this graph represent a polynomial of second order, therefore the values r/D can be replaced by a function, f (r/D).

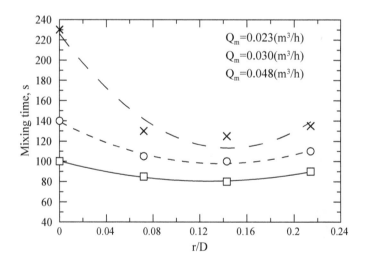

Fig. 11.24 Effect of the radial position on mixing time (h/D = 0.143). Ref. [32]

$$f\left(\frac{r}{D}\right) = a + b\left(\frac{r}{D}\right) + c\left(\frac{r}{D}\right)^2$$

As shown in the graph, the values of the constants, a, b and c, change with Q. The equation that describes the highest Q in the graph is shown below:

$$f\left(\frac{r}{D}\right) = 100 - 311\left(\frac{r}{D}\right) + 1223\left(\frac{r}{D}\right)^2$$

This is equivalent to:

$$f\left(\frac{r}{D}\right) = 1 - 3.11\left(\frac{r}{D}\right) + 12.2\left(\frac{r}{D}\right)^2$$

Pan et al. used the following relationship:

$$f\left(\frac{r}{D}\right) = 1 - \left(\frac{r}{D}\right) + 3.33\left(\frac{r}{D}\right)^2$$

It was found that any of the expressions that describe the functional relationship can be used without a significant effect on the final outcome of the dimensional formula that summarizes the whole experimental work. Table 11.23 summarized the experimental data from Pan et al. [32].

Applying the statistical tools of excel the following values are obtained: $\ln C = -3.61$, $x_1 = 7.051$, $x_2 = 0.9706$, $x_3 = -1.7690$. The regression coefficient (R^2) is 0.88. Replacing these values in the general expression we obtain the following solution:

$$\left(\frac{\tau^2 g}{D}\right) = 242 \times 10^{-6}\left(1 - \frac{r}{D} + 3.33\left(\frac{r}{D}\right)^2\right)^{7.051}\left(\frac{h}{D}\right)^{0.971}\left(\frac{Q\mu}{D^2\sigma}\right)^{-1.769}$$

The prediction capability of this equation is shown in Fig. 11.25.

The comparison of calculated and experimental data on mixing time is shown in Fig. 11.26.

Pan et al. [32] compared fluid flow patterns from a conventional ladle furnace (LMF) what they called ladle bottom blow stirring (LBS) and the CAS process and found important differences:

(i) In LBS, the flow pattern remains similar in a wide range of H/D ratios. The results shown in Fig. 11.27 are for the same gas flow rate, central gas injection and compares two H/D ratios; 1 and 1.4. In this range of the aspect ratio, the flow pattern in the LBS process remains steady, on the contrary, in the CAS process is unsteady, producing a different flow pattern when the H/D ratio is above 1.1–1.2, however, this behavior is beneficial for the CAS process because by increasing the H/D ratio decreases mixing time.

(ii) The snorkel increases mixing time in the CAS process. As the depth of immersion of the snorkel increases (h), it also increases mixing time.

Table 11.23 Experimental dimensionless groups using data from Pan et al. [32]

$\frac{\tau^2 g}{D}$	$\frac{Q\mu}{D^2\sigma}$	h/D	f (r/D)	log π_1	log π_2	log π_3	log π_4
2.19E+06	5.53E−07	0.072	1.000	6.340	−6.257	−1.143	0.000
6.48E+06	5.53E−07	0.143	1.000	6.811	−6.257	−0.845	0.000
2.65E+06	5.53E−07	0.143	0.945	6.423	−6.257	−0.845	−0.024
1.55E+06	5.53E−07	0.072	0.925	6.189	−6.257	−1.143	−0.034
2.55E+06	5.53E−07	0.143	0.925	6.407	−6.257	−0.845	−0.034
3.15E+06	5.53E−07	0.214	0.925	6.499	−6.257	−0.670	−0.034
2.75E+06	5.53E−07	0.143	0.939	6.439	−6.257	−0.845	−0.028
6.87E+05	1.11E−06	0.072	1.000	5.837	−5.956	−1.143	0.000
1.85E+06	1.11E−06	0.143	1.000	6.268	−5.956	−0.845	0.000
5.92E+05	1.11E−06	0.143	0.945	5.772	−5.956	−0.845	−0.024
5.47E+05	1.11E−06	0.143	0.925	5.738	−5.956	−0.845	−0.034
6.39E+05	1.11E−06	0.143	0.939	5.805	−5.956	−0.845	−0.028
1.97E+05	1.46E−06	0.072	1.000	5.295	−5.837	−1.143	0.000
6.87E+05	1.46E−06	0.143	1.000	5.837	−5.837	−0.845	0.000
2.37E+06	1.46E−06	0.214	1.000	6.374	−5.837	−0.670	0.000
3.86E+05	1.46E−06	0.143	0.945	5.587	−5.837	−0.845	−0.024
3.16E+05	1.46E−06	0.072	0.925	5.500	−5.837	−1.143	−0.034
3.50E+05	1.46E−06	0.143	0.925	5.545	−5.837	−0.845	−0.034
5.05E+05	1.46E−06	0.214	0.925	5.703	−5.837	−0.670	−0.034
4.24E+05	1.46E−06	0.143	0.939	5.627	−5.837	−0.845	−0.028
3.50E+05	2.33E−06	0.143	1.000	5.545	−5.633	−0.845	0.000
5.92E+05	2.33E−06	0.214	1.000	5.772	−5.633	−0.670	0.000
2.53E+05	2.33E−06	0.143	0.945	5.403	−5.633	−0.845	−0.024
1.97E+05	2.33E−06	0.072	0.925	5.295	−5.633	−1.143	−0.034
2.24E+05	2.33E−06	0.143	0.925	5.351	−5.633	−0.845	−0.034
3.50E+05	2.33E−06	0.214	0.925	5.545	−5.633	−0.670	−0.034
2.84E+05	2.33E−06	0.143	0.939	5.453	−5.633	−0.845	−0.028
8.76E+04	2.91E−06	0.143	1.000	4.942	−5.536	−0.845	0.000
4.24E+05	2.91E−06	0.214	1.000	5.627	−5.536	−0.670	0.000

(iii)   Mixing time decreases and then increases when the radial position increases.
The minimum mixing time was reported for a radial position at about r/D =
0.14. The behavior of the radial position is different to the immersion depth
and gas flow rate. Decreasing the snorkel immersion depth and increasing
gas flow rate always decreases mixing time. This difference suggested to the
authors the use of a polynomial function for the radial position.

(iv)   Mixing time between LBS and CAS can be compared. LBS is equivalent to
h/D = 0. It was found that mixing time for LBS is shorter compared to CAS.

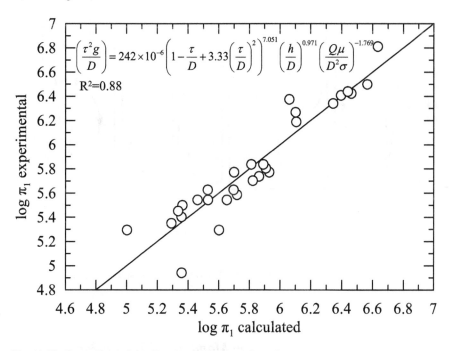

Fig. 11.25 Comparison of calculated and experimental results

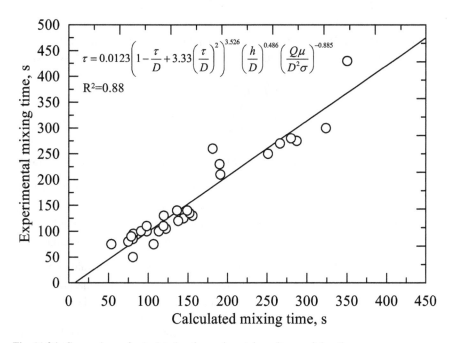

Fig. 11.26 Comparison of calculated and experimental results on mixing time

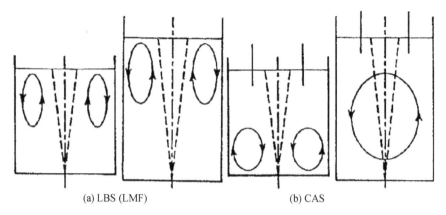

(a) LBS (LMF)   (b) CAS

Fig. 11.27 Comparison of flow patterns from **a** LBS and **b** CAS

Mandal and Mazumdar [33] criticized this work. This discussion is explained in the following section.

11.3.2 Experimental Data from Mandal and Mazumdar [33, 34]

Mandal and Mazumdar correctly pointed out that a system that includes the following 8 variables and 3 fundamental dimensions is described by 5 dimensionless groups:

$$\tau = f(Q, D, h, r, \mu, \sigma, g)$$
$$\frac{\tau^2 g}{D} = f\left(\frac{Q^2}{gD^5}\right)\left(\frac{Dg\mu^2}{\sigma^2}\right)\left(\frac{h}{D}\right)\left(\frac{r}{D}\right)$$

Actually, in the paper by Mandal and Mazumdar, the group $\left(\frac{Dg\mu^2}{\sigma^2}\right)$ is erroneously reported as $\left(\frac{Dg\mu}{\sigma}\right)$. The term $\left(\frac{Dg\mu}{\sigma}\right)$ is not a dimensionless group. The important thing is that this group contains only constants because of the use of one model and only one fluid (water). On that basis, the previous equation can be reduced. The implications of this simplification is that the Froude number disappears $\left(\frac{Q^2}{gD^5}\right)$ and is replaced by another group, $\left(\frac{Q\mu}{D^2\sigma}\right)$, that depends on the physical properties of the fluid rather than on the inertial and gravitational forces. To confirm that is the Froude number the main dimensionless group that controls dynamic similarity, they used experimental data for the CAS process from Pan et al. [32] and also from Mazumdar and Guthrie reported in Mandal's thesis [34]. It is important to study the results with a relationship

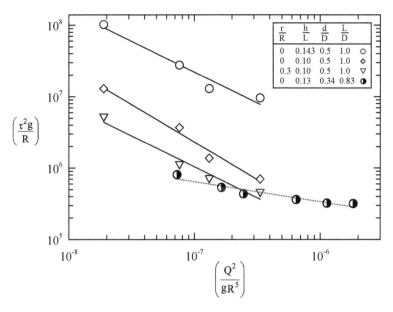

Fig. 11.28 Effect of Froude number on the dimensionless mixing time. After Mandal

that includes only two dimensionless groups. In Fig. 11.28 it is observed different lines, corresponding to different ratios: r/R, h/L, d/D and L/D. This is an indication that the sole relationship between two variables cannot include all the experimental data.

$$\pi'_1 = f\left(\pi'_2\right)$$
$$\frac{\tau^2 g}{R} = C\left(\frac{Q^2}{gR^5}\right)^x$$

Dimensional analysis

Another point raised by Mandal and Mazumdar was the absence of the height of the liquid in the initial set of variables considering that Pan et al. found that it has an important effect. Therefore, Mandal and Mazumdar [33] suggested to eliminate the physical properties of the fluid (μ, σ) and replace these variables by L and d.

$$\tau = f(Q, D, h, r, d, L, g)$$

where: τ is the mixing time, Q is the bottom gas flow rate (STP), D is inner diameter of the ladle, h is the submerged depth of slag baffle into the liquid bath, r is the radial distance between the center point of ladle bottom and that of the porous plug, d is the diameter of snorkel, L is the depth of liquid in ladle.

Variable	Symbol	Units	Dimensions
Mixing time	τ	s	[T]
Ladle inner diameter	D	m	[L]
Gas flow rate (STP)	Q	m³/s	$[L^3 T^{-1}]$
Baffle depth	h	m	[L]
Radial position	r	m	[L]
Gravitational acceleration	g	m/s²	$[LT^{-2}]$
Diameter of snorkel	d	m	[L]
Depth of liquid in ladle	L	m	[L]

The problem involves 8 variables and 2 fundamental dimensions. The number of dimensionless groups is six (n − k = 8 − 2 = 6). Using the method of repeating variables, we choose D and g as the repeating variables.

$$\pi_1 = \tau(D)^{a_1}(g)^{b_1}$$
$$\pi_2 = Q(D)^{a_2}(g)^{b_2}$$
$$\pi_3 = h(D)^{a_3}(g)^{b_3}$$
$$\pi_4 = d(D)^{a_4}(g)^{b_4}$$
$$\pi_5 = r(D)^{a_5}(g)^{b_5}$$
$$\pi_6 = L(D)^{a_6}(g)^{b_6}$$

$$\pi_1 = f(\pi_2)(\pi_3)(\pi_4)(\pi_5)(\pi_6)$$

First π-group:

$$\pi_1 = \tau(D)^{a_1}(g)^{b_1}$$
$$\pi_1 = (T)(L)^{a_1}\left(LT^{-2}\right)^{b_1} = T^0 L^0$$

Balance:
   L: $a_1 + b_1 = 0.$
   T: $1 - 2b_1 = 0.$

Solution:
   $b_1 = \frac{1}{2}$
   $a_1 = -\frac{1}{2}$

Substitution of these values on the π_1-group yields:

$$\pi_1 = \frac{\tau g^{1/2}}{D^{1/2}} = \frac{\tau^2 g}{D}$$

Second π-group:

$$\pi_2 = Q(D)^{a_2} (g)^{b_2}$$
$$\pi_2 = (L^3 T^{-1})(L)^{a_2}(LT^{-2})^{b_2} = T^0 L^0$$

Balance:
  L: $3 + a_2 + b_2 = 0$.
  T: $-1 - 2b_2 = 0$.

Solution:
  $b_2 = -\frac{1}{2}$
  $a_2 = -5/2$.

Substituting values:

$$\pi_2 = \frac{Q}{g^{1/2}D^{\frac{5}{2}}} = \frac{Q^2}{gD^5}$$

Third π-group:

$$\pi_3 = h(D)^{a_3} (g)^{b_3}$$
$$\pi_3 = (L)(L)^{a_3}(LT^{-2})^{b_3} = T^0 L^0$$

Balance:
  L: $1 + a_3 + b_3 = 0$.
  T: $-2b_3 = 0$.

Solution:
  $b_3 = 0$.
  $a_3 = -1$.

Substituting values:

$$\pi_3 = \frac{h}{D}$$

Fourth π-group:

$$\pi_4 = d(D)^{a_4} (g)^{b_4}$$
$$\pi_4 = (L)(L)^{a_4}(LT^{-2})^{b_4} = T^0 L^0$$

Balance:
  L: $1 + a_4 + b_4 = 0$.
  T: $-2b_4 = 0$.

Solution:

 $b_4 = 0.$

 $a_4 = -1.$

Substituting values:

$$\pi_4 = \frac{d}{D}$$

Fifth π-group:

$$\pi_5 = r(D)^{a_5}(g)^{b_5}$$
$$\pi_5 = (L)(L)^{a_5}(LT^{-2})^{b_5} = M^0 T^0 L^0$$

Balance:

 L: $1 + a_5 + b_5 = 0.$

 T: $-2b_5 = 0.$

Solution:

 $b_5 = 0.$

 $a_5 = -1.$

Substituting values:

$$\pi_5 = \frac{r}{D}$$

Sixth π-group:

$$\pi_6 = L(D)^{a_6}(g)^{b_6}$$
$$\pi_6 = (L)(L)^{a_6}(LT^{-2})^{b_6} = M^0 T^0 L^0$$

Balance:

 L: $1 + a_6 + b_6 = 0.$

 T: $-2b_6 = 0.$

Solution:

 $b_6 = 0.$

 $a_6 = -1.$

Substituting values:

$$\pi_6 = \frac{L}{D}$$

Substituting into the general expression:

$$\pi_1 = f(\pi_2)(\pi_3)(\pi_4)(\pi_5)(\pi_6)$$

$$\frac{\tau^2 g}{D} = f\left(\frac{Q^2}{gD^5}\right)\left(\frac{h}{D}\right)\left(\frac{d}{D}\right)\left(\frac{r}{D}\right)\left(\frac{L}{D}\right)$$

This expression can be reduced:

$$\pi'_3 = \frac{\pi_3}{\pi_6} = \frac{h}{L}$$

$$\pi'_4 = \frac{\pi_4}{\pi_5} = \frac{d}{r}$$

Then, the expression is reduced to 4 dimensionless groups:

$$\frac{\tau^2 g}{D} = f\left(\frac{Q^2}{gD^5}\right)\left(\frac{h}{L}\right)\left(\frac{d}{r}\right)$$

The experimental database used by Mandal and Mazumdar [34] is indicated in Table 11.24

Table 11.24 Experimental data base employed by Mandal and Mazumdar [34]

	t,s	Q, Nm³/s	π'_1	π'_2	π_1	π_2	π'_3	π'_4
Pan et al. R = 0.14 D = 0.28 m	1210	3.16E−06	1.02E+08	1.89E−08	5.12E+07	5.92E−10	0.143	1.00E−06
	630	6.33E−06	2.78E+07	7.60E−08	1.39E+07	2.38E−09	0.143	1.00E−06
	430	8.33E−06	1.29E+07	1.32E−07	6.47E+06	4.11E−09	0.143	1.00E−06
	370	1.33E−05	9.58E+06	3.36E−07	4.79E+06	1.05E−08	0.143	1.00E−06
	430	3.16E−06	1.29E+07	1.89E−08	6.47E+06	5.92E−10	0.102	1.00E−06
	230	6.33E−06	3.70E+06	7.60E−08	1.85E+06	2.38E−09	0.102	1.00E−06
	140	8.33E−06	1.37E+06	1.32E−07	6.86E+05	4.11E−09	0.102	1.00E−06
	100	1.33E−05	7.00E+05	3.36E−07	3.50E+05	1.05E−08	0.102	1.00E−06
	270	3.16E−06	5.10E+06	1.89E−08	2.55E+06	5.92E−10	0.102	2.86E−01
	125	6.33E−06	1.09E+06	7.60E−08	5.47E+05	2.38E−09	0.102	2.86E−01
	100	8.33E−06	7.00E+05	1.32E−07	3.50E+05	4.11E−09	0.102	2.86E−01
	80	1.33E−05	4.48E+05	3.36E−07	2.24E+05	1.05E−08	0.102	2.86E−01
Mazumdar and Guthrie R = 0.56 D = 1.12 m	215	1.98E−04	8.09E+05	7.26E−08	4.04E+05	2.32E−06	0.13	1.00E−06
	175	2.98E−04	5.36E+05	1.65E−07	2.68E+05	5.27E−06	0.13	1.00E−06
	144	5.91E−04	3.63E+05	6.47E−07	1.81E+05	2.07E−05	0.13	1.00E−06
	136	7.81E−04	3.24E+05	1.13E−06	1.62E+05	3.62E−05	0.13	1.00E−06
	158	3.65E−04	4.37E+05	2.47E−07	2.18E+05	7.90E−06	0.13	1.00E−06
	135	9.94E−04	3.19E+05	1.83E−06	1.59E+05	5.86E−05	0.13	1.00E−06

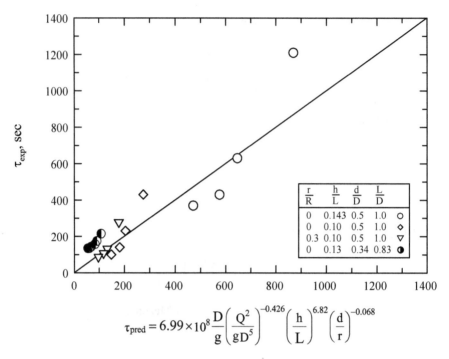

$$\tau_{pred} = 6.99 \times 10^{8} \frac{D}{g} \left(\frac{Q^2}{gD^5} \right)^{-0.426} \left(\frac{h}{L} \right)^{6.82} \left(\frac{d}{r} \right)^{-0.068}$$

Fig. 11.29 Predicted versus experimental data on mixing time using four dimensionless groups. After Mandal [34]

Using 4 dimensionless groups, covers all the variables. the results are now shown in Fig. 11.29. It is observed that most experimental data can be predicted reasonably well.

11.4   Entrainment of Iron Droplets at Liquid/Liquid Interfaces Due to Rising Gas Bubbles

When a gas is injected and the thickness of the upper phase is large, bubbles rise to the top and drag material from the lower phase into the upper phase, in the form of a film around the bubble and also a tail is formed when crossing the interface, as shown in Fig. 11.30. Entrainment of the steel droplets in a top slag layer contribute to enhance mass transfer.

Dimensional analysis (Droplet residence time):

Based on experimental measurements, droplet residence time (t_d) is inversely proportional to droplet radius. The average residence time of a droplet from the lower phase in the upper phase, in sec. is expressed in terms of the following variables: r_d is the

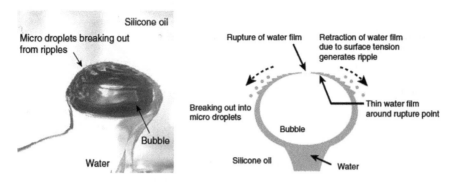

Fig. 11.30 Ruptured water film around a rising bubble. *Source* Uemura et al. Europhys Lett 92:34,004, 2010

average droplet radius, in cm. ρ_d is the density of the lower phase in g/cm^3, ρ_s is the density of the upper phase in g/cm^3, μ_s is the viscosity of the upper phase in g/cm·s and g is the acceleration of gravity in cm/s^2.

$$t_d = f(r_d, \rho_d, \rho_s, \mu_s, g)$$

Variable	Symbol	Units	Dimensions
Average residence time	t_d	s	$[T]$
Average droplet radius	r_d	cm	$[L]$
Density of lower phase	ρ_d	g/cm^3	$[ML^{-3}]$
Density of upper phase	ρ_s	g/cm^3	$[ML^{-3}]$
Viscosity of upper phase	μ_s	g/cm s	$[ML^{-1}T^{-1}]$
Gravitational acceleration	g	m/s^2	$[LT^{-2}]$

The problem involves 6 variables and 3 fundamental dimensions. The number of dimensionless groups is six (n − k = 6 − 3 = 3). Using the method of repeating variables, we choose D and g as the repeating variables.

$$\pi_1 = t_d(\rho_s)^{a_1} (g)^{b_1} (r_d)^{c_1}$$
$$\pi_2 = \mu_s(\rho_s)^{a_2} (g)^{b_2} (r_d)^{c_2}$$
$$\pi_3 = \rho_d(\rho_s)^{a_3} (g)^{b_3} (r_d)^{c_3}$$

First π-group:

$$\pi_1 = t_d(\rho_s)^{a_1} (g)^{b_1} (r_d)^{c_1}$$
$$\pi_1 = (T)(ML^{-3})^{a_1} (LT^{-2})^{b_1} (L)^{c_1} = M^0 T^0 L^0$$

Balance:

  M: $a_1 = 0$
  L: $-3a_1 + b_1 + c_1 = 0$
  T: $1 - 2b_1 = 0$

Solution:

  $a_1 = 0$
  $b_1 = \frac{1}{2}$
  $c_1 = 3a_1 - b_1 = -\frac{1}{2}$

Substitution of these values on the π_1-group yields:

$$\pi_1 = t_d \sqrt{\frac{g}{r_d}}$$

Second π-group:

$$\pi_2 = \mu_s (\rho_s)^{a_2} (g)^{b_2} (r_d)^{c_2}$$
$$\pi_1 = \left(ML^{-1}T^{-1}\right)\left(ML^{-3}\right)^{a_2}\left(LT^{-2}\right)^{b_2}(L)^{c_2} = M^0T^0L^0$$

Balance:

  M: $1 + a_2 = 0$
  L: $-1 - 3a_2 + b_2 + c_2 = 0$
  T: $-1 - 2b_2 = 0$

Solution:

$a_2 = -1.$
  $b_2 = -\frac{1}{2}$
  $c_2 = 1 + 3a_2 - b_2 \therefore c_2 = 1 - 3 + \frac{1}{2} = -3/2$

Substituting values:

$$\pi_2 = \frac{\mu_s}{\rho_s g^{1/2} r_d^{3/2}} \equiv \frac{\rho_s g^{1/2} r_d^{3/2}}{\mu_s} = \frac{\rho_s^2 g r_d^3}{\mu_s^2}$$

Third π-group

$$\pi_3 = \rho_d (\rho_s)^{a_3} (g)^{b_3} (r_d)^{c_3}$$
$$\pi_3 = \left(ML^{-3}\right)\left(ML^{-3}\right)^{a_3}\left(LT^{-2}\right)^{b_3}(L)^{c_3} = M^0T^0L^0$$

Balance:

  M: $1 + a_3 = 0$
  L: $-3 - 3a_3 + b_3 + c_3 = 0$

T: $-2b_3 = 0$

Solution:

$a_3 = -1$

$b_3 = 0$

$c_3 = 3 + 3a_3 - b_3 = 3 - 3 = 0$

Substituting values:

$$\pi_3 = \frac{\rho_d}{\rho_s}$$

$$\pi_1 = f(\pi_2)(\pi_3)$$

$$t_d\sqrt{\frac{g}{r_d}} = f\left(\frac{\rho_s^2 g r_d^3}{\mu_s^2}\right)\left(\frac{\rho_d}{\rho_s}\right)$$

The final general function to describe the average residence time of droplet from the lower phase in the upper phase is expressed as follows:

$$t_d\sqrt{\frac{g}{r_d}} = C\left(\frac{\rho_s^2 g r_d^3}{\mu_s^2}\right)^{x_1}\left(\frac{\rho_d}{\rho_s}\right)^{x_2}$$

Reiter and Schwerdtfeger [35] reported a slightly different expression. They included the density difference in the second dimensionless group, as follows:

$$t_d\sqrt{\frac{g}{r_d}} = C\left(\frac{\rho_s \Delta\rho g r_d^3}{\mu_s^2}\right)^{x_1}\left(\frac{\rho_s}{\rho_d}\right)^{x_2}$$

Experimental work:

Reiter and Schwerdtfeger [35, 36] used high speed photography to study the passage of single bubbles of nitrogen through liquid–liquid interfaces, comparing three systems: water/cyclohexane, mercury/water and mercury/silicon oil.

The experimental set up consisted of a vessel with a square section with dimensions $175 \times 175 \times 500$ mm. The height of the lower phase was about 160 mm and the upper phase 120 mm. A device was employed to control each bubble volume using a cup inside a cylinder. A scale marked in mm was placed on the back of one wall. Illumination was controlled with 4 lamps of 1 kW each.

The residence time of a droplet in the upper phase (t_d) was measured as the time from its formation until reached the lower phase. The initial time of formation was not accurate for all droplets. The time at which the first droplet was visible was taken as a common time of formation. This error is small because the time difference for the formation of droplets is small compared with its residence time.

	T, °C	ρ, g/cm^3	μ, g/cm s	σ, g/s^2	v, cm^2/s		T, °C	σ_{ij}, g/s^2
Water	20	1	0.01	72	0.01	Hg/Water	20	375
Cyclohexane	22	0.78	0.0093	25.3	0.012	Hg/Si-oil	20	425

(continued)

(continued)

	T, °C	ρ, g/cm^3	μ, g/cm s	σ, g/s^2	ν, cm^2/s		T, °C	σ_{ij}, g/s^2
Mercury	20	13.6	0.0155	475	0.0011	W/C-Hex	20	45
Silicon oil	20	0.96	0.48	20.8	0.50			

It was observed that increasing the volume of the bubbles, the mass of the lower phase carried into the upper phase increases. The mass of the lower phase carried into the upper phase increases and the number of droplets decreases in systems with lower interfacial energy and density difference. Due to the high interfacial tension when mercury is involved, the bubbles have a large residence time at the interface before crossing to the upper phase. The film thickness in the Hg/silicon oil system was in the order of 10 microns. It was observed a general tendency of higher droplet's residence time as the droplet radius decreased. It was also observed that in general the bubble size has a negligible effect on the droplet residence time, with the exception of mercury and small bubbles.

The effect of interfacial energies, viscosity of the lower phase and the bubble size was neglected in the dimensional analysis of the droplet residence time. Table 11.25 shows the experimental data.

Applying the statistical tools of excel the following values are obtained: $\log C = 1.393$, $x_1 = 0.1068$, $x_2 = -0.406$. The regression coefficient (R^2) is 0.658. Replacing

Table 11.25 Experimental data on droplet's residence time [35, 36]

	r_d, cm	t_d, s	g, cm/s^2	μ_s, g/cm s	ρ_s, g/cm^3	ρ_d, g/cm^3	π_1	π_2	π_3
Mercury-silicon oil	0.049	0.066	981	0.00048	0.96	13.6	9.3	461,655	14.17
	0.05	0.471	981	0.00048	0.96	13.6	66.0	490,500	14.17
	0.052	0.345	981	0.00048	0.96	13.6	47.4	551,746	14.17
	0.053	0.378	981	0.00048	0.96	13.6	51.4	584,193	14.17
Mercury-water	0.035	0.046	981	0.01	1.00	13.6	7.7	421	13.60
	0.039	0.144	981	0.01	1.00	13.6	22.8	582	13.60
	0.043	0.159	981	0.01	1.00	13.6	24.0	780	13.60
	0.049	0.137	981	0.01	1.00	13.6	19.4	1154	13.60
Water-cyclohexane	0.21	1.73	981	0.0093	0.78	1	118.2	63,907	1.28
	0.23	0.96	981	0.0093	0.78	1	62.7	83,961	1.28
	0.18	1.54	981	0.0093	0.78	1	113.7	40,245	1.28
	0.31	1.61	981	0.0093	0.78	1	90.6	205,578	1.28
	0.26	0.97	981	0.0093	0.78	1	59.6	121,286	1.28
	0.23	1	981	0.0093	0.78	1	65.3	83,961	1.28
	0.34	1.38	981	0.0093	0.78	1	74.1	271,225	1.28
	0.29	1.35	981	0.0093	0.78	1	78.5	168,301	1.28
	0.29	1.27	981	0.0093	0.78	1	73.9	168,301	1.28
	0.38	1.19	981	0.0093	0.78	1	60.5	378,654	1.28
	0.42	2.21	981	0.0093	0.78	1	106.8	511,258	1.28

these values in the general expression which describes the average residence time of droplet from the lower phase in the upper phase is expressed as follows:

$$t_d \sqrt{\frac{g}{r_d}} = 24.7 \left(\frac{\rho_s^2 g r_d^3}{\mu_s^2} \right)^{0.10} \left(\frac{\rho_d}{\rho_s} \right)^{-0.406}$$

This result indicates that the droplet residence time increases by increasing the droplet radius size. They also observed a higher velocity as the droplet size increased, therefore it would be expected a shorter residence time for larger droplets. The data base reported by Reiter and Schwerdtfeger [35] gives only partial experimental results. In their dimensional analysis they considered the density difference as an additional density term. Using their dimensionless groups, the coefficients for the π-groups are; $x_1 = 0.107$, $x_2 = -0.575$. The regression coefficient is low; $R^2 = 0.65$. A possible explanation of this contradictory results is probably due to the uncertainty in the experimental measurements on the residence time. The authors are aware of the error in the measurements but claimed to be small. Their analysis describing the effect of the droplet radius on the droplet residence time shows a high dispersion of the data.

According with the experimental data, the authors found a stronger relationship between droplet radius and bubble size. This analysis is carried below.

Dimensional analysis (droplet radius):

The average droplet radius (r_d), in cm. is expressed in terms of the following variables: bubble radius (r_b) in cm, density of the lower phase (ρ_d) in g/cm^3, density of the upper phase (ρ_s) in g/cm^3, interfacial tension between two liquids in g/s^2, interfacial tension gas/lower phase, in g/s^2, and acceleration of gravity (g), in cm/s^2. The viscosities of phases involved was neglected.

$$r_d = f(r_b, \rho_d, \rho_s, \sigma_{sd}, \sigma_{gd}, g)$$

Variable	Symbol	Units	Dimensions
Average droplet radius	r_d	cm	[L]
Average bubble radius	r_b	cm	[L]
Density of lower phase	ρ_d	g/cm^3	$[ML^{-3}]$
Density of upper phase	ρ_s	g/cm^3	$[ML^{-3}]$
Interfacial tension between two liquids	σ_{sd}	g/s^2	$[MT^{-2}]$
Interfacial tension gas/lower phase	σ_{gd}	g/s^2	$[MT^{-2}]$
Gravitational acceleration	g	m/s^2	$[LT^{-2}]$

The problem involves 7 variables and 3 fundamental dimensions. The number of dimensionless groups is four $(n - k = 7 - 3 = 4)$. Using the method of repeating variables, we choose r_b, ρ_s and σ_{gd} as the repeating variables.

$$\pi_1 = r_d(r_b)^{a_1} (\rho_s)^{b_1} \left(\sigma_{gd}\right)^{c_1}$$
$$\pi_2 = \sigma_{sd}(r_b)^{a_2} (\rho_s)^{b_2} \left(\sigma_{gd}\right)^{c_2}$$
$$\pi_3 = \rho_d(r_b)^{a_3} (\rho_s)^{b_3} \left(\sigma_{gd}\right)^{c_3}$$
$$\pi_4 = g \ (r_b)^{a_4} (\rho_s)^{b_4} \left(\sigma_{gd}\right)^{c_4}$$

First π-group:

$$\pi_1 = r_d(r_b)^{a_1} (\rho_s)^{b_1} \left(\sigma_{gd}\right)^{c_1}$$
$$\pi_1 = (L)(L)^{a_1} (ML^{-3})^{b_1} (MT^{-2})^{c_1} = M^0 T^0 L^0$$

Balance:
  M: $b_1 + c_1 = 0$
  L: $1 + a_1 + b_1 + c_1 = 0$
  T: $-2c_1 = 0$

Solution:
  $c_1 = 0$
  $b_1 = 0.$
  $a_1 = -1.$

Substitution of these values on the π_1-group yields:

$$\pi_1 = \frac{r_d}{r_b}$$

Second π-group:

$$\pi_2 = \sigma_{sd}(r_b)^{a_2} (\rho_s)^{b_2} \left(\sigma_{gd}\right)^{c_2}$$
$$\pi_1 = (MT^{-2})(L)^{a_2} (ML^{-3})^{b_2} (MT^{-2})^{c_2} == M^0 T^0 L^0$$

Balance:
  M: $1 + b_2 + c_2 = 0$
  L: $a_2 - 3b_2 = 0$
  T: $-2 - 2c_2 = 0$

Solution:$c_2 = -1.$
  $b_2 = 0$
  $a_2 = 0$

Substituting values:

$$\pi_2 = \frac{\sigma_{sd}}{\sigma_{gd}}$$

Third π-group

$$\pi_3 = \rho_d (r_b)^{a_3} (\rho_s)^{b_3} (\sigma_{gd})^{c_3}$$
$$\pi_3 = \left(ML^{-3}\right)(L)^{a_3} \left(ML^{-3}\right)^{b_3} \left(MT^{-2}\right)^{c_3} = M^0 T^0 L^0$$

Balance:

M: $1 + b_3 + c_3 = 0$
L: $-3 + a_3 - 3b_3 = 0$
T: $-2c_3 = 0$

Solution:

$c_3 = 0$
$b_3 = -1$
$c_3 = 0$

Substituting values:

$$\pi_3 = \frac{\rho_d}{\rho_s}$$

Fourth π-group:

$$\pi_4 = g \, (r_b)^{a_4} (\rho_s)^{b_4} (\sigma_{gd})^{c_4}$$
$$\pi_1 = \left(LT^{-2}\right)(L)^{a_4} \left(ML^{-3}\right)^{b_4} \left(MT^{-2}\right)^{c_4} = M^0 T^0 L^0$$

Balance:

M: $b_4 + c_4 = 0$
L: $1 + a_4 - 3b_4 = 0$
T: $-2 - 2c_4 = 0$

Solution:

$c_4 = -1$
$b_4 = 1.$
$a_4 = 2.$

Substitution of these values on the π_4-group yields:

$$\pi_4 = \frac{g \rho_s r_b^2}{\sigma_{gd}}$$

Substituting values:

$$\pi_1 = f(\pi_2)(\pi_3)(\pi_4)$$

$$\frac{r_d}{r_b} = f\left(\frac{\sigma_{sd}}{\sigma_{gd}}\right)\left(\frac{\rho_d}{\rho_s}\right)\left(\frac{g\rho_s r_b^2}{\sigma_{gd}}\right)$$

The final general function to describe the average droplet radius size in the upper phase is expressed as follows:

$$\frac{r_d}{r_b} = C\left(\frac{\sigma_{sd}}{\sigma_{gd}}\right)^{x_1}\left(\frac{\rho_d}{\rho_s}\right)^{x_2}\left(\frac{g\rho_s r_b^2}{\sigma_{gd}}\right)^{x_3}$$

The fourth dimensionless group is the Eötvos or Bond number which is used to characterize the shape of bubbles or drops, defines a ratio of gravitational to surface tension forces.

$$\frac{r_d}{r_b} = C\left(\frac{\sigma_{sd}}{\sigma_{gd}}\right)^{x_1}\left(\frac{\rho_d}{\rho_s}\right)^{x_2}(Eo_b)^{x_3}$$

Experimental data:

The interfacial tension gas liquid, σ_{gd}, was not clarified in the original papers by Reiter and Schwerdtfeger [35, 36]. The values employed correspond to the gas–liquid interfacial tensions for the systems mercury/nitrogen and water/nitrogen reported by Hills [37] and Franck [38], respectively. The experimental data on the average droplet's radius and the dimensionless groups is shown in Table 11.26.

Applying the statistical tools of excel the following values are obtained: log C = 0.307, $x_1 = 2.212$, $x_2 = 1.265$, $x_3 = -0.196$. The regression coefficient (R^2) is 0.988. Replacing these values in the general expression which describes the average droplet radius size in the upper phase is expressed as follows:

$$\frac{r_d}{r_b} = 2.03\left(\frac{\sigma_{sd}}{\sigma_{gd}}\right)^{2.21}\left(\frac{\rho_d}{\rho_s}\right)^{1.26}(Eo_b)^{-0.196}$$

The final result is quite similar to that reported by the authors. The main difference in the exponent of the second pi-group, 2.21 versus 1.60 which can be attributed to different values for the gas–liquid interfacial tensions (σ_{gd}). Figure 11.31 shows the predicted values for r_d using the previous relationship.

11.5 Tundish Open Slag Eye (TOSE)

The ladle shroud is a refractory tube employed to transfer molten steel from ladle to tundish at the continuous casting plant, as shown in Fig. 11.32. The historical evolution of the ladle shrouds has been described in detail by Zhang et al. [39], the following introduction on ladle shrouds is based on this work. The ladle shroud is a

Table 11.26 Experimental data on the average droplet radius

r_d, cm	r_b, cm	ρ_s, g/cm^3	ρ_d, g/cm^3	σ_{sd}, g/s^2	σ_{gd}, g/s^2	π_1	π_2	π_3	π_4
0.049	0.91	0.96	13.6	425	475	0.05	0.89	0.07	1.66
0.05	0.98	0.96	13.6	425	475	0.05	0.89	0.07	1.92
0.052	1.11	0.96	13.6	425	475	0.05	0.89	0.07	2.46
0.053	1.39	0.96	13.6	425	475	0.04	0.89	0.07	3.82
0.035	0.94	1.00	13.6	375	475	0.04	0.79	0.07	1.83
0.039	1.07	1.00	13.6	375	475	0.04	0.79	0.07	2.35
0.043	1.18	1.00	13.6	375	475	0.04	0.79	0.07	2.88
0.049	1.34	1.00	13.6	375	475	0.04	0.79	0.07	3.69
0.21	0.49	0.78	1	45	72	0.43	0.63	0.78	2.58
0.23	0.49	0.78	1	45	72	0.47	0.63	0.78	2.58
0.18	0.52	0.78	1	45	72	0.34	0.63	0.78	2.91
0.31	0.58	0.78	1	45	72	0.54	0.63	0.78	3.52
0.26	0.66	0.78	1	45	72	0.39	0.63	0.78	4.62
0.23	0.73	0.78	1	45	72	0.32	0.63	0.78	5.59
0.34	0.89	0.78	1	45	72	0.38	0.63	0.78	8.51
0.29	0.94	0.78	1	45	72	0.31	0.63	0.78	9.43
0.29	0.96	0.78	1	45	72	0.30	0.63	0.78	9.78
0.38	1.10	0.78	1	45	72	0.35	0.63	0.78	12.74
0.42	1.38	0.78	1	45	72	0.30	0.63	0.78	20.23

component of the teeming process during continuous casting. Initially, the teeming process was carried out by ingot casting but eventually continuous casting has reached a share of more than 96%. In the early 1970s the stream of liquid steel was poured into the tundish without protection against the atmosphere. The first record to apply a protective shroud was in the late 1970s in the USA, which decreased steel reoxidation, however a perfect sealing between the ladle shroud and the slide gate-nozzle system is not yet possible, leading to air infiltration. Air infiltration increases when the flow rate of liquid steel increases. This problem can be decreased improving the sealing of all junctions but in addition to that solution, argon gas has also been injected. The bubble size of argon depends on the gas flow rate, decreasing its size if the argon gas flow rate decreases. Small bubbles, lower than about 1 mm follow the stream of liquid steel but larger bubbles rise to the top and open the slag layer, creating a tundish open slag eye (TSOE) which increases not only absorption of air but also heat losses.

Guthrie et al. [40–42] have defined a low gas flow rate (0.2 l/min) in a water model to produce small bubbles, lower than 1.5 mm. Small bubbles contribute to remove non-metallic inclusions and furthermore, can suppress the formation of a ladle eye around the ladle shroud because the small bubbles are dispersed in the bulk liquid and

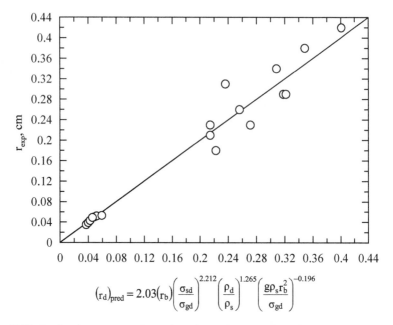

$$(r_d)_{pred} = 2.03(r_b)\left(\frac{\sigma_{sd}}{\sigma_{gd}}\right)^{2.212}\left(\frac{\rho_d}{\rho_s}\right)^{1.265}\left(\frac{g\rho_s r_b^2}{\sigma_{gd}}\right)^{-0.196}$$

Fig. 11.31 Predicted versus experimental results on droplet radius using dimensional analysis. After Reiter and Schwerdtfeger [35, 36]

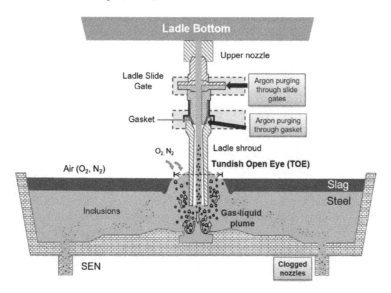

Fig. 11.32 Shroud arrangement (taken from Ma et al. [54])

rise slowly to the top, leaving the liquid without the formation of a permanent slag eye. Formation of microbubbles under steelmaking conditions remains a challenge today due to the large differences between a water model and the steel/slag system.

Chattopadhyay et al. [43–54] has studied in detail the formation of the slag eye in the tundish.

The slag eye area increases, increasing the gas flow rate [43]. A gas fraction of 6% has been suggested to decrease the slag eye area [45]. The numerical prediction of slag eye can be done using the Discrete Particle Model (DPM) to treat a two phase gas–liquid system [45], however, for multiphase flow requiring the presence of the slag phase, the Volume of Fluid (VOF) model is coupled with DPM. In this case, VOF is used to track the steel/slag interface and DPM the bubble/liquid interface [46, 53]. Numerical predictions of slag eye area as a function of the gas flow rate, can be described by the following relationship [53]:

$$A_{tose} = 0.017Q - 0.159$$

where: A_{tose} is the tundish open slag eye area in m^2 and Q is the gas flow rate in Nl/min.

Increasing submergence depth, turbulence decreases and also decreases slag emulsification [44].

The prediction capability of the slag eye area can be increased applying Artificial Neural Networks [54].

Numerical modelling reported by Chatterjee et al. [49] for a tundish of 63 ton. and a height of liquid steel of 1.22 m, a maximum surface velocity without gas injection was reported as 0.08 m/s. With a gas flow rate of 3.2 Nl/min the maximum surface velocity increased to 0.5 m/s, defining the onset of slag eye formation. As the gas flow rate increases, also the slag eye and air absorption increase. These authors reported an increase in oxygen absorption from 1 g/min to 15 g/min, when the slag eye area increased from 0.088 to 0.13 m^2, and a diagram called "eye diagram" was proposed which describes the effect of slag eye and exposure time to the atmosphere on total oxygen pick up. The previous results indicate that steel reoxidation due to formation of the slag eye are a potential source of non-metallic inclusions which in turn also affect nozzle clogging [50]. In addition to this they also reported a higher population density of inclusions in the tundish central region, which corresponds to the region of the slag eye. These findings suggest a change of sampling position, from the conventional position close to the stopper rods to the region of the slag eye area [55].

Ladle shroud misalignment from a vertical position promotes eccentric slag eye [51]. This phenomena should be prevented because in turn promotes asymmetrical turbulent kinetic energy contours and slag entrainment in the bulk of liquid steel [44, 45].

The bubble–liquid interaction and the corresponding formation of slag eye in both a ladle and a tundish is different. In the ladle, the gas is injected through the bottom and rises to the top. In the case of the tundish, the gas is injected from the top and moves to the bottom, then reverses to the top when the buoyancy forces become equal to the inertial forces. Increasing the gas flow rate, the buoyancy forces also increase. If the buoyancy effects overcome the inertial forces, flow reversal of the bubble plume takes place at a higher level [49].

It has been speculated that once argon leaves the tundish or the ladle, a protective layer of inert gas can be formed around the naked slag eye. Chattopadhyay et al. [52] reported data on the densities of argon an air at room and high temperatures. The density of argon at room temperature is higher than air, 1.784 versus 1.275 kg/m^3, respectively. At room temperature argon would form a protective shield, however at high temperatures, say 1550 °C, the density of argon is 0.267 kg/m^3. Air near the tundish can reach temperatures in the region from 500–700 °C, having an average density of 0.4–0.5 kg/m^3, clearly higher than argon, therefore the protective inert gas layer would not be possible. These authors calculated an argon mole fraction of about 0.05–0.1 above the slag eye. Oxygen pick-up was also calculated as a function of the slag eye area. Oxygen pick-up increased from 15 to 35 g/min for a slag eye area of 0.12 and 0.38 m^2, respectively.

The following table summarizes those differences. As a consequence of these differences, the velocity of the plume, fluid flow in the two-phase zone and finally its effect on the slag eye area is expected to be different.

	Ladle open slag eye (LOSE)	Tundish open slag eye (TOSE)
Vessel geometry	Cylinder	Rectangular box
Gas injection	Bottom	Top
Height of liquid	Higher	Lower
Injection device	Porous plugs or nozzles	Pipe
Plume shape	Cylindrical or conical	Cylindrical
Slag eye shape	Near circular	Irregular

Chattopadhyay and Chatterjee [46] applied correlations to estimate the slag eye area, developed for ladle conditions, to estimate the value for tundish conditions and found errors from 75 to 100%, which clearly indicates the need to re-evaluate in a separate way the formation of slag eye in the tundish. In this work they employed water and mineral oil to simulate liquid steel and slag, respectively. Slag thickness was about 6% with respect to the height of liquid in the tundish.

Dimensional analysis of the problem

The problem is the definition of one equation to describe the effect of process variables on the Tundish Open Slag Eye (TOSE) area. The slag eye area (A_e), based on the previous analysis, depends on the following variables: plume velocity (U_p), height of primary liquid (H), oil thickness (h_s), density of water (ρ_1), density difference oil–water ($\Delta\rho$), oil's kinematic viscosity (ν) and gravity (g).

$$A_e = f\left(U_p, H, h_s, \rho_1, \Delta\rho, g\right)$$

Variable	Symbol	Units	Dimensions
Slag eye	A_e	m^2	$[L^2]$
Plume velocity	U_p	m/s	$[LT^{-1}]$
Height of liquid	H	m	$[L]$
Slag thickness	h_s	m	$[L]$
Kinematic viscosity	ν	m^2/s	$[L^2T^{-1}]$
Density of liquid	ρ	kg/m^3	$[ML^{-3}]$
Density difference	$\Delta\rho$	kg/m^3	$[ML^{-3}]$
Gravitational acceleration	g	m/s^2	$[LT^{-2}]$

The problem involves 8 variables and 3 fundamental dimensions. The number of dimensionless groups is five ($n - k = 8 - 3 = 5$).

This problem is similar to a previous one, described in Sect. 11.2.3. The only precaution here is to consider that h in this problem correspond to slag thickness.

$$\pi_1 = f(\pi_2)(\pi_3)(\pi_4)(\pi_5)$$
$$\frac{A_e}{h^2} = C\left(\frac{U_p^2}{gh}\right)^{x_1}\left(\frac{\Delta\rho}{\rho_1}\right)^{x_2}\left(\frac{H}{h}\right)^{x_3}\left(\frac{\nu_s}{hU_p}\right)^{x_4}$$

If the values of H and h remain constant or the exponent is assumed to be close to unity the previous expression reduces to:

$$\frac{A_e}{Hh} = C\left(\frac{U_p^2}{gh}\right)^{x_1}\left(\frac{\Delta\rho}{\rho_1}\right)^{x_2}\left(\frac{\nu_s}{hU_p}\right)^{x_4}$$

Because of the differences previously discussed in bubble behavior in ladles and tundishes, new expressions for the bubble size and plume velocity under tundish conditions are required in order to complete the dimensional analysis for the slag eye area.

The velocity of the plume (U_p) is affected by the following variables: bubble diameter (d_b), height of primary liquid (H), density of water (ρ_1), density of gas (ρ_b), viscosity of water (μ_1) and gravity (g).

$$U_p = f(H, d_b, \rho_1, \rho_b, \mu_1, g)$$

Variable	Symbol	Units	Dimensions
Plume velocity	U_p	m/s	$[LT^{-1}]$
Bubble diameter	d_b	m	$[L]$
Height of liquid	H	m	$[L]$
Dynamic viscosity of water	μ_1	kg/(m × s)	$[ML^{-1}T^{-1}]$

(continued)

(continued)

Variable	Symbol	Units	Dimensions
Density of water	ρ_l	kg/m^3	$[ML^{-3}]$
Density of gas	ρ_b	kg/m^3	$[ML^{-3}]$
Gravitational acceleration	g	m/s^2	$[LT^{-2}]$

The problem involves 7 variables and 3 fundamental dimensions. The number of dimensionless groups is four ($n - k = 7 - 3 = 4$). Using the method of repeating variables, we choose d_b, ρ_l and μ_l as the repeating variables.

$$\pi_1 = U_p(d_b)^{a_1}(\rho_1)^{b_1}(\mu_1)^{c_1}$$
$$\pi_2 = g(d_b)^{a_2}(\rho_1)^{b_2}(\mu_1)^{c_2}$$
$$\pi_3 = \rho_b(d_b)^{a_3}(\rho_1)^{b_3}(\mu_1)^{c_3}$$
$$\pi_4 = H(d_b)^{a_4}(\rho_1)^{b_4}(\mu_1)^{c_4}$$
$$\pi_1 = f(\pi_2)(\pi_3)(\pi_4)$$

First π-group:

$$\pi_1 = U_p(d_b)^{a_1}(\rho_1)^{b_1}(\mu_1)^{c_1}$$
$$\pi_1 = \left(LT^{-1}\right)(L)^{a_1}\left(ML^{-3}\right)^{b_1}\left(ML^{-1}T^{-1}\right)^{c_1} = M^0T^0L^0$$

Balance:
  M: $b_1 + c_1 = 0$.
  L: $1 + a_1 - 3b_1 - c_1 = 0$
  T: $-1 - c_1 = 0$

Solution:
  $c_1 = -1$
  $b_1 = -c_1 = 1$
  $a_1 = 1$.

Substitution of these values on the π_1-group yields:

$$\pi_1 = \frac{d_b \rho_1 U_p}{\mu_1}$$

Second π-group:

$$\pi_2 = g(d_b)^{a_2}(\rho_1)^{b_2}(\mu_1)^{c_2}$$
$$\pi_2 = \left(LT^{-2}\right)(L)^{a_2}\left(ML^{-3}\right)^{b_2}\left(ML^{-1}T^{-1}\right)^{c_2} = M^0T^0L^0$$

Balance:

M: $b_2 + c_2 = 0$.

L: $1 + a_2 - 3b_2 - c_2 = 0$

T: $-2 - c_2 = 0 \therefore c_2 = -2$

Solution:

$c_2 = -2$.

$b_2 = 2$.

$a_2 = 3$

Substituting values:

$$\pi_2 = \frac{g\rho_l^2 d_b^3}{\mu_l^2}$$

Third π-group:

$$\pi_3 = \rho_b (d_b)^{a_3} (\rho_l)^{b_3} (\mu_l)^{c_3}$$
$$\pi_3 = \left(ML^{-3}\right)(L)^{a_3} \left(ML^{-3}\right)^{b_3} \left(ML^{-1}T^{-1}\right)^{c_3} = M^0 T^0 L^0$$

Balance:

M: $1 + b_3 + c_3 = 0$.

L: $-3 + a_3 - 3b_3 - c_3 = 0$

T: $-c_3 = 0$

Solution:

$c_3 = 0$

$b_3 = -1$

$a_3 = 0$

Substituting values:

$$\pi_3 = \frac{\rho_b}{\rho_l}$$

Fourth π-group:

$$\pi_4 = H(d_b)^{a_4} (\rho_l)^{b_4} (\mu_l)^{c_4}$$
$$\pi_4 = (L)(L)^{a_4} \left(ML^{-3}\right)^{b_4} \left(ML^{-1}T^{-1}\right)^{c_4} = M^0 T^0 L^0$$

Balance:

M: $b_4 + c_4 = 0$.

L: $1 + a_4 - 3b_4 - c_4 = 0$

T: $-c_4 = 0$

Solution:

$$c_4 = 0$$
$$b_4 = 0$$
$$a_4 = -1$$

Substituting values:

$$\pi_4 = \frac{H}{d_b}$$

Therefore,

$$\pi_1 = f(\pi_2)(\pi_3)(\pi_4)$$
$$\frac{d_b \rho_l U_p}{\mu_l} = f\left(\frac{g\rho_l^2 d_b^3}{\mu_l^2}\right)\left(\frac{\rho_b}{\rho_l}\right)\left(\frac{H}{d_b}\right)$$

The final general function to describe plume velocity in the tundish is expressed as follows:

$$\frac{d_b \rho_l U_p}{\mu_l} = C\left(\frac{g\rho_l^2 d_b^3}{\mu_l^2}\right)^{y_1}\left(\frac{\rho_b}{\rho_l}\right)^{y_2}\left(\frac{H}{d_b}\right)^{y_3}$$

It should be noticed that the original work by Chatterjee and Chattopadhyay [47] has a mistake in the computation of the second dimensionless group. They reported the following result:

$$\pi_2 = \frac{g d_b^3}{\mu_l^2 \rho_l^2}$$

An important step in dimensional analysis is to make sure that each dimensionless group is dimensionless:

$$\pi_2 = \frac{g d_b^3}{\mu_l^2 \rho_l^2} = \frac{L}{T^2}\frac{L^3}{}\frac{L^2 T^2}{M^2}\frac{L^6}{M^2}$$

Clearly, is non-dimensional. The expression developed is dimensionless:

$$\pi_2 = \frac{g\rho_l^2 d_b^3}{\mu_l^2} = \frac{L}{T^2}\frac{M^2}{L^6}\frac{L^3}{}\frac{L^2 T^2}{M^2} = -$$

The relevant variables that affect bubble size depend on the regime of bubble formation; at low gas flow rates (<1 ml/s) the bubble size depends on the nozzle diameter and properties of the liquid, from 1 to 1000 ml/s the bubble size depends on the gas flow rate and nozzle diameter and, at very high gas flow rates is the jetting regime. Both ladle and tundish operate in the middle range.

Since there is no nozzle diameter in the case of the tundish, the dimension in this case is the height of the liquid. Therefore, the bubble diameter (d_b) depends on the following variables: gas flow rate (Q), height of the liquid (H) and gravity force (g).

$$d_b = f(Q, H, g)$$

Variable	Symbol	Units	Dimensions
Bubble diameter	d_b	m	[L]
Height of liquid steel	H	m	[L]
Gas flow rate (STP)	Q	m³/s	$[L^3T^{-1}]$
Gravitational acceleration	g	m/s²	$[LT^{-2}]$

The problem involves 4 variables and 2 fundamental dimensions. The number of dimensionless groups is two $(n - k = 4 - 2 = 4)$. Using the method of repeating variables, we choose H and g as the repeating variables.

$$\pi_1 = d_b(H)^{a_1}(g)^{b_1}$$
$$\pi_2 = Q(H)^{a_2}(g)^{b_2}$$
$$\pi_1 = f(\pi_2)$$

First π-group:

$$\pi_1 = d_b(H)^{a_1}(g)^{b_1}$$
$$\pi_1 = (L)(L)^{a_1}(LT^{-2})^{b_1} = T^0L^0$$

Balance:
L: $1 + a_1 + b_1 = 0.$
T: $-2b_1 = 0.$

Solution:
$b_1 = 0.$
$a_1 = -1.$

Substitution of these values on the π_1-group yields:

$$\pi_1 = \frac{d_b}{H}$$

Second π-group:

$$\pi_2 = Q(H)^{a_2}(g)^{b_2}$$
$$\pi_2 = \left(L^3T^{-1}\right)(L)^{a_2}(LT^{-2})^{b_2} = T^0L^0$$

Balance:

$L: 3 + a_2 + b_2 = 0.$

$T: -1 - 2b_2 = 0.$

Solution:

$b_2 = -\frac{1}{2}$

$a_2 = -5/2.$

Substituting values:

$$\pi_2 = \frac{Q}{g^{1/2}H^{\frac{5}{2}}} = \frac{Q^2}{gH^5}$$

$$\pi_1 = f(\pi_2)$$

$$\frac{d_b}{H} = f\left(\frac{Q^2}{gH^5}\right)$$

The general function to describe the bubble size as a function of Q, H and g is the following:

$$\frac{d_b}{H} = C\left(\frac{Q^2}{gH^5}\right)^n$$

Therefore, in order to define the form of the function for TOSE, a first group of experimental data between bubble diameter and operational variables is required. Then, the plume velocity as a function of; H, d_b, ρ_l, ρ_b, μ_l, g, is obtained using experimental data, and finally with relationships for bubble size and plume velocity under tundish conditions, a relationship for the slag eye is established. This work is carried out in the following section.

Experimental data from Chattopadhyay and Chatterjee [47, 56]:

Two water models with scales 1:1 and 1:3 were built from a 12 ton. delta shaped, four-strand, billet casting tundish built in the Rio Tinto Iron and Titanium (RTIT) plant at Sorel-Tracy in Canada. A square tank with a height of 3 m and base of 1 m was used to represent the ladle in the full-scale model. The dimensions of the full scale tundish are: height 1.02 m, length 4.38 m, short width 0.51 m and long width 1.02 m. The flow rate of water was 0.17 m³/min which maintained a steady-state height of water of 0.5 m. The immersion depth of the ladle shroud was 60 mm. The slag phase was simulated using polyethylene beds (PE) with a density of 920 kg/m³, with a diameter of 2.5–3 mm. the thickness of the slag layer was 0.02 m. The flow rate of water in the one third scale model was calculated using Froude similarity, corresponding to 0.01 m³/min.

$$Q_m = \lambda^{2.5} \cdot Q_p$$

The slag layer was simulated using three different oils: mineral oil, Mazola oil and machine oil. Their physical properties and the range of the variables are shown below. The slag thickness in the full-scale model is equivalent to 4–6% of the height of the liquid and for the one third scale model from 6 to 18%. Compressed air was employed as the gas phase, to replace argon.

	ρ, kg/m^3	μ, Pa s	ν, m^2/s
Water	998	1×10^{-3}	1×10^{-6}
Mazola oil	925	5×10^{-2}	5.41×10^{-5}
Mineral oil	870	1.7×10^{-2}	1.95×10^{-5}
Machine oil	910	4.55×10^{-2}	5×10^{-5}
PE	920	1×10^{-2}	1.09×10^{-5}

	Q, Nl/min	H, m	h_s, m
Full scale model	3.4–46.8	0.5, 0.6, 0.7	0.03
1:3 scale model	0.2–1.7	0.167	0.01, 0.02, 0.03

Measurements: High-definition video photography was used to measure both bubble size and slag eye areas. The bubble diameter was measured using the full-scale model.

Table 11.27 shows the experimental data about the average bubble diameter as a function of gas flow rate and height of the liquid. Notice that the bubble diameter ranges from 4 to 12 mm.

Applying the statistical tools of excel the following values are obtained: $\log C = 0.08$, $n = 0.2908$. The regression coefficient (R^2) is 0.876. Replacing these values in the general expression we obtain the following solution:

$$\frac{d_b}{H} = 1.2 \left(\frac{Q^2}{gH^5} \right)^{0.2908}$$

This equation is used to compute the bubble diameter in the one third scale model. Figure 11.33 describes the previous relationship.

Now the experimental data of plume velocity as a function of bubble diameter, height of liquid and other variables is analyzed. Since only one type of gas is injected (air), the ratio of gas density to liquid density was constant in the experimental work, therefore this dimensionless group was included in the constant, as follows:

$$\frac{d_b \rho_l U_p}{\mu_l} = C \left(\frac{g\rho_l^2 d_b^3}{\mu_l^2} \right)^{y_1} \left(\frac{H}{d_b} \right)^{y_2}$$

Table 11.28 shows the experimental data, including the bubble diameter predictions for the one third scale model.

Table 11.27 Experimental data on bubble diameter

H, m	Q, m³ s⁻¹	d_b, m	$\frac{d_b}{H}$	$\frac{Q^2}{gH^5}$	$\log\left(\frac{d_b}{H}\right)$	$\log\left(\frac{Q^2}{gH^5}\right)$
0.5	5.67E−05	0.004	0.0070	1.05E−08	−2.155	−7.979
0.5	1.13E−04	0.004	0.0070	4.17E−08	−2.155	−7.380
0.5	1.70E−04	0.004	0.0080	9.43E−08	−2.097	−7.026
0.5	2.27E−04	0.005	0.0100	1.68E−07	−2.000	−6.774
0.5	2.83E−04	0.005	0.0100	2.61E−07	−2.000	−6.583
0.5	3.40E−04	0.008	0.0160	3.77E−07	−1.796	−6.424
0.5	3.97E−04	0.008	0.0160	5.14E−07	−1.796	−6.289
0.5	4.53E−04	0.010	0.0200	6.69E−07	−1.699	−6.174
0.5	5.10E−04	0.010	0.0200	8.48E−07	−1.699	−6.071
0.5	5.67E−04	0.012	0.0240	1.05E−06	−1.620	−5.979
0.5	6.23E−04	0.012	0.0240	1.27E−06	−1.620	−5.898
0.5	6.80E−04	0.012	0.0240	1.51E−06	−1.620	−5.822
0.5	7.37E−04	0.012	0.0240	1.77E−06	−1.620	−5.752
0.6	5.67E−05	0.004	0.0058	4.21E−09	−2.234	−8.375
0.6	1.13E−04	0.004	0.0058	1.67E−08	−2.234	−7.776
0.6	1.70E−04	0.004	0.0067	3.79E−08	−2.176	−7.422
0.6	2.27E−04	0.005	0.0083	6.76E−08	−2.079	−7.170
0.6	2.83E−04	0.005	0.0083	1.05E−07	−2.079	−6.979
0.6	3.40E−04	0.008	0.0133	1.52E−07	−1.875	−6.819
0.6	3.97E−04	0.008	0.0133	2.07E−07	−1.875	−6.685
0.6	4.53E−04	0.010	0.0167	2.69E−07	−1.778	−6.570
0.6	5.10E−04	0.010	0.0167	3.41E−07	−1.778	−6.467
0.6	5.67E−04	0.012	0.0200	4.21E−07	−1.699	−6.375
0.6	6.23E−04	0.012	0.0200	5.09E−07	−1.699	−6.293
0.6	6.80E−04	0.012	0.0200	6.06E−07	−1.699	−6.217
0.6	7.37E−04	0.012	0.0200	7.12E−07	−1.699	−6.147
0.7	6.00E−05	0.004	0.0050	2.18E−09	−2.301	−8.661
0.7	1.20E−04	0.004	0.0050	8.73E−09	−2.301	−8.059
0.7	1.80E−04	0.004	0.0057	1.97E−08	−2.243	−7.707
0.7	2.40E−04	0.005	0.0071	3.49E−08	−2.146	−7.457
0.7	3.00E−04	0.005	0.0071	5.46E−08	−2.146	−7.263
0.7	3.60E−04	0.008	0.0114	7.86E−08	−1.942	−7.105
0.7	4.20E−04	0.009	0.0129	1.07E−07	−1.891	−6.971
0.7	4.80E−04	0.010	0.0143	1.40E−07	−1.845	−6.855
0.7	5.40E−04	0.012	0.0164	1.77E−07	−1.784	−6.752
0.7	6.00E−04	0.012	0.0171	2.18E−07	−1.766	−6.661

(continued)

Table 11.27 (continued)

H, m	Q, m³ s⁻¹	d_b, m	$\frac{d_b}{H}$	$\frac{Q^2}{gH^5}$	$\log\left(\frac{d_b}{H}\right)$	$\log\left(\frac{Q^2}{gH^5}\right)$
0.7	6.60E−04	0.012	0.0171	2.64E−07	−1.766	−6.578
0.7	7.20E−04	0.012	0.0171	3.14E−07	−1.766	−6.502
0.7	7.80E−04	0.014	0.0197	3.69E−07	−1.705	−6.433

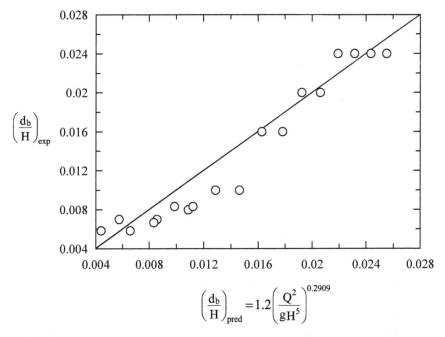

$$\left(\frac{d_b}{H}\right)_{pred} = 1.2\left(\frac{Q^2}{gH^5}\right)^{0.2909}$$

Fig. 11.33 Predicted versus experimental results on the dimensionless bubble diameter by dimensional analysis. Adapted from Chattopadhyay and Chatterjee [47, 56]

Applying the statistical tools of excel the following values are obtained: $\log C = 0.232$, $y_1 = 0.568$, $y_2 = 0.0179$. The regression coefficient (R^2) is 0.995. Replacing these values in the general expression we obtain the following solution:

$$\frac{d_b \rho_l U_p}{\mu_l} = 1.709\left(\frac{g\rho_l^2 d_b^3}{\mu_l^2}\right)^{0.568}\left(\frac{H}{d_b}\right)^{0.018}$$

This equation is employed to compute the velocity of the plume for all the experiments, both full scale and one third scale models. The main difference of this relationship with that reported by Chatterjee and Chattopadhyay [47] is in the constant. These authors reported the following values: $C = 1.257 \times 10^7$, $y_1 = 0.587$, $y_2 = 0.046$. In spite of the large differences in the constant, comparing the predictions

Table 11.28 Experimental data on plume velocities

Scale	U_p, ms^{-1}	H, m	d_b, m	ρ_l kg/m^3	μ_l Pa s	g m/s^2	$\dfrac{d_b\rho_l U_p}{\mu_l}$	$\dfrac{g\rho_l^2 d_b^3}{\mu_l^2}$	$\dfrac{H}{d_b}$
1:3	0.29	0.167	0.0006	998	0.001	9.81	171.2	2.12E−09	278
1:3	0.37	0.167	0.0012	998	0.001	9.81	443.5	1.70E−08	139
1:3	0.46	0.167	0.0017	998	0.001	9.81	772.1	4.82E−08	98
1:3	0.54	0.167	0.0021	998	0.001	9.81	1131.1	9.09E−08	80
1:3	0.63	0.167	0.0025	998	0.001	9.81	1558.0	1.53E−07	67
1:3	0.71	0.167	0.0029	998	0.001	9.81	2051.9	2.39E−07	58
1:3	0.80	0.167	0.0032	998	0.001	9.81	2535.4	3.21E−07	52
1:3	0.88	0.167	0.0034	998	0.001	9.81	2981.1	3.86E−07	49
1:1	0.73	0.5	0.0035	998	0.001	9.81	2547.6	4.21E−07	143
1:1	0.85	0.5	0.0035	998	0.001	9.81	2980.2	4.21E−07	143
1:1	0.98	0.5	0.0040	998	0.001	9.81	3897.2	6.28E−07	125
1:1	1.10	0.5	0.0050	998	0.001	9.81	5491.0	1.23E−06	100
1:1	1.23	0.5	0.0050	998	0.001	9.81	6108.5	1.23E−06	100
1:1	1.35	0.5	0.0080	998	0.001	9.81	10762.5	5.02E−06	63
1:1	1.47	0.5	0.0080	998	0.001	9.81	11745.7	5.02E−06	63
1:1	1.60	0.5	0.0100	998	0.001	9.81	15924.2	9.81E−06	50
1:1	1.72	0.5	0.0100	998	0.001	9.81	17156.2	9.81E−06	50
1:1	1.85	0.5	0.0120	998	0.001	9.81	22071.9	1.70E−05	42
1:1	1.97	0.5	0.0120	998	0.001	9.81	23560.0	1.70E−05	42
1:1	2.10	0.5	0.0120	998	0.001	9.81	25046.8	1.70E−05	42
1:1	2.22	0.5	0.0120	998	0.001	9.81	26524.1	1.70E−05	42
1:1	0.73	0.6	0.0035	998	0.001	9.81	2545.5	4.21E−07	171
1:1	0.85	0.6	0.0035	998	0.001	9.81	2980.2	4.21E−07	171
1:1	0.98	0.6	0.0040	998	0.001	9.81	3900.0	6.28E−07	150
1:1	1.10	0.6	0.0050	998	0.001	9.81	5490.0	1.23E−06	120
1:1	1.23	0.6	0.0050	998	0.001	9.81	6106.0	1.23E−06	120
1:1	1.35	0.6	0.0080	998	0.001	9.81	10762.5	5.02E−06	75
1:1	1.47	0.6	0.0080	998	0.001	9.81	11750.5	5.02E−06	75
1:1	1.60	0.6	0.0100	998	0.001	9.81	15923.2	9.81E−06	60
1:1	1.72	0.6	0.0100	998	0.001	9.81	17162.2	9.81E−06	60
1:1	1.85	0.6	0.0120	998	0.001	9.81	22070.7	1.70E−05	50
1:1	1.97	0.6	0.0120	998	0.001	9.81	23560.0	1.70E−05	50
1:1	2.10	0.6	0.0120	998	0.001	9.81	25049.2	1.70E−05	50
1:1	2.22	0.6	0.0120	998	0.001	9.81	26530.1	1.70E−05	50
1:1	0.73	0.7	0.0035	998	0.001	9.81	2545.1	4.21E−07	200

(continued)

Table 11.28 (continued)

Scale	U_p, ms^{-1}	H, m	d_b, m	ρ_l kg/m^3	μ_l Pa s	g m/s^2	$\frac{d_b\rho_l U_p}{\mu_l}$	$\frac{g\rho_l^2 d_b^3}{\mu_l^2}$	$\frac{H}{d_b}$
1:1	0.85	0.7	0.0035	998	0.001	9.81	2977.1	4.21E−07	200
1:1	0.98	0.7	0.0040	998	0.001	9.81	3897.6	6.28E−07	175
1:1	1.10	0.7	0.0050	998	0.001	9.81	5492.5	1.23E−06	140
1:1	1.23	0.7	0.0050	998	0.001	9.81	6108.5	1.23E−06	140
1:1	1.35	0.7	0.0080	998	0.001	9.81	10756.9	5.02E−06	88
1:1	1.47	0.7	0.0090	998	0.001	9.81	13219.3	7.15E−06	78
1:1	1.60	0.7	0.0100	998	0.001	9.81	15917.2	9.81E−06	70
1:1	1.85	0.7	0.0115	998	0.001	9.81	21145.4	1.49E−05	61
1:1	1.97	0.7	0.0120	998	0.001	9.81	23557.6	1.70E−05	58
1:1	2.10	0.7	0.0120	998	0.001	9.81	25040.8	1.70E−05	58
1:1	2.22	0.7	0.0120	998	0.001	9.81	26532.5	1.70E−05	58

using their relationship and our result, the difference in velocities is less than 10%. Figure 11.34 compares the predicted and experimental values for the dimensionless plume velocity.

The final step is the relationship to describe TOSE, under the assumption that the exponent in the ratio of gas density/liquid density is unity.

$$\pi_1 = f\,(\pi_2)(\pi_3)(\pi_4)$$
$$\frac{A_e}{Hh} = C\left(\frac{U_p^2}{gh}\right)^{x_1}\left(\frac{\Delta\rho}{\rho_l}\right)^{x_2}\left(\frac{\nu_s}{hU_p}\right)^{x_4}$$

It should be noticed that data corresponding to slag eyes were taken from the plots reported by Chatterjee and Chattopadhyay [47]. There is a small error compared with the real value, on average 5%. Table 11.29 contains the experimental database to compute the ladle eye by dimensionless analysis.

Applying the statistical tools of excel for multi-linear regression analysis and transforming the expression for slag eye in log space, the following values are obtained: log C = 2.48, $x_1 = 1.649$, $x_2 = 1.605$, $x_3 = 0.0688$. The regression coefficient (R^2) is 0.962. Replacing these values in the general expression we obtain the following solution:

$$\log\frac{A_e}{Hh} = \log C + x_1\log\left(\frac{U_p^2}{gH}\right) + x_2\log\left(\frac{\Delta\rho}{\rho_l}\right) + x_3\log\left(\frac{\nu_s}{HU_p}\right)$$

Replacing the constant and the exponents:

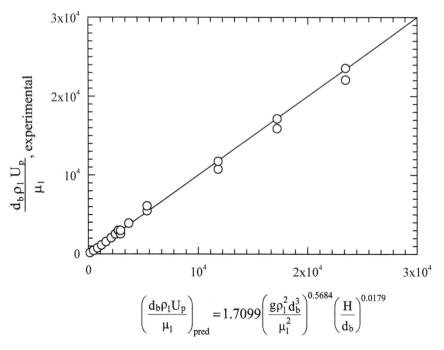

$$\left(\frac{d_b\rho_1 U_p}{\mu_1}\right)_{pred} = 1.7099\left(\frac{g\rho_1^2 d_b^3}{\mu_1^2}\right)^{0.5684}\left(\frac{H}{d_b}\right)^{0.0179}$$

Fig. 11.34 Predicted versus experimental results on the dimensionless plume velocity by dimensional analysis. Adapted from Chattopadhyay and Chatterjee [47, 56]

$$\frac{A_e}{Hh} = 308.2\left(\frac{U_p^2}{gH}\right)^{1.649}\left(\frac{\Delta\rho}{\rho_l}\right)^{1.605}\left(\frac{v_s}{HU_p}\right)^{0.0688}$$

Comparison of slag eye calculated with the previous expression and experimental data is shown in Fig. 11.35.

11.6   Liquid Steel Discharge from a Ladle to the Tundish Using a Protective Shroud [57–62]

In continuous casting the transfer of liquid steel from the ladle to the tundish is carried out by opening a slide-gate at the bottom of the ladle. In this way liquid steel is discharged. If the stream of liquid steel is not protected during this transfer operation, oxygen and nitrogen from the surrounding atmosphere interact with the stream of liquid steel. In many applications this reoxidation should be fully prevented. This is done using a shroud partially immersed in the tundish and injecting a protective gas inside the shroud.

Table 11.29 Experimental database to compute ladle eye by dimensionless analysis

Oil type	h, m	H, m	db, m	U_p, m/s	A, m^2	Q, m^3/s	π_1	π_2	π_3	π_4
PE	0.03	0.5	0.0035	0.840	0.11	5.7E−05	7.6	2.4	0.078	4.3E−04
PE	0.03	0.5	0.0035	0.840	0.16	1.1E−04	10.4	2.4	0.078	4.3E−04
PE	0.03	0.5	0.004	0.921	0.25	1.7E−04	17.0	2.9	0.078	3.9E−04
PE	0.03	0.5	0.005	1.073	0.33	2.3E−04	22.2	3.9	0.078	3.4E−04
PE	0.03	0.5	0.005	1.073	0.57	2.8E−04	37.8	3.9	0.078	3.4E−04
PE	0.03	0.5	0.008	1.483	0.78	3.4E−04	52.2	7.5	0.078	2.5E−04
PE	0.03	0.5	0.008	1.483	1.00	4.0E−04	66.5	7.5	0.078	2.5E−04
PE	0.03	0.5	0.01	1.729	1.27	4.5E−04	84.8	10.2	0.078	2.1E−04
PE	0.03	0.5	0.01	1.729	2.03	5.1E−04	135.7	10.2	0.078	2.1E−04
PE	0.03	0.5	0.012	1.959	2.47	5.7E−04	164.3	13.0	0.078	1.9E−04
PE	0.03	0.5	0.012	1.959	2.78	6.2E−04	185.2	13.0	0.078	1.9E−04
PE	0.03	0.5	0.012	1.959	3.21	6.8E−04	213.9	13.0	0.078	1.9E−04
PE	0.03	0.5	0.012	1.959	**3.80**	7.4E−04	253.0	13.0	0.078	1.9E−04
PE	0.03	0.6	0.0035	0.843	0.12	5.7E−05	6.5	2.4	0.078	4.3E−04
PE	0.03	0.6	0.0035	0.843	0.21	1.1E−04	11.8	2.4	0.078	4.3E−04
PE	0.03	0.6	0.004	0.924	0.41	1.7E−04	22.6	2.9	0.078	3.9E−04
PE	0.03	0.6	0.005	1.077	0.58	2.3E−04	32.4	3.9	0.078	3.4E−04
PE	0.03	0.6	0.005	1.077	0.88	2.8E−04	48.6	3.9	0.078	3.4E−04
PE	0.03	0.6	0.008	1.488	1.23	3.4E−04	68.2	7.5	0.078	2.4E−04
PE	0.03	0.6	0.008	1.488	1.60	4.0E−04	88.8	7.5	0.078	2.4E−04
PE	0.03	0.6	0.01	1.734	1.77	4.5E−04	98.5	10.2	0.078	2.1E−04
PE	0.03	0.6	0.01	1.734	2.83	5.1E−04	157.3	10.2	0.078	2.1E−04
PE	0.03	0.6	0.012	1.966	4.05	5.7E−04	224.7	13.1	0.078	1.8E−04
PE	0.03	0.6	0.012	1.966	4.75	6.2E−04	263.9	13.1	0.078	1.8E−04
PE	0.03	0.6	0.012	1.966	5.20	6.8E−04	288.8	13.1	0.078	1.8E−04
PE	0.03	0.6	0.012	1.966	6.41	7.4E−04	356.3	13.1	0.078	1.8E−04
PE	0.03	0.7	0.0035	0.845	0.12	5.7E−05	5.7	2.4	0.078	4.3E−04
PE	0.03	0.7	0.0035	0.845	0.34	1.2E−04	16.1	2.4	0.078	4.3E−04
PE	0.03	0.7	0.004	0.926	0.46	1.8E−04	21.8	2.9	0.078	3.9E−04
PE	0.03	0.7	0.005	1.080	0.67	2.4E−04	32.1	4.0	0.078	3.4E−04
PE	0.03	0.7	0.005	1.080	1.09	3.0E−04	51.7	4.0	0.078	3.4E−04
PE	0.03	0.7	0.008	1.492	1.32	3.6E−04	63.0	7.6	0.078	2.4E−04
PE	0.03	0.7	0.009	1.618	1.85	4.2E−04	88.2	8.9	0.078	2.2E−04
PE	0.03	0.7	0.01	1.739	2.11	4.8E−04	100.4	10.3	0.078	2.1E−04
PE	0.03	0.7	0.0115	1.914	3.24	5.4E−04	154.3	12.5	0.078	1.9E−04

(continued)

Table 11.29 (continued)

Oil type	h, m	H, m	db, m	U_p, m/s	A, m^2	Q, m^3/s	π_1	π_2	π_3	π_4
PE	0.03	0.7	0.012	1.971	4.47	6.0E−04	212.9	13.2	0.078	1.8E−04
PE	0.03	0.7	0.012	1.971	5.41	6.6E−04	257.6	13.2	0.078	1.8E−04
PE	0.03	0.7	0.012	1.971	6.72	7.2E−04	319.9	13.2	0.078	1.8E−04
PE	0.03	0.7	0.0138	2.170	8.20	7.8E−04	390.5	16.0	0.078	1.7E−04
Mineral oil	0.01	0.167	0.00061	0.248	0.01	3.5E−06	7.1	0.6	0.128	7.9E−03
Mineral oil	0.01	0.167	0.00116	0.386	0.03	7.1E−06	15.1	1.5	0.128	5.1E−03
Mineral oil	0.01	0.167	0.00167	0.495	0.05	1.1E−05	29.6	2.5	0.128	3.9E−03
Mineral oil	0.01	0.167	0.00212	0.584	0.09	1.4E−05	51.7	3.5	0.128	3.3E−03
Mineral oil	0.01	0.167	0.00252	0.657	0.13	1.8E−05	77.0	4.4	0.128	3.0E−03
Mineral oil	0.01	0.167	0.0028	0.707	0.17	2.1E−05	104.2	5.1	0.128	2.8E−03
Mineral oil	0.01	0.167	0.00317	0.769	0.28	2.5E−05	166.0	6.0	0.128	2.5E−03
Mineral oil	0.01	0.167	0.00343	0.812	0.34	2.8E−05	204.9	6.7	0.128	2.4E−03
Mazola oil	0.01	0.167	0.0006	0.245	0.01	3.5E−06	3.5	0.6	0.073	2.2E−02
Mazola oil	0.01	0.167	0.00116	0.386	0.01	7.1E−06	5.7	1.5	0.073	1.4E−02
Mazola oil	0.01	0.167	0.00166	0.493	0.02	1.1E−05	14.4	2.5	0.073	1.1E−02
Mazola oil	0.01	0.167	0.00212	0.584	0.04	1.4E−05	22.7	3.5	0.073	9.3E−03
Mazola oil	0.01	0.167	0.00252	0.657	0.08	1.8E−05	45.9	4.4	0.073	8.2E−03
Mazola oil	0.01	0.167	0.00288	0.720	0.08	2.1E−05	49.6	5.3	0.073	7.5E−03
Mazola oil	0.01	0.167	0.00317	0.769	0.11	2.5E−05	64.6	6.0	0.073	7.0E−03
Mazola oil	0.01	0.167	0.00343	0.812	0.12	2.8E−05	71.6	6.7	0.073	6.7E−03
Mazola oil	0.02	0.167	0.00061	0.248	0.00	3.5E−06	0.9	0.3	0.073	1.1E−02
Mazola oil	0.02	0.167	0.00116	0.386	0.00	7.1E−06	1.4	0.8	0.073	7.0E−03
Mazola oil	0.02	0.167	0.00167	0.495	0.01	1.1E−05	3.8	1.3	0.073	5.5E−03
Mazola oil	0.02	0.167	0.00212	0.584	0.02	1.4E−05	5.5	1.7	0.073	4.6E−03
Mazola oil	0.02	0.167	0.00253	0.659	0.05	1.8E−05	14.5	2.2	0.073	4.1E−03
Mazola oil	0.02	0.167	0.0029	0.724	0.06	2.1E−05	17.2	2.7	0.073	3.7E−03
Mazola oil	0.02	0.167	0.00318	0.771	0.07	2.5E−05	21.2	3.0	0.073	3.5E−03
Mazola oil	0.02	0.167	0.00343	0.812	0.08	2.8E−05	25.0	3.4	0.073	3.3E−03
Machine oil	0.01	0.167	0.00061	0.248	0.01	3.5E−06	3.7	0.6	0.088	2.0E−02
Machine oil	0.01	0.167	0.00116	0.386	0.01	7.1E−06	7.4	1.5	0.088	1.3E−02
Machine oil	0.01	0.167	0.00167	0.495	0.03	1.1E−05	16.9	2.5	0.088	1.0E−02
Machine oil	0.01	0.167	0.00212	0.584	0.05	1.4E−05	28.0	3.5	0.088	8.6E−03

(continued)

Table 11.29 (continued)

Oil type	h, m	H, m	db, m	U_p, m/s	A, m²	Q, m³/s	π_1	π_2	π_3	π_4
Machine oil	0.01	0.167	0.00252	0.657	0.08	1.8E−05	50.5	4.4	0.088	7.6E−03
Machine oil	0.01	0.167	0.00288	0.720	0.11	2.1E−05	64.8	5.3	0.088	6.9E−03
Machine oil	0.01	0.167	0.00318	0.771	0.12	2.5E−05	73.1	6.1	0.088	6.5E−03
Machine oil	0.01	0.167	0.00343	0.812	0.14	2.8E−05	84.9	6.7	0.088	6.2E−03
Machine oil	0.03	0.167	0.0006	0.245	0.00	3.5E−06	0.3	0.2	0.088	6.8E−03
Machine oil	0.03	0.167	0.00116	0.386	0.00	7.1E−06	0.9	0.5	0.088	4.3E−03
Machine oil	0.03	0.167	0.00167	0.495	0.01	1.1E−05	2.8	0.8	0.088	3.4E−03
Machine oil	0.03	0.167	0.00212	0.584	0.02	1.4E−05	4.0	1.2	0.088	2.9E−03
Machine oil	0.03	0.167	0.00253	0.659	0.04	1.8E−05	7.5	1.5	0.088	2.5E−03
Machine oil	0.03	0.167	0.0029	0.724	0.05	2.1E−05	10.8	1.8	0.088	2.3E−03
Machine oil	0.03	0.167	0.00318	0.771	0.08	2.5E−05	15.3	2.0	0.088	2.2E−03
Machine oil	0.03	0.167	0.00343	0.812	0.09	2.8E−05	18.2	2.2	0.088	2.1E−03

Additional data: kinematic viscosity of oils, oils density and density differences oil-water

A conventional ladle shroud system includes a slide gate, a collector nozzle and the shroud proper. Some features are shown in Fig. 11.36. The shroud proper is a refractory tube used to transfer liquid steel from the ladle to the tundish, avoiding direct contact with the surrounding atmosphere. The slide gate opens during teeming from 30 to 100% and discharges liquid steel by gravity into the tundish. A perfect sealing between the collector nozzle and the shroud inhibits air ingress, however under industrial conditions, due to misalignment and wearing of materials air ingress is possible. It can be estimated based on the increment of nitrogen which usually ranges from 3 to 5 ppm but can be as high as 12 ppm [61]. The extent of nitrogen pick-up gives a measure of oxygen ingress and potential reoxidation of liquid steel.

There are three ways to introduce the protective gas into the shroud; in two cases the gas is injected in the vicinity of the shroud-collector nozzle joint and a third case injected the gas across the shroud cross section via a circular groove running on the inside of the shroud wall. Singh and Mazumdar [61] reported large variations on the conditions of argon injection through the shroud in the continuous casting of both

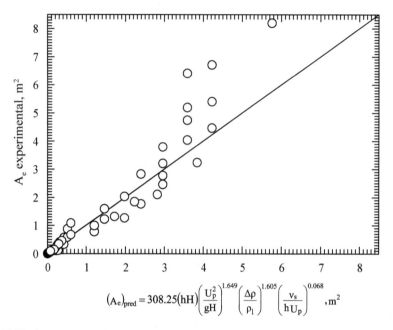

$$(A_e)_{pred} = 308.25(hH)\left(\frac{U_p^2}{gH}\right)^{1.649}\left(\frac{\Delta\rho}{\rho_1}\right)^{1.605}\left(\frac{v_s}{hU_p}\right)^{0.068}, m^2$$

Fig. 11.35 Comparison predicted versus experimental data on TOSE. Adapted from Chattopadhyay and Chatterjee [47, 56]

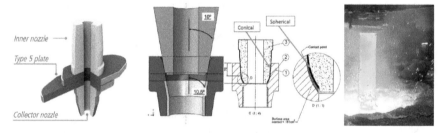

Fig. 11.36 Illustrations showing the slide gate, collector nozzle and shroud arrangements. Taken from Vesuvius [57]

blooms and slabs and found that the gas flow rate has large effects on the teeming process. Due to opacity of liquid steel what happens inside of a shroud cannot be visualized. In order to visualize the fluidynamics inside a shroud it is necessary to build a scaled physical model.

There is a large variety of shroud designs. Wen et al. [57] reported some benefits using the trumpet-shaped ladle shroud (TLS). The swirling ladle shroud (SLS) has been reported to decrease the impact velocity at the tundish bottom, about 1/3 compared with the conventional design [63]. A variation of SLS is the Dissipative Ladle shroud (DLS) [64]. The conventional ladle shroud (CLS) is a straight pipe. Figure 11.37 shows the previous shroud designs.

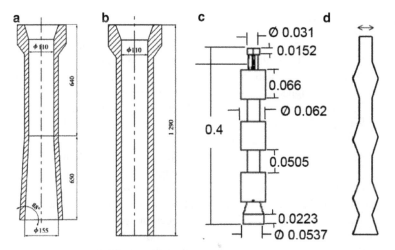

Fig. 11.37 Ladle shroud designs. **a** TLS, **b** CLS, **c** SLS, **d** DLS

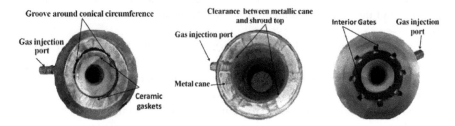

Fig. 11.38 Gas injection modes in the shroud. Taken from Singh [58]

The way the gas is injected into the shroud also has some variations. Figure 11.38 shows three modes of gas injection. In the first mode the gas is distributed across the shroud cross section using a circular groove. In the other modes the gas is released in the vicinity of the shroud-collector nozzle joint.

The interaction liquid steel-refractory is critical to steel cleanliness. Currently the most common type of refractories for shrouds are alumina-graphite. The mixture is sintered using phenolic resin as a bonding agent. Typical chemical compositions (wt%) for two plants are shown in the following table.

Component	A	B
Al_2O_3	53	39
Carbon	31	32
SiO_2	13	28
Ani-oxidants	3	1

Experimental work:

One of the objectives of Singh's work was to define a single dimensionless equation capable to describe the height of the jet for different types of shrouds employed in bloom casting (BCS) and slab casting (SCS) as a function the relevant variables. A water model was designed with a geometric scale $\lambda = 1:1$. This scale allows to fully satisfy dynamic similarity. There is no slide gate to control the discharge of water into the shroud. This is equivalent to the case of 100% opening in the real case.

Figure 11.39a shows a water model illustrating the flow rate of the liquid stream (Q_L), the gas flow rate (Q_g), shroud diameter (D_s), shroud's depth of immersion (d_s) and the height of the free jet (H_m). The total length of the shroud is L_s.

The range in the variables investigated is summarized in Table 11.30. It can be observed that the dimensions of the shroud for slab casting are higher because of the larger flow rate of liquid steel. The gas injection port diameter in all cases was 6.35 mm. The gas flow rate is referred to 1 atm and 1600 °C.

Preliminary experiments were carried out to observe the behavior of the free jet.

In order to read the height of the free jet, the exact position of the liquid/gas interphase also called meniscus position, is required, however this surface is highly dynamic and therefore a minimum of 5–7 readings are needed to take an average value.

It was found a drastic effect of the gas flow rate on the height of the confined jet, as can be shown from Fig. 11.40 reported by Singh and Mazumdar [61]. This figure shows a progressive increase in the ratio Q_g/Q_L, starting from the initial point when liquid is discharged. Cases (a) and (b) represents the initial teeming process or a change of a ladle. Due to a decrease in the level of steel in the tundish, the tip of the shroud can be exposed to the atmosphere. The initial teeming is carried out in this condition because if the shroud is immersed backflow of liquid can occur and then leakage through the shroud-collector joint. Case (c) represents the case when the shroud is immersed and there is no gas injection. Case (d) represent the case of a low Q_g/Q_L ratio. In cases (e) and (f) the ratio Q_g/Q_L is increased. From these experiments it is clear that the ratio Q_g/Q_L has a large effect on the height of the confined jet.

At extremely low gas flow rates (case d) the height of the free jet is zero and the gas dissolves in the liquid but once a critical minimum gas flow rate is reached there is the formation of a free jet.

The effect of the immersion depth of the shroud was investigated; The experimental results for BCS are shown in Table 11.31. It was observed that the immersion depth has practically no effect on the height of the confined jet. The values for the height reported represent an average of 7 measurements. This result was attributed to similar pressures at the tip of the shroud for different immersion depths, taking into account that at the top of the surface of the liquid inside the shroud, the pressure is atmospheric for both cases. Due to density differences in the case of the liquid steel case, a small effect of the immersion depth can be expected [61].

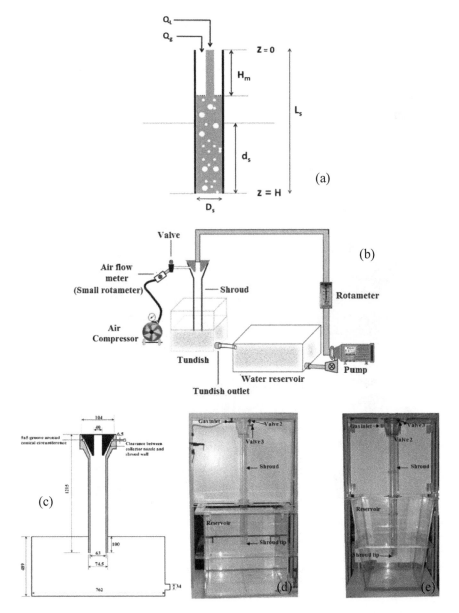

Fig. 11.39 Discharge of a liquid stream through a shroud; **a** schematic, **b** general experimental set up, **c** dimensions bloom casting shroud (BCS), **d** BCS model, **e** SCS model. After Singh [58]

Table 11.30 Variables in the prototype and the water model

	BCS		SCS	
	Prototype	WM	Prototype	WM
L_S (mm)	1215	1215	1640	1640
D_S (mm)	65	62.8	110	95
D_{CN} (mm)	40.45	40	70	70
Q_L (l/min)	98–140	98–140	350–460	350–460
Q_g (l/min)	12.4–24.8	12.4–24.8	78–96	78–96
d_S (mm)	100–300	100–300	100–400	100–400

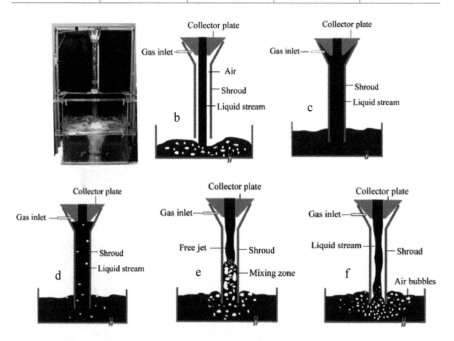

Fig. 11.40 Effect of gas flow rate on the height of the confined jet. After [61]

Table 11.31 Effect of immersion depth. After [58]

$Q_L \times 10^4$, m^3/s	$Q_g \times 10^4$, m^3/s	d_s, mm	$\overline{H}_m$, mm
16.67	3.33	50	377
		100	376
		150	378
		200	381
		250	376

Dimensional analysis I: Based on the previous knowledge of the variables that affect the height of the confided jet, the following 5 variables were considered: gas flow rate, liquid flow rate, shroud diameter and gravity,

$$H_m = f\left(Q_g, Q_L, D_s, g\right)$$

The dimensionless group are defined using the matrix method.

$$H_m = f(Q_g)^a (Q_L)^b (D_s)^c (g)^d$$

The variables and their dimensions are related as follows:

	H_m	Q_g	Q_L	D_s	g
L	1	3	3	1	1
T	−1	−1	−1	0	−2

The dimensional matrix is expressed as follows:

$$D = \begin{pmatrix} 1 & 3 & 3 & 1 & 1 \\ 0 & -1 & -1 & 0 & -2 \end{pmatrix}$$

The augmented matrix becomes:

$$\begin{pmatrix} 1 & 0 & 0 & 0 & 0 \\ 0 & 1 & 0 & 0 & 0 \\ 0 & 0 & 1 & 0 & 0 \\ 1 & 3 & 3 & 1 & 1 \\ 0 & -1 & -1 & 0 & -2 \end{pmatrix}$$

The square (S) and remaining (R) matrices are:

$$S = \begin{pmatrix} 1 & 1 \\ 0 & -2 \end{pmatrix}, R = \begin{pmatrix} 1 & 3 & 3 \\ 0 & -1 & -1 \end{pmatrix}$$

The solution requires S^{-1} and $-S^{-1}R$:

$$\begin{pmatrix} I & 0 \\ -S^{-1}R & S^{-1} \end{pmatrix} = \begin{pmatrix} 1 & 0 & 0 & 0 & 0 \\ 0 & 1 & 0 & 0 & 0 \\ 0 & 0 & 1 & 0 & 0 \\ -1 & -\frac{5}{2} & -\frac{5}{2} & 1 & \frac{1}{2} \\ 0 & -\frac{1}{2} & -\frac{1}{2} & 0 & -\frac{1}{2} \end{pmatrix}$$

The π-groups are defined from the previous solution, as follows:

$$
\begin{array}{ccccccc}
 & \pi_1 & \pi_2 & \pi_3 & & & \\
H_m & 1 & 0 & 0 & 0 & 0 \\
Q_g & 0 & 1 & 0 & 0 & 0 \\
Q_L & 0 & 0 & 1 & 0 & 0 \\
D_s & -1 & -\frac{5}{2} & -\frac{5}{2} & 1 & \frac{1}{2} \\
g & 0 & -\frac{1}{2} & -\frac{1}{2} & 0 & -\frac{1}{2}
\end{array}
$$

$$
\pi_1 = \frac{H_m}{D_s}
$$

$$
\pi_2 = \left(\frac{Q_g^2}{gD_s^5} \right)
$$

$$
\pi_3 = \left(\frac{Q_L^2}{gD_s^5} \right)
$$

$$
\pi_1 = f(\pi_2, \pi_3)
$$

$$
\frac{H_m}{D_s} = C \left(\frac{Q_g^2}{gD_s^5} \right)^a \left(\frac{Q_L^2}{gD_s^5} \right)^b
$$

Equivalent new groups; π_2-group dividing π_2/π_3 and π_3-group $1/\pi_3$:

$$
\frac{H_m}{D_s} = K \left(\frac{Q_g}{Q_L} \right)^a \left(\frac{gD_s^5}{Q_L^2} \right)^b
$$

The previous equation can be recast in logarithmic form, as follows.

$$
\ln \left(\frac{H_m}{D_s} \right) = \ln K + a \ln \left(\frac{Q_g}{Q_L} \right) + b \ln \left(\frac{gD_s^5}{Q_L^2} \right)
$$

Performing a multilinear regression analysis with the experimental data, the constants, $\ln K$, a and b can be found. If the regression coefficient is higher than about 0.95 it will be an indication that the previous dimensionless groups correctly describe the dimensionless height of the free jet. When the whole set of results were evaluated, a comparison between the experimental data and the predicted value reported a large standard error. However, if the experimental data is analyzed on a separate basis for each type of shroud the resulting equation satisfactorily predicts the experimental data, as will be shown below. Table 11.32 shows the experimental data for the slab caster and Table 11.33 the experimental data for the bloom caster. In the last three columns the units involve only meters and seconds.

The multilinear regression analysis was carried out in excel. Once the analysis is completed, the values for the intercept and constants are provided on a separate sheet.

Table 11.32 Experimental results for SCS, $D_S = 95$ mm, $D_{CN} = 70$ mm

Exp data			Data processing		
$Q_{(l)}$ l/min	$Q_{(g)}$ l/min	$H_{(m)}$, mm	$\ln (H_m/D_s)$	$\ln (Q_g/Q_l)$	$\ln \frac{gD_s^5}{Q_l^2}$
200	10	385	1.40	−3.00	1.92
200	20	850	2.19	−2.30	1.92
200	30	1145	2.49	−1.90	1.92
200	40	1385	2.68	−1.61	1.92
200	50	1532	2.78	−1.39	1.92
250	12.5	329	1.24	−3.00	1.48
250	25	651	1.92	−2.30	1.48
250	37.5	1187	2.53	−1.90	1.48
250	50	1426	2.71	−1.61	1.48
250	62.5	1561	2.80	−1.39	1.48
300	15	255	0.99	−3.00	1.11
300	30	569	1.79	−2.30	1.11
300	45	1185	2.52	−1.90	1.11
300	60	1500	2.76	−1.61	1.11
300	75	1642	2.85	−1.39	1.11
350	17.5	181	0.64	−3.00	0.80
350	35	655	1.93	−2.30	0.80
350	52.5	1200	2.54	−1.90	0.80
350	70	1513	2.77	−1.61	0.80
350	87.5	1646	2.85	−1.39	0.80

The results for the two types of shrouds, BCS and SCS, give the following final form of the dimensionless equations:

For BCS: ($R^2 = 0.94$)

$$\frac{H_m}{D_s} = 24.13 \left(\frac{Q_g}{Q_L}\right)^{1.042} \left(\frac{gD_s^5}{Q_L^2}\right)^{0.334}$$

For SCS: ($R^2 = 0.94$)

$$\frac{H_m}{D_s} = 75.56 \left(\frac{Q_g}{Q_L}\right)^{1.128} \left(\frac{gD_s^5}{Q_L^2}\right)^{0.146}$$

The comparison between experimental data and predicted values of the free jet with the previous equations is indicated in Figs. 11.41 and 11.42

In the previous equations it is shown that the height of the free jet is inversely proportional to the liquid flow rate and directly proportional to the gas flow rate.

Table 11.33 Experimental results for BCS, $D_S = 62.8$ mm, $D_{CN} = 40$ mm

Exp data			Data processing		
$Q_{(l)}$ l/min	$Q_{(g)}$ l/min	$H_{(m)}$, mm	$\ln (H_m/D_s)$	$\ln (Q_g/Q_l)$	$\ln \frac{gD_s^5}{Q_l^2}$
75	3.75	152	0.884	−2.996	1.814
75	7.5	288	1.523	−2.303	1.814
75	11.25	402	1.856	−1.897	1.814
75	15	491	2.056	−1.609	1.814
75	18.75	626	2.299	−1.386	1.814
100	5	100	0.465	−2.996	1.238
100	10	179	1.047	−2.303	1.238
100	15	243	1.353	−1.897	1.238
100	20	343	1.698	−1.609	1.238
100	25	562	2.192	−1.386	1.238
115	5.75	108	0.542	−2.996	0.959
115	11.5	179	1.047	−2.303	0.959
115	17.25	238	1.332	−1.897	0.959
115	23	347	1.709	−1.609	0.959
115	28.75	579	2.221	−1.386	0.959
120	6	100	0.465	−2.996	0.874
120	12	176	1.031	−2.303	0.874
120	18	239	1.337	−1.897	0.874
120	24	354	1.729	−1.609	0.874
120	30	597	2.252	−1.386	0.874
130	6.5	75	0.178	−2.996	0.714
130	13	152	0.884	−2.303	0.714
130	19.5	281	1.498	−1.897	0.714
130	26	425	1.912	−1.609	0.714
130	32.5	642	2.325	−1.386	0.714

Dimensional analysis II: The results from both BCS and SCS were not unified with one single relationship using five variables. It means that one or more variables are missing. One variable not accounted for before was the diameter of the water jet. This parameter is defined by the diameter of the collector nozzle (D_{CN}). In the previous experiments it was noticed that both BCS and SCS have different values.

The new dimensionless analysis includes *6 variables*:

$$H_m = f\left(Q_g, Q_L, D_s, D_{cn}, g\right)$$

The dimensionless groups will be obtained applying the matrix method. The number of equations is defined depending on the number of dimensions, in this case

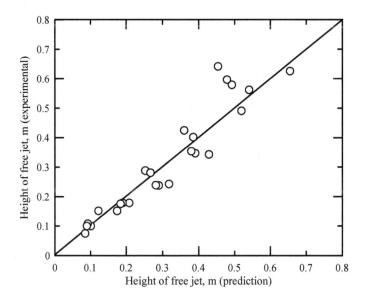

Fig. 11.41 Comparison between experimental data and predicted values for BCS

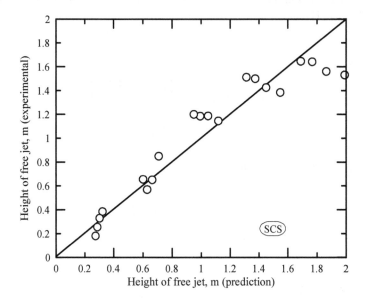

Fig. 11.42 Comparison between experimental data and predicted values for SCS

[L] and [T]. Each equation is of the following form:

$$H_m = f(Q_g)^{x_1}(Q_L)^{x_2}(D_s)^{x_3}(D_{cn})^{x_4}(g)^{x_5}$$
$$[L^1 T^0] = f\left[L^3 T^{-1}\right]^{x_1}\left(L^3 T^{-1}\right)^{x_2}(L)^{x_3}(L)^{x_4}\left(LT^{-2}\right)^{x_5}$$

The system of equations indicates five unknowns (x_1 to x_5):

$$
\begin{pmatrix}
 & H_m & Q_g & Q_L & D_s & D_{cn} & g \\
[L] & 1 & 3 & 3 & 1 & 1 & 1 \\
[T] & 0 & -1 & -1 & 0 & 0 & -2
\end{pmatrix}
$$

The dimensional matrix is defined as D:

$$
D = \begin{pmatrix}
1 & 3 & 3 & 1 & 1 & 1 \\
0 & -1 & -1 & 0 & 0 & -2
\end{pmatrix}
$$

The system of equations involves 6 variables and we have 2 equations, then, the size of the identity matrix should be $6 - 2 = 4$. It means 4 coefficients are given the value of unity. The augmented matrix is then expressed as follows:

$$
\begin{pmatrix}
1 & 0 & 0 & 0 & 0 & 0 \\
0 & 1 & 0 & 0 & 0 & 0 \\
0 & 0 & 1 & 0 & 0 & 0 \\
0 & 0 & 0 & 1 & 0 & 0 \\
1 & 3 & 3 & 1 & 1 & 1 \\
0 & -1 & -1 & 0 & 0 & -2
\end{pmatrix}
\begin{pmatrix}
k_1 \\ k_2 \\ k_3 \\ k_4 \\ k_5 \\ k_6
\end{pmatrix}
=
\begin{pmatrix}
k_1 \\ k_2 \\ k_3 \\ k_4 \\ b_1 \\ b_2
\end{pmatrix}
$$

The square (S) and remaining (R) matrices are:

$$
R = \begin{pmatrix}
1 & 3 & 3 & 1 \\
0 & -1 & -1 & 0
\end{pmatrix}
$$

$$
S = \begin{pmatrix}
1 & 1 \\
0 & -2
\end{pmatrix}
$$

The solution requires S^{-1} and $-S^{-1}R$:

$$
S^{-1} = \begin{pmatrix} 1 & 1 \\ 0 & -2 \end{pmatrix}^{-1} = \begin{pmatrix} 1 & 0.5 \\ 0 & -0.5 \end{pmatrix}
$$

$$
S^{-1}R = \begin{pmatrix} 1 & 1 \\ 0 & -2 \end{pmatrix}^{-1} \begin{pmatrix} 1 & 3 & 3 & 1 \\ 0 & -1 & -1 & 0 \end{pmatrix} = \begin{pmatrix} 1 & 2.5 & 2.5 & 1 \\ 0 & 0.5 & 0.5 & 0 \end{pmatrix}
$$

Solution:

$$
\begin{pmatrix} k_1 \\ k_2 \\ k_3 \\ k_4 \\ k_5 \end{pmatrix} = \begin{pmatrix} 1 & 0 & 0 & 0 & 0 & 0 \\ 0 & 1 & 0 & 0 & 0 & 0 \\ 0 & 0 & 1 & 0 & 0 & 0 \\ 0 & 0 & 0 & 1 & 0 & 0 \\ 1 & 2.5 & 2.5 & 1 & 1 & 0.5 \\ 0 & 0.5 & 0.5 & 0 & 0 & -0.5 \end{pmatrix} \begin{pmatrix} k_1 \\ k_2 \\ 0 \\ 0 \\ 0 \\ 0 \end{pmatrix}
$$

The values of the dimensionless groups is based on the initial arrangement of the variables:

$$
\pi_1 \equiv \begin{array}{cccccc} H_m & Q_g & Q_L & D_s & D_{cn} & g \\ 1 & 0 & 0 & 0 & 1 & 0 \end{array}
$$

$$
\pi_1 = \frac{H_m}{D_{cn}}
$$

In a similar way:

$$
\pi_2 = \frac{Q_g}{D_{cn}^{2.5} g^{0.5}}
$$

$$
\pi_3 = \frac{Q_L}{D_{cn}^{2.5} g^{0.5}} = \frac{Q_L^2}{D_{cn}^5 g}
$$

$$
\pi_4 = \frac{D_s}{D_{cn}}
$$

The general dimensionless relationship is:

$$
\frac{H_m}{D_{cn}} = K \left(\frac{Q_g^2}{g D_{cn}^5} \right)^a \left(\frac{Q_L^2}{g D_{cn}^5} \right)^b \left(\frac{D_s}{D_{cn}} \right)^c
$$

Which can be recast as follows:

$$
\frac{H_m}{D_s} = K \left(\frac{Q_g}{Q_L} \right)^a \left(\frac{g D_s^5}{Q_L^2} \right)^b \left(\frac{D_{cn}}{D_s} \right)^c
$$

Modified groups are obtained as follows: $\pi_1 = \frac{\pi_1}{\pi_5}$, $\pi_2 = \frac{\pi_2}{\pi_3}$, $\pi_4 = \frac{1}{\pi_4}$

The previous expression is employed to study the whole experimental results for BCS and SCS with different shroud and collector nozzle diameters. The data base is given in Table 11.34.

The multilinear regression analysis was carried out in excel. The following values are obtained: $\log C = 2.03$, $x_1 = 0.99$, $x_2 = 0.19$, $x_3 = 2.46$. The regression coefficient (R^2) was 0.86 and standard error of 13%. Replacing these values, the particular solution is:

Table 11.34 Experimental results for whole set of experimental data (After Ref)

	Q_l l/min	Q_g l/min	H_m, mm	D_s, mm	D_{cn}, mm	log π_1	log π_2	log π_3	log π_4
BCS 40	75	3.8	145	62.8	40	0.363	−1.301	0.788	−0.196
	75	7.5	303	62.8	40	0.684	−1.000	0.788	−0.196
	75	11.3	376	62.8	40	0.777	−0.824	0.788	−0.196
	75	15.0	495	62.8	40	0.897	−0.699	0.788	−0.196
	75	18.8	624	62.8	40	0.997	−0.602	0.788	−0.196
	75	22.5	618	62.8	40	0.993	−0.523	0.788	−0.196
	75	26.3	793	62.8	40	1.102	−0.456	0.788	−0.196
	75	30.0	888	62.8	40	1.150	−0.398	0.788	−0.196
	75	33.8	911	62.8	40	1.161	−0.347	0.788	−0.196
	75	37.5	979	62.8	40	1.193	−0.301	0.788	−0.196
	100	5.0	83	62.8	40	0.123	−1.301	0.538	−0.196
	100	10.0	164	62.8	40	0.417	−1.000	0.538	−0.196
	100	15.0	260	62.8	40	0.617	−0.824	0.538	−0.196
	100	20.0	378	62.8	40	0.779	−0.699	0.538	−0.196
	100	25.0	586	62.8	40	0.970	−0.602	0.538	−0.196
	100	30.0	676	62.8	40	1.032	−0.523	0.538	−0.196
	100	35.0	833	62.8	40	1.123	−0.456	0.538	−0.196
	100	40.0	931	62.8	40	1.171	−0.398	0.538	−0.196
	100	45.0	987	62.8	40	1.196	−0.347	0.538	−0.196
	110	5.5	81	62.8	40	0.108	−1.301	0.455	−0.196
	110	11.0	163	62.8	40	0.413	−1.000	0.455	−0.196
	110	16.5	268	62.8	40	0.631	−0.824	0.455	−0.196
	110	22.0	407	62.8	40	0.812	−0.699	0.455	−0.196
	110	27.5	523	62.8	40	0.920	−0.602	0.455	−0.196
	110	33.0	712	62.8	40	1.055	−0.523	0.455	−0.196
	110	38.5	866	62.8	40	1.139	−0.456	0.455	−0.196
	110	44.0	953	62.8	40	1.181	−0.398	0.455	−0.196
	110	49.5	1011	62.8	40	1.207	−0.347	0.455	−0.196
	120	6.0	79	62.8	40	0.100	−1.301	0.379	−0.196
	120	12.0	163	62.8	40	0.415	−1.000	0.379	−0.196
	120	18.0	281	62.8	40	0.650	−0.824	0.379	−0.196
	120	24.0	367	62.8	40	0.767	−0.699	0.379	−0.196
	120	30.0	562	62.8	40	0.952	−0.602	0.379	−0.196
	120	36.0	737	62.8	40	1.070	−0.523	0.379	−0.196
	120	42.0	895	62.8	40	1.154	−0.456	0.379	−0.196
	120	48.0	997	62.8	40	1.201	−0.398	0.379	−0.196

(continued)

Table 11.34 (continued)

	Q_l l/min	Q_g l/min	H_m, mm	D_s, mm	D_{cn}, mm	log π_1	log π_2	log π_3	log π_4
	130	6.5	76	62.8	40	0.080	−1.301	0.310	−0.196
	130	13.0	165	62.8	40	0.419	−1.000	0.310	−0.196
	130	19.5	282	62.8	40	0.652	−0.824	0.310	−0.196
	130	26.0	368	62.8	40	0.768	−0.699	0.310	−0.196
	130	32.5	576	62.8	40	0.962	−0.602	0.310	−0.196
	130	39.0	770	62.8	40	1.088	−0.523	0.310	−0.196
	130	45.5	942	62.8	40	1.176	−0.456	0.310	−0.196
	130	52.0	999	62.8	40	1.202	−0.398	0.310	−0.196
SCS 70	200	10.0	385	95	70	0.608	−1.301	0.835	−0.133
	200	20.0	860	95	70	0.957	−1.000	0.835	−0.133
	200	30.0	1145	95	70	1.081	−0.824	0.835	−0.133
	200	40.0	1385	95	70	1.164	−0.699	0.835	−0.133
	200	50.0	1532	95	70	1.207	−0.602	0.835	−0.133
	200	60.0	1744	95	70	1.264	−0.523	0.835	−0.133
	250	12.5	329	95	70	0.540	−1.301	0.641	−0.133
	250	25.0	651	95	70	0.836	−1.000	0.641	−0.133
	250	37.5	1187	95	70	1.097	−0.824	0.641	−0.133
	250	50.0	1426	95	70	1.176	−0.699	0.641	−0.133
	250	62.5	1561	95	70	1.216	−0.602	0.641	−0.133
	250	75.0	1645	95	70	1.238	−0.523	0.641	−0.133
	300	15.0	255	95	70	0.429	−1.301	0.482	−0.133
	300	30.0	569	95	70	0.778	−1.000	0.482	−0.133
	300	45.0	1185	95	70	1.096	−0.824	0.482	−0.133
	300	60.0	1500	95	70	1.198	−0.699	0.482	−0.133
	300	75.0	1642	95	70	1.238	−0.602	0.482	−0.133
	300	90.0	1751	95	70	1.265	−0.523	0.482	−0.133
	350	17.5	181	95	70	0.280	−1.301	0.348	−0.133
	350	35.0	655	95	70	0.839	−1.000	0.348	−0.133
	350	52.5	1200	95	70	1.101	−0.824	0.348	−0.133
	350	70.0	1513	95	70	1.202	−0.699	0.348	−0.133
	350	87.5	1646	95	70	1.239	−0.602	0.348	−0.133
	350	105.0	1774	95	70	1.271	−0.523	0.348	−0.133
SCS 60	200	10.0	335	95	60	0.547	−1.300	0.835	−0.200
	200	20.0	565	95	60	0.774	−1.000	0.835	−0.200
	200	30.0	855	95	60	0.954	−0.824	0.835	−0.200
	200	40.2	1248	95	60	1.118	−0.697	0.835	−0.200

(continued)

Table 11.34 (continued)

Q_l l/min	Q_g l/min	H_m, mm	D_s, mm	D_{cn}, mm	log π_1	log π_2	log π_3	log π_4
200	50.0	1375	95	60	1.161	−0.602	0.835	−0.200
200	60.0	1487	95	60	1.195	−0.523	0.835	−0.200
227	12.0	372	95	60	0.593	−1.277	0.725	−0.200
227	23.0	514	95	60	0.733	−0.995	0.725	−0.200
227	34.0	912	95	60	0.982	−0.824	0.725	−0.200
227	45.0	1266	95	60	1.125	−0.703	0.725	−0.200
227	57.0	1438	95	60	1.180	−0.600	0.725	−0.200
250	12.5	258	95	60	0.434	−1.302	0.641	−0.200
250	25.0	398	95	60	0.622	−1.000	0.641	−0.200
250	37.5	828	95	60	0.940	−0.824	0.641	−0.200
250	50.0	1280	95	60	1.129	−0.699	0.641	−0.200
250	62.5	1420	95	60	1.175	−0.602	0.641	−0.200
250	75.0	1372	95	60	1.160	−0.523	0.641	−0.200
255	13.0	360	95	60	0.579	−1.292	0.624	−0.200
255	26.0	526	95	60	0.743	−0.992	0.624	−0.200
255	38.0	972	95	60	1.010	−0.827	0.624	−0.200
255	51.0	1340	95	60	1.149	−0.699	0.624	−0.200
255	64.0	1458	95	60	1.186	−0.600	0.624	−0.200
282	14.0	249	95	60	0.418	−1.305	0.536	−0.200
282	28.0	406	95	60	0.631	−1.003	0.536	−0.200
282	42.0	892	95	60	0.973	−0.827	0.536	−0.200
282	56.0	1342	95	60	1.150	−0.702	0.536	−0.200
282	70.0	1507	95	60	1.200	−0.605	0.536	−0.200
300	15.0	223	95	60	0.371	−1.301	0.482	−0.200
300	30.0	375	95	60	0.596	−1.000	0.482	−0.200
300	45.0	857	95	60	0.955	−0.824	0.482	−0.200
300	60.0	1370	95	60	1.159	−0.699	0.482	−0.200
300	75.0	1505	95	60	1.200	−0.602	0.482	−0.200
310	16.0	223	95	60	0.371	−1.287	0.454	−0.200
310	31.0	410	95	60	0.635	−1.000	0.454	−0.200
310	46.0	932	95	60	0.992	−0.828	0.454	−0.200
310	62.0	1398	95	60	1.168	−0.699	0.454	−0.200
310	78.0	1515	95	60	1.203	−0.599	0.454	−0.200
340	17.0	192	95	60	0.306	−1.302	0.374	−0.200
340	34.0	413	95	60	0.638	−1.000	0.374	−0.200
340	51.0	935	95	60	0.993	−0.824	0.374	−0.200

(continued)

Table 11.34 (continued)

Q_l l/min	Q_g l/min	H_m, mm	D_s, mm	D_{cn}, mm	log π_1	log π_2	log π_3	log π_4
340	68.0	1429	95	60	1.177	−0.699	0.374	−0.200
350	17.5	197	95	60	0.317	−1.301	0.349	−0.200
350	35.0	368	95	60	0.588	−1.000	0.349	−0.200
350	52.5	968	95	60	1.008	−0.824	0.349	−0.200
350	70.0	1426	95	60	1.176	−0.699	0.349	−0.200
368	19.0	180	95	60	0.278	−1.287	0.305	−0.200
368	37.0	527	95	60	0.744	−0.997	0.305	−0.200
368	56.0	1108	95	60	1.067	−0.818	0.305	−0.200
368	74.0	1493	95	60	1.196	−0.697	0.305	−0.200

$$\frac{H_m}{D_s} = 108 \left(\frac{Q_g}{Q_L}\right)^{0.99} \left(\frac{gD_s^5}{Q_L^2}\right)^{0.19} \left(\frac{D_{cn}}{D_s}\right)^{2.5}$$

The previous equation is valid for the following conditions: (1) $0.005 < Q_g/Q_L <$ 0.50, $12.5 \times 10^{-4} < Q_L < 58 \times 10^{-4}$ m³/s, (2) BCS: $D_{cn} = 40$, $D_s = 62.8$ mm, (3) SCS: $D_{cn} = 60$ and 70 mm $D_s = 95$ mm.

Figure 11.43 compares the experimental and predicted values for the height of the free jet.

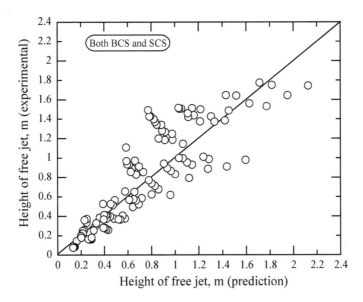

Fig. 11.43 Comparison between experimental and predicted values for the height of the free jet using the database reported by Singh

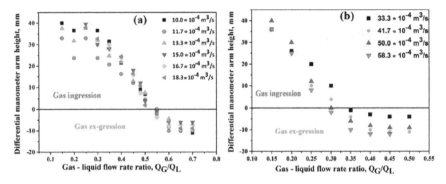

Fig. 11.44 Evaluation of air ingress: **a** BCS, **b** SCS. After Singh [58]

The effect of gas injection and air ingress was evaluated using two additional ports on the top of the shroud, open to the atmosphere. The two ports were connected to manometers. The gas flow rate was increased and air ingress was related with the manometric pressure. Positive values indicate air ingress and negative values represent argon gas sealing. The results are shown in Fig. 11.44. It can be shown that there is a critical ratio Q_g/Q_L that impedes air ingress. Volumetric gas flow rates below $0.5Q_L$ for BCS and $0.35Q_L$ for SCS have the potential of air ingress. This is an important contribution from Singh and Mazumdar [61] because even if argon is injected but the flow rate is below those values, air ingress is possible if the seal between connector nozzle and shroud is not perfect. As the gas flow rate increases the problem of air ingress is eliminated, the height of the confined jet also increases and erosion of the refractory decreases but also turbulence increases inside the tundish and the size of the ladle eye.

11.7   Teeming of Liquid Steel

Teeming is the pouring or drainage of liquid steel from one vessel to another, commonly using a slide gate at the bottom of the first vessel. This operation is common in steelmaking in both the integrated and mini-mill plants. During the teeming process, due to tangential forces, a vortex is formed. This phenomenon is the cause of slag entrainment into the second vessel and formation of exogeneous inclusions.

Carry-Over Slag (COS) from the BOF or EAF into a ladle creates significant problems. First of all, the slag from a BOF or EAF is oxidizing, containing a high concentration of iron oxide (20–40%). The slag in the next process, secondary metallurgy, is a reducing slag (FeO should be lower than 2%) in order to remove Sulphur. Any amount of COS is detrimental in ladle metallurgy:

- The amount of deoxidizers increases, increasing the amount of Non-Metallic Inclusions (NMI). NMI affect the subsequent continuous casting process due to nozzle clogging, formation of cracks and other defects in billets and slabs as well as lower mechanical properties of the final product
- The reducing conditions in the ladle furnace promote P reversion
- The rate and extent of desulphurization decreases, largely affecting productivity
- A higher FeO decreases slag basicity requiring higher additions of fluxes which in turn increases energy consumption and slag volume.

11.7.1 Experimental Work from Kuwana [65]

Kuwana et al. [65] investigated the onset of vortex formation (a critical height of the liquid) as a function of the tangential velocity using two geometric scale vessels with different discharge nozzle diameters.

Dimensional analysis

The critical height (H_{cr}) depends on the following 8 variables: liquid's physical properties; density (ρ), viscosity (μ), surface tension (σ); ladle dimensions; nozzle diameter (d), ladle inner diameter (D), initial height of the liquid (H_0); tangential velocity ($U_{\theta 0}$) and, gravity force (g).

$$H_{cr} = f(U_{\theta 0}, H_0, D, d, \mu, \sigma, \rho, g)$$

Variable	Symbol	Units	Dimensions
Critical height	H_{cr}	m	[L]
Ladle inner diameter	D	m	[L]
Initial height of liquid	H_0	m	[L]
Nozzle diameter	d	m	[L]
Surface tension	σ	N/m	$[MT^{-2}]$
Dynamic viscosity	μ	kg/(m × s)	$[ML^{-1}T^{-1}]$
Density	ρ	kg/m³	$[ML^{-3}]$
Initial tangential velocity	$U_{\theta 0}$	m/s	$[LT^{-1}]$
Gravity	g	m/s²	$[LT^{-2}]$

There are 9 variables and 3 dimensions. According with the π-theorem this yields 6 dimensionless groups. As non-repeating variables we choose; $U_{\theta 0}$, ρ and d. Next, we define the values for the six dimensionless groups as follows:

$$\pi_1 = H_{cr}(U_{\theta 0})^{a_1}(\rho)^{b_1}(d)^{c_1}$$
$$\pi_2 = H_0(U_{\theta 0})^{a_2}(\rho)^{b_2}(d)^{c_2}$$

$$\pi_3 = D(U_{\theta 0})^{a_3}(\rho)^{b_3}(d)^{c_3}$$
$$\pi_4 = \mu(U_{\theta 0})^{a_4}(\rho)^{b_4}(d)^{c_4}$$
$$\pi_5 = \sigma(U_{\theta 0})^{a_5}(\rho)^{b_5}(d)^{c_5}$$
$$\pi_6 = g(U_{\theta 0})^{a_6}(\rho)^{b_6}(d)^{c_6}$$

First π-group:

$$\pi_1 = H_{cr}(U_{\theta 0})^{a_1}(\rho)^{b_1}(d)^{c_1}$$
$$\pi_1 = (L)(LT^{-1})^{a_1}(ML^{-3})^{b_1}(L)^{c_1} = M^0 T^0 L^0$$

Balance:
  M: $b_1 = 0$
  L: $1 + a_1 - 3b_1 + c_1 = 0$
  T: $-a_1 = 0$

Solution: $a_1 = 0$.
  $b_1 = 0$
  $c_1 = -1 - a_1 + 3b_1 = -1$

Substituting values:

$$\pi_1 = \frac{H_{cr}}{d}$$

Second π-group:

$$\pi_2 = H_0(U_{\theta 0})^{a_2}(\rho)^{b_2}(d)^{c_2}$$
$$\pi_2 = (L)(LT^{-1})^{a_2}(ML^{-3})^{b_2}(L)^{c_2} = M^0 T^0 L^0$$

Balance:
  M: $b_2 = 0$
  L: $1 + a_2 - 3b_2 + c_2 = 0$
  T: $-a_2 = 0$

Solution:

$a_2 = 0$.
  $b_2 = 0$
  $c_2 = -1 - a_2 + 3b_2 \therefore c_2 = -1$

Substituting values:

$$\pi_2 = \frac{H_0}{d}$$

Third π-group:

$$\pi_3 = D(U_{\theta 0})^{a_3}(\rho)^{b_3}(d)^{c_3}$$
$$\pi_3 = (L)\left(LT^{-1}\right)^{a_3}\left(ML^{-3}\right)^{b_3}(L)^{c_3} = M^0 T^0 L^0$$

Balance:
  M: $b_3 = 0$
  L: $1 + a_3 - 3b_3 + c_3 = 0$
  T: $-a_3 = 0$

Solution:

$a_3 = 0.$
  $b_3 = 0$
  $c_3 = -1 - a_3 + 3b_3 \therefore c_3 = -1$

Substituting values:

$$\pi_3 = \frac{D}{d}$$

Fourth π-group:

$$\pi_4 = \mu(U_{\theta 0})^{a_4}(\rho)^{b_4}(d)^{c_4}$$
$$\pi_4 = \left(ML^{-1}T^{-1}\right)\left(LT^{-1}\right)^{a_4}\left(ML^{-3}\right)^{b_4}(L)^{c_4} = M^0 T^0 L^0$$

Balance:
  M: $1 + b_4 = 0$
  L: $-1 + a_4 - 3b_4 + c_4 = 0$
  T: $-1 - a_4 = 0$

Solution:

$a_4 = -1.$
  $b_4 = -1$
  $c_4 = 1 - a_4 + 3b_4 \therefore c_4 = -1$

Substituting values:

$$\pi_4 = \frac{\mu}{\rho U_{\theta 0} d}$$

Fifth π-group:

$$\pi_5 = \sigma(U_{\theta 0})^{a_5}(\rho)^{b_5}(d)^{c_5}$$
$$\pi_5 = \left(MT^{-2}\right)\left(LT^{-1}\right)^{a_5}\left(ML^{-3}\right)^{b_5}(L)^{c_5} = M^0 T^0 L^0$$

Balance:
   M: $1 + b_5 = 0$
   L: $a_5 - 3b_5 + c_5 = 0$
   T: $-2 - a_5 = 0$

Solution:

$a_5 = -2.$
   $b_5 = -1$
   $c_5 = -a_5 + 3b_5 \therefore c_3 = -1$

Substituting values:

$$\pi_5 = \frac{\sigma}{\rho U_{\theta 0}^2 d}$$

Sixth π-group:

$$\pi_6 = g(U_{\theta 0})^{a_6} (\rho)^{b_6} (d)^{c_6}$$
$$\pi_6 = (LT^{-2})(LT^{-1})^{a_6} (ML^{-3})^{b_6} (L)^{c_6} = M^0 T^0 L^0$$

Balance:
   M: $b_6 = 0$
   L: $1 + a_6 - 3b_6 + c_6 = 0$
   T: $-2 - a_6 = 0$

Solution:

$a_6 = -2.$
   $b_6 = 0$
   $c_6 = -1 - a_6 + 3b_6 \therefore c_6 = 1$

Substituting values:

$$\pi_6 = \frac{gd}{U_{\theta 0}^2}$$

The final functional relationship is expressed as follows:

$$\pi_1 = f(\pi_2)(\pi_3)(\pi_4)(\pi_5)(\pi_6)$$
$$\frac{H_{cr}}{d} = f\left(\frac{H_0}{d}\right)\left(\frac{D}{d}\right)\left(\frac{\mu}{\rho U_{\theta 0} d}\right)\left(\frac{\sigma}{\rho U_{\theta 0}^2 d}\right)\left(\frac{gd}{U_{\theta 0}^2}\right)$$

The dimensionless groups, π_4, π_5, π_6 are written in the inverse form to recognize familiar dimensionless groups. The general form is:

$$\frac{H_{cr}}{d} = C\left(\frac{H_0}{d}\right)^{x_1}\left(\frac{D}{d}\right)^{x_2}\left(\frac{\rho U_{\theta 0}d}{\mu}\right)^{x_3}\left(\frac{\rho U_{\theta 0}^2 d}{\sigma}\right)^{x_4}\left(\frac{U_{\theta 0}^2}{gd}\right)^{x_5}$$

The dimensionless groups, π_4, π_5, π_6 are Re, We and Fr numbers.

$$\frac{H_{cr}}{d} = C\left(\frac{H_0}{d}\right)^{x_1}\left(\frac{D}{d}\right)^{x_2}(Re)^{x_3}(We)^{x_4}(Fr)^{x_5}$$

Experimental data from Kuwana et al. [65]

Kuwana et al. [65] carried out physical and mathematical modelling work to define the effect of the tangential speed on the critical height of the liquid. The critical height is defined when the tip of the vortex reaches the bottom of the vessel. A critical height has a corresponding initial tangential velocity.

The physical modeling work was carried out comparing two cylindrical water models. The tangential speed was measured using Particle Image Velocimetry (PIV). To measure the tangential velocity the laser beam was directed in the horizontal direction at a height corresponding to the initial height of the liquid and a camera capturing images from the top of the ladle and to measure the critical height of the liquid the positions of the laser beam and camera were inverted. The initial tangential velocity was imparted manually with a wooden stirrer using various stirring speeds. Once the initial tangential speed reached a certain value the nozzle was opened. For each ladle, the variables are the critical height and the initial tangential speed.

Kuwana et al. [65] found for initial tangential velocities ($U_{\theta 0}$) close to zero, less than 1 mm/s, a critical height of the liquid also close to zero. Increasing $U_{\theta 0}$ also increased the critical height of the liquid. When $U_{\theta 0}$ was higher than 3 mm/s H_{cr} reached H_0, a condition indicating formation of a funnel-shaped vortex as soon as the nozzle is opened. Therefore, to reduce vortex formation the tangential velocities should be decreased. The ladle dimensions and water properties are indicated in Table 11.35 and Fig. 11.45.

The experimental data consists of information on the critical height as a function of the tangential velocity for the two vessels, shown in Table 11.36.

Applying the statistical tools of excel the following values are obtained: $\log C = 2.76$, $x_1 = 0$, $x_2 = 0$, $x_3 = 0$, $x_4 = -0.204$, $x_5 = 0.6187$. The regression coefficient (R^2) was 0.57. Replacing these values in the general expression:

Table 11.35 Ladle dimensions and water properties

	H0, mm	D, mm	d, mm	μ, Pa s	ρ, g/mm^3	σ, g/s^2
Small ladle	127	254	3.81			
Large ladle	127	254	7.62			
Water				1×10^{-3}	1×10^{-3}	73

Fig. 11.45 Water models employed by Kuwana et al. [65] to study vortex formation during the drainage of liquid from a ladle

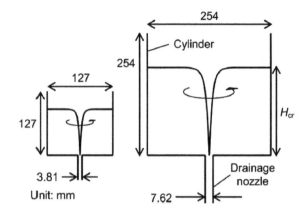

$$\frac{H_{cr}}{d} = C\left(\frac{H_0}{d}\right)^{x_1}\left(\frac{D}{d}\right)^{x_2}(Re)^{x_3}(We)^{x_4}(Fr)^{x_5}$$

$$\frac{H_{cr}}{d} = 578\left(\frac{\rho U_{\theta 0}^2 d}{\sigma}\right)^{-0.204}\left(\frac{U_{\theta 0}^2}{gd}\right)^{0.619}$$

The regression analysis indicates the initial height (H_0) and ladle diameter (D) as well as the Re number are not relevant in the definition of the critical height for vortex formation (H_{cr}/d). Plotting by separate the effect of Re, Fr and We numbers in Fig. 11.46, it is shown that the effect of those dimensionless groups increases very rapid for low values but then the dimensionless critical height (similar results for $H_{cr}/H0$ and H_{cr}/d) remains constant.

Figure 11.47 shows the results when the general expression in terms of both Fr and We numbers is applied.

11.7.2   Experimental Work from Kamaraj et al. [66]

Kamaraj et al. [67] compared vortex formation, simulating teeming from a BOF, with and without an oil layer and reported that the oil layer, in general, decreased the critical height for vortex formation.

A water model, with a geometric scale 1:7 was made with transparent perplex. Figure 11.48 and Table 11.37 show the dimensions in cm. of the model and prototype. The experiments were made with the half-BOF Perspex model, Fig. 11.48b. Table 11.38 indicates the range of the variables evaluated.

During one experiment the model was filled with water with an initial liquid height (H_0). Waiting time is the time elapsed from the moment the vessel is filled with water until the start of tapping. This time accounts for the motion in the liquids before tapping. The flow rate was controlled with a digital flowmeter. Two cameras, one external and one underwater, were used to observe the height of the liquid as a

Table 11.36 Experimental data base and experimental dimensionless groups

U mm/s	H_{cr} mm	H0, mm	D, mm	d, mm	π_1	π_2	π_3	π_4	π_5	π_6
0.48	25.6	127	127	3.81	6.73	33.33	33.33	1.815	1.2E−05	0.00001
0.72	24.5	127	127	3.81	6.44	33.33	33.33	2.729	2.7E−05	0.00001
1.09	29.3	127	127	3.81	7.69	33.33	33.33	4.147	6.2E−05	0.00003
1.35	40.3	127	127	3.81	10.57	33.33	33.33	5.157	9.6E−05	0.00005
1.59	30.7	127	127	3.81	8.07	33.33	33.33	6.074	1.3E−04	0.00007
1.78	39.9	127	127	3.81	10.47	33.33	33.33	6.780	1.7E−04	0.00008
3.13	99.6	127	127	3.81	26.15	33.33	33.33	11.929	5.1E−04	0.00026
4.35	107.3	127	127	3.81	28.16	33.33	33.33	16.592	9.9E−04	0.00051
4.46	111.1	127	127	3.81	29.15	33.33	33.33	16.994	1.0E−03	0.00053
4.78	108.0	127	127	3.81	28.35	33.33	33.33	18.215	1.2E−03	0.00061
5.07	118.3	127	127	3.81	31.04	33.33	33.33	19.327	1.3E−03	0.00069
6.06	116.8	127	127	3.81	30.65	33.33	33.33	23.081	1.9E−03	0.00098
6.43	109.0	127	127	3.81	28.62	33.33	33.33	24.504	2.2E−03	0.00111
8.93	111.9	127	127	3.81	29.37	33.33	33.33	34.038	4.2E−03	0.00214
9.44	112.3	127	127	3.81	29.46	33.33	33.33	35.965	4.7E−03	0.00238
13.35	117.6	127	127	3.81	30.88	33.33	33.33	50.874	9.3E−03	0.00477
14.18	116.9	127	127	3.81	30.68	33.33	33.33	54.019	1.0E−02	0.00538
15.00	116.5	127	127	3.81	30.58	33.33	33.33	57.164	1.2E−02	0.00602
0.45	1.1	254	254	7.62	0.15	33.33	33.33	3.447	2.1E−05	0.00000
0.88	1.5	254	254	7.62	0.19	33.33	33.33	6.693	8.1E−05	0.00001
1.26	69.4	254	254	7.62	9.11	33.33	33.33	9.626	1.7E−04	0.00002
1.64	69.0	254	254	7.62	9.06	33.33	33.33	12.466	2.8E−04	0.00004
1.93	68.7	254	254	7.62	9.01	33.33	33.33	14.698	3.9E−04	0.00005
1.94	121.5	254	254	7.62	15.95	33.33	33.33	14.816	3.9E−04	0.00005
2.45	121.5	254	254	7.62	15.95	33.33	33.33	18.670	6.3E−04	0.00008
3.69	162.7	254	254	7.62	21.35	33.33	33.33	28.138	1.4E−03	0.00018
4.01	174.1	254	254	7.62	22.84	33.33	33.33	30.555	1.7E−03	0.00022
4.14	207.1	254	254	7.62	27.18	33.33	33.33	31.516	1.8E−03	0.00023
4.57	174.4	254	254	7.62	22.89	33.33	33.33	34.814	2.2E−03	0.00028
5.07	180.7	254	254	7.62	23.71	33.33	33.33	38.659	2.7E−03	0.00034
7.76	206.8	254	254	7.62	27.14	33.33	33.33	59.107	6.3E−03	0.00080
9.81	201.0	254	254	7.62	26.37	33.33	33.33	74.738	1.0E−02	0.00129
10.74	198.6	254	254	7.62	26.06	33.33	33.33	81.839	1.2E−02	0.00154
11.57	200.2	254	254	7.62	26.28	33.33	33.33	88.128	1.4E−02	0.00179
20.00	220.6	254	254	7.62	28.94	33.33	33.33	152.407	4.2E−02	0.00535
23.65	220.2	254	254	7.62	28.90	33.33	33.33	180.202	5.8E−02	0.00748
24.21	220.3	254	254	7.62	28.90	33.33	33.33	184.462	6.1E−02	0.00784

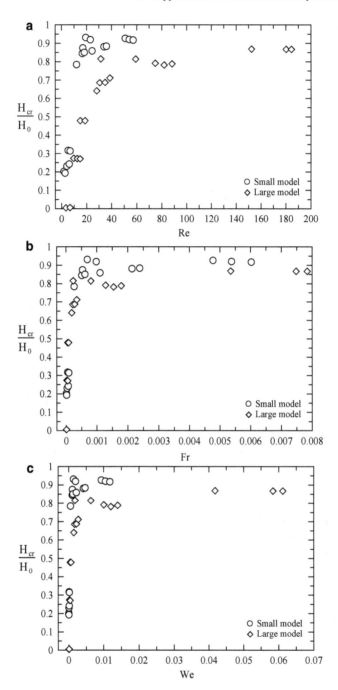

Fig. 11.46 Effect of different dimensionless groups on the dimensionless critical height for vortex formation: **a** Re, **b** Fr and **c** We. Experimental data from Kuwana et al. [28]

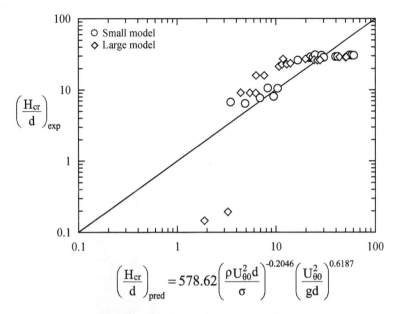

$$\left(\frac{H_{cr}}{d}\right)_{pred} = 578.62\left(\frac{\rho U_{\theta 0}^2 d}{\sigma}\right)^{-0.2046}\left(\frac{U_{\theta 0}^2}{gd}\right)^{0.6187}$$

Fig. 11.47 Comparison between predicted and experimental data on the critical for vortex formation during draining of a liquid, using dimensional analysis. Experimental data from Kuwana et al. [28]

function of time. The beginning of a vortex is the formation of a dimple at the interface air/water or oil/water. As the teeming process continues, the dimple extends into the liquid forming a vortex tail. Once the vortex tail reaches the tap hole is defined the critical height for vortex formation. At the end of the tapping process the vortex tail decreases and forms a drain sink for the overlying phase. As soon as the drain of the overlying phase starts, the tapping process is suspended. The collected mixture of air and water is settled for some time to separate the liquids and then their volumes are measured. Each experiment was repeated three times.

Two groups of experiments with and without a top oil layer were carried out.

The tapping process should be suspended at the onset of the drain sink to avoid massive carry over slag or air entrainment.

Yield loss was estimated based on the amount of water left in the BOF vessel after the tapping process.

Dimensional analysis without oil layer:

Vortex formation without an oil layer is also relevant to study because represents the case when air is the overlying phase. Reaching a critical height for vortex formation implies air infiltration into the tapping stream.

The critical vortex height (H_v) depends on the following 8 variables: liquid's physical properties; density (ρ), dynamic viscosity (μ), surface tension (σ); BOF dimensions; taphole diameter (d), distance from rear end of BOF vessel to the taphole (D); liquid's velocity (U), initial height of the primary liquid (H_0), oil thickness (h)

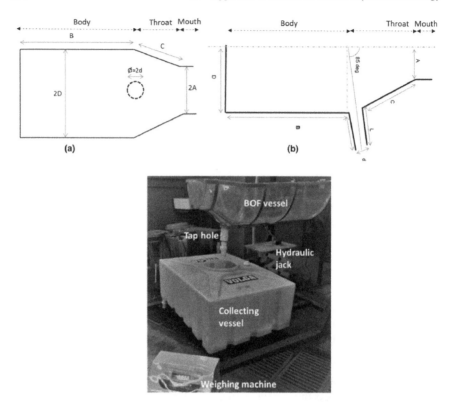

Fig. 11.48 Water model **a** general dimensions, **b** half model, **c** overall experimental set up (Taken from Kamaraj et al. [66])

Table 11.37 Dimensions of prototype and water model

Scale	D	B	C	A	L	d
1:1	246	665	154	147	60	
1:7	35	95	22	21	9	

Table 11.38 Range of variables evaluated

		I	II	III	IV
Initial liquid height, cm	H0	8, 11, 14		13	12.5, 13, 13.5
Tap hole diameter, cm	d	1.05, 1.44, 1.77, 2.14			1.44, 21.4
Initial filling flow rate, l/min	Q	20, 40	20, 30, 40	20	20
Dwell/waiting time, s	t	30, 90	30, 60, 90	90	90
Oil type	–	–	–	A, B, C	A, B
Oil thickness, cm	h	–	–	1	0.5, 1, 1.5
Number of experiments		16	108	12	12

dwell time (t) and and gravity force (g). The dwell or waiting time is the time elapsed after filling the vessel.

$$H_v = f(H0, D, d, U, g, \mu, \rho, t)$$

Variable	Symbol	Units	Dimensions
Critical vortex height	Hv	m	[L]
Initial height of liquid	H0	m	[L]
Characteristic length	D	m	[L]
Taphole diameter	d	m	[L]
Tapping velocity	U	m/s	$[LT^{-1}]$
Dynamic viscosity	μ	kg/(m × s)	$[ML^{-1}T^{-1}]$
Density	ρ	kg/m^3	$[ML^{-3}]$
Dwell/waiting time	t	s	[T]
Gravity acceleration	g	m/s^2	$[LT^{-2}]$

There are 9 variables and 3 dimensions. According with the π-theorem this yields 6 dimensionless groups. As non-repeating variables we choose; H_0, g and D, which contain the three dimensions. Next, we define the values for the seven dimensionless groups as follows:

$$\pi_1 = H_v(H_0)^{a_1}(g)^{b_1}(\mu)^{c_1}$$
$$\pi_2 = d(H_0)^{a_2}(g)^{b_2}(\mu)^{c_2}$$
$$\pi_3 = D(H_0)^{a_3}(g)^{b_3}(\mu)^{c_3}$$
$$\pi_4 = U(H_0)^{a_4}(g)^{b_4}(\mu)^{c_4}$$
$$\pi_5 = t(H_0)^{a_5}(g)^{b_5}(\mu)^{c_5}$$
$$\pi_6 = \rho(H_0)^{a_6}(g)^{b_6}(\mu)^{c_6}$$

First π-group:

$$\pi_1 = H_v(H_0)^{a_1}(g)^{b_1}(\mu)^{c_1}$$
$$\pi_1 = (L)(L)^{a_1}(LT^{-2})^{b_1}(ML^{-1}T^{-1})^{c_1} = M^0T^0L^0$$

Balance:
M: $c_1 = 0$
L: $1 + a_1 + b_1 - c_1 = 0$
T: $-2b_1 - c_1 = 0$

Solution:

$c_1 = 0.$
  $b_1 = 0$
  $a_1 = -1$

Substituting values:

$$\pi_1 = \frac{H_v}{H_0}$$

Second π-group:

$$\pi_2 = d\,(H_0)^{a_2}\,(g)^{b_2}\,(\mu)^{c_2}$$
$$\pi_2 = (L)(L)^{a_2}\left(LT^{-2}\right)^{b_2}\left(ML^{-1}T^{-1}\right)^{c_2} = M^0T^0L^0$$

Balance:
  M: $c_2 = 0$
  L: $1 + a_2 + b_2 - c_2 = 0$
  T: $-2b_2 - c_2 = 0$

Solution:

$c_1 = 0.$
  $b_2 = 0$
  $a_2 = -1$

Substituting values:

$$\pi_2 = \frac{d}{H_0}$$

Third π-group:

$$\pi_3 = D\,(H_0)^{a_3}\,(g)^{b_3}\,(\mu)^{c_3}$$
$$\pi_3 = (L)(L)^{a_3}\left(LT^{-2}\right)^{b_3}\left(ML^{-1}T^{-1}\right)^{c_3} = M^0T^0L^0$$

Balance:
  M: $c_3 = 0$
  L: $1 + a_3 + b_3 - c_3 = 0$
  T: $-2b_3 - c_3 = 0$

Solution:

$c_3 = 0.$
  $b_3 = 0$
  $a_3 = -1$

Substituting values:

$$\pi_3 = \frac{D}{H_0}$$

Fourth π-group:

$$\pi_4 = U(H_0)^{a_4}(g)^{b_4}(\mu)^{c_4}$$
$$\pi_4 = \left(LT^{-1}\right)(L)^{a_4}\left(LT^{-2}\right)^{b_4}\left(ML^{-1}T^{-1}\right)^{c_4} = M^0T^0L^0$$

Balance:

M: $c_4 = 0$
L: $1 + a_4 + b_4 - c_4 = 0$
T: $-1 - 2b_4 - c_4 = 0$

Solution:

$c_4 = 0.$
$\quad b_4 = -\frac{1}{2}$
$\quad a_4 = -1 - b_4 + c_4 \therefore a_4 = -\frac{1}{2}$

Substituting values:

$$\pi_4 = \frac{U}{\sqrt{gH_0}} = \frac{U^2}{gH_0}$$

Fifth π-group:

$$\pi_5 = t(H_0)^{a_5}(g)^{b_5}(\mu)^{c_5}$$
$$\pi_5 = (T)(L)^{a_5}\left(LT^{-2}\right)^{b_5}\left(ML^{-1}T^{-1}\right)^{c_5} = M^0T^0L^0$$

Balance:

M: $c_5 = 0$
L: $a_5 + b_5 - c_5 = 0$
T: $1 - 2b_5 - c_5 = 0$

Solution:

$c_5 = 0.$
$\quad b_5 = \frac{1}{2}$
$\quad a_5 = -b_5 + c_5 \therefore a_5 = -\frac{1}{2}$

Substituting values:

$$\pi_5 = \sqrt{\frac{g}{H_0}}\,t = \frac{gt^2}{H_0}$$

Sixth π-group:

$$\pi_6 = \rho(H_0)^{a_6}(g)^{b_6}(\mu)^{c_6}$$
$$\pi_6 = \left(ML^{-3}\right)(L)^{a_6}\left(LT^{-2}\right)^{b_6}\left(ML^{-1}T^{-1}\right)^{c_6} = M^0T^0L^0$$

Balance:
  M: $1 + c_6 = 0$
  L: $-3 + a_6 + b_6 - c_6 = 0$
  T: $-2b_6 - c_6 = 0$

Solution:

$c_6 = -1.$
  $b_6 = \frac{1}{2}$
  $a_6 = 3 - b_6 + c_6 \therefore a_6 = 3/2$

Substituting values:

$$\pi_6 = \frac{\rho g^{\frac{1}{2}} H_0^{\frac{3}{2}}}{\mu}$$

The final functional relationship is expressed as follows:

$$\pi_1 = f(\pi_2)(\pi_3)(\pi_4)(\pi_5)(\pi_6)$$

$$\frac{H_v}{H_0} = f\left(\frac{d}{H_0}\right)\left(\frac{D}{H_0}\right)\left(\frac{U^2}{gH_0}\right)\left(\frac{gt^2}{H_0}\right)\left(\frac{\rho g^{\frac{1}{2}} H_0^{\frac{3}{2}}}{\mu}\right)$$

Two dimensionless groups can be simplified:

$$\frac{\pi_2}{\pi_3} = \frac{d}{D}$$

$$\pi_6 = \frac{\rho g^{\frac{1}{2}} H_0^{\frac{3}{2}}}{\mu} = \frac{\rho g^{\frac{1}{2}} H_0^{\frac{1}{2}} H_0}{\mu} = \frac{\sqrt{gH_0}}{U}\frac{U\rho H_0}{\mu} = \frac{gH_0}{U^2}\frac{U\rho H_0}{\mu}$$

Therefore, the form of the general function is as follows:

$$\frac{H_v}{H_0} = C\left(\frac{d}{D}\right)^{x_1}\left(\frac{U^2}{gH_0}\right)^{y_1}\left(\frac{gH_0}{U^2}\right)^{y_2}\left(\frac{gt^2}{H_0}\right)^{x_3}\left(\frac{U\rho H_0}{\mu}\right)^{x_4}$$

In the general function, the numerator and denominator in each group are interchangeable, therefore we can collect the two terms containing the Fr number:

$$\frac{H_v}{H_0} = C\left(\frac{d}{D}\right)^{x_1}\left(\frac{U^2}{g H_0}\right)^{x_2}\left(\frac{gt^2}{H_0}\right)^{x_3}\left(\frac{U\rho H_0}{\mu}\right)^{x_4}$$

It can be noticed that π_4 and π_6 are the Fr and Re numbers, respectively;

$$\frac{H_v}{H_0} = C\left(\frac{d}{D}\right)^{x_1}(Fr)^{x_2}\left(\frac{gt^2}{H_0}\right)^{x_3}(Re)^{x_4}$$

The drain velocity, U, can be defined from an energy balance, applying Bernoulli's equation:

$$\frac{P_1}{\rho_1} + \frac{U_1^2}{2} + gz_1 = \frac{P_2}{\rho_2} + \frac{U_2^2}{2} + gz_2$$

The free surface and tapping stream are exposed to the atmosphere, then $P_1 = P_2 = P_{atm}$. If z_2, the tapping hole, is the datum line, then $z_2 = 0$ and velocity at the free surface, $U_1 \cong 0$, then, the drain velocity (U_2 in this case) is given by the following equation:

$$U_2 = \sqrt{2gz_1}$$

Actually, this result can also be obtained applying dimensional analysis: the exit velocity (U_2) is affected by gravity (g) and height of the liquid (z_1):

$$U_2 = f(g, z_1)$$

This problem involves 3 variables and 2 dimensions, which results in one dimensionless group:

$$\pi_1 = U_2(g)^{a_1}(z_1)^{b_1}$$
$$\pi_1 = (LT^{-1})(LT^{-2})^{a_1}(L)^{b_1} = M^0 T^0 L^0$$

Solution:
$a_1 = \frac{1}{2}$
$b_1 = \frac{1}{2}$
Since there is only one dimensionless group:

$$\pi_1 = \frac{U_2}{\sqrt{gz_1}} = \text{Constant}$$
$$U_2 = C\sqrt{gz_1}$$

These terms form the Froude number:

$$\frac{U_2^2}{gz_1} = C$$

Experimental data:

The initial hydrodynamic conditions in the liquid, before tapping, are defined by the filling rate of the BOF vessel. The flow rate and diameter of the pipe employed to fill the BOF with water can be used to define the filling velocity (U_1). On the other hand, the tapping or draining velocity (U_2) of the BOF vessel depends on the variable height of the liquid. This value changes during the tapping process. For each experiment only the initial tapping velocity can be defined as a constant.

In the previous dimensional analysis both Fr and Re numbers contain the liquid's velocity. Kamaraj et al. [68] choose the filling velocity to compute the Fr number on the basis that the drain velocity was already included in the product gH_0, and used the initial drain velocity in the calculation of the Re number. In this way both velocities are considered in this problem.

$$\pi_3 = Fr = \frac{U^2}{gH_0} = \frac{U_1^2}{gH_0}$$
$$\pi_5 = Re = \frac{\rho U H_0}{\mu} = \frac{\rho U_2 H_0}{\mu}$$

Table 11.39 indicates the experimental database and the values of the experimental dimensionless groups.

Applying the statistical tools of excel the following values are obtained: $\log C = 5.739$, $x_1 = 1.786$, $x_2 = -0.044$, $x_3 = 0.076$, $x_4 = -0.633$. The regression coefficient (R^2) was 0.893. Replacing these values in the general equation:

$$\frac{H_v}{H_0} = 5.487 \times 10^5 \left(\frac{d}{D}\right)^{1.786} (Fr)^{-0.044} \left(\frac{gt^2}{H_0}\right)^{0.076} (Re)^{-0.633}$$

The expression is described in Fig. 11.49. It is very similar to the expression reported by Kamaraj et al. [66].

$$\frac{H_v}{H_0} = 4.4 \times 10^5 \left(\frac{d}{D}\right)^{1.79} \left(\frac{1}{Fr}\right)^{0.05} \left(\frac{H_0}{gt^2}\right)^{-0.07} \left(\frac{1}{Re}\right)^{-0.63}$$

11.8   Rate of Consumption of Mold Fluxes

Once liquid steel is produced, it is solidified through several processes; ingot casting, continuous casting and sand casting. Continuous casting (CC) of steel is the primary route with approximately 96.6% worldwide in 2019 [69]. The evolution of this process has been described by Wolf [70] and Smil [71]. The CC ratio increased

Table 11.39 Experimental data base and experimental dimensionless groups

H_v, Cm	Q, l/min	t, s	H_0, cm	d, cm	U_1, m/s	U_2, m/s	π_1	π_2	π_3	π_4	π_5
1.80	40	30	14.1	1.05	3.65	1.66	0.13	0.01	9.65	6.3E+04	2.6E+05
1.87	40	60	14.0	1.05	3.65	1.66	0.13	0.01	9.71	2.5E+05	2.5E+05
1.70	40	90	14.0	1.05	3.65	1.66	0.12	0.01	9.71	5.7E+05	2.5E+05
1.87	40	30	11.2	1.05	3.65	1.48	0.17	0.01	12.14	7.9E+04	1.8E+05
1.97	40	60	11.2	1.05	3.65	1.48	0.18	0.01	12.14	3.2E+05	1.8E+05
1.77	40	90	11.2	1.05	3.65	1.48	0.16	0.01	12.14	7.1E+05	1.8E+05
1.67	40	30	8.0	1.05	3.65	1.25	0.21	0.01	17.00	1.1E+05	1.1E+05
1.63	40	90	8.1	1.05	3.65	1.26	0.20	0.01	16.79	9.8E+05	1.1E+05
4.58	40	30	14.2	1.41	3.65	1.67	0.32	0.01	9.58	6.2E+04	2.6E+05
4.50	40	60	14.0	1.41	3.65	1.66	0.32	0.01	9.71	2.5E+05	2.5E+05
4.37	40	90	14.0	1.41	3.65	1.66	0.31	0.01	9.71	5.7E+05	2.5E+05
4.23	40	30	11.0	1.41	3.65	1.47	0.38	0.01	12.36	8.0E+04	1.8E+05
4.37	40	60	11.0	1.41	3.65	1.47	0.40	0.01	12.36	3.2E+05	1.8E+05
4.57	40	90	11.0	1.41	3.65	1.47	0.42	0.01	12.36	7.2E+05	1.8E+05
3.70	40	30	8.0	1.41	3.65	1.25	0.46	0.01	17.00	1.1E+05	1.1E+05
5.07	40	60	8.0	1.41	3.65	1.25	0.63	0.01	17.00	4.4E+05	1.1E+05
5.17	40	90	8.0	1.41	3.65	1.25	0.65	0.01	17.00	9.9E+05	1.1E+05
5.50	40	30	14.0	1.77	3.65	1.66	0.39	0.02	9.71	6.3E+04	2.5E+05
6.35	40	60	14.1	1.77	3.65	1.66	0.45	0.02	9.65	2.5E+05	2.6E+05
7.68	40	90	14.1	1.77	3.65	1.66	0.54	0.02	9.65	5.6E+05	2.6E+05
5.55	40	30	11.2	1.77	3.65	1.48	0.50	0.02	12.14	7.9E+04	1.8E+05
6.18	40	60	11.0	1.77	3.65	1.47	0.56	0.02	12.36	3.2E+05	1.8E+05
7.63	40	90	11.3	1.77	3.65	1.49	0.67	0.02	12.04	7.0E+05	1.8E+05
5.35	40	30	8.1	1.77	3.65	1.26	0.66	0.02	16.79	1.1E+05	1.1E+05
5.90	40	60	8.0	1.77	3.65	1.25	0.74	0.02	17.00	4.4E+05	1.1E+05
6.57	40	90	8.1	1.77	3.65	1.26	0.81	0.02	16.79	9.8E+05	1.1E+05
5.47	40	30	14.0	2.14	3.65	1.66	0.39	0.02	9.71	6.3E+04	2.5E+05
6.10	40	60	13.9	2.14	3.65	1.65	0.44	0.02	9.78	2.5E+05	2.5E+05
6.20	40	90	14.0	2.14	3.65	1.66	0.44	0.02	9.71	5.7E+05	2.5E+05
5.70	40	30	11.0	2.14	3.65	1.47	0.52	0.02	12.36	8.0E+04	1.8E+05
5.90	40	60	11.0	2.14	3.65	1.47	0.54	0.02	12.36	3.2E+05	1.8E+05
7.95	40	90	11.0	2.14	3.65	1.47	0.72	0.02	12.36	7.2E+05	1.8E+05
5.13	40	30	8.1	2.14	3.65	1.26	0.63	0.02	16.79	1.1E+05	1.1E+05
6.20	40	60	8.0	2.14	3.65	1.25	0.78	0.02	17.00	4.4E+05	1.1E+05
6.63	40	90	8.0	2.14	3.65	1.25	0.83	0.02	17.00	9.9E+05	1.1E+05

(continued)

Table 11.39 (continued)

H_v, Cm	Q, l/min	t, s	H_0, cm	d, cm	U_1, m/s	U_2, m/s	π_1	π_2	π_3	π_4	π_5
1.77	30	30	14.0	1.05	2.74	1.66	0.13	0.01	5.46	6.3E+04	2.5E+05
1.70	30	60	14.0	1.05	2.74	1.66	0.12	0.01	5.46	2.5E+05	2.5E+05
1.77	30	90	14.0	1.05	2.74	1.66	0.13	0.01	5.46	5.7E+05	2.5E+05
1.87	30	30	11.2	1.05	2.74	1.48	0.17	0.01	6.83	7.9E+04	1.8E+05
1.70	30	60	11.0	1.05	2.74	1.47	0.15	0.01	6.95	3.2E+05	1.8E+05
1.70	30	90	11.1	1.05	2.74	1.48	0.15	0.01	6.89	7.2E+05	1.8E+05
1.87	30	30	8.3	1.05	2.74	1.28	0.22	0.01	9.22	1.1E+05	1.2E+05
1.93	30	60	8.0	1.05	2.74	1.25	0.24	0.01	9.56	4.4E+05	1.1E+05
2.00	30	90	8.0	1.05	2.74	1.25	0.25	0.01	9.56	9.9E+05	1.1E+05
4.35	30	30	14.2	1.41	2.74	1.67	0.31	0.01	5.39	6.2E+04	2.6E+05
4.58	30	60	14.0	1.41	2.74	1.66	0.33	0.01	5.46	2.5E+05	2.5E+05
4.73	30	90	14.0	1.41	2.74	1.66	0.34	0.01	5.46	5.7E+05	2.5E+05
4.93	30	30	11.0	1.41	2.74	1.47	0.45	0.01	6.95	8.0E+04	1.8E+05
5.05	30	60	11.0	1.41	2.74	1.47	0.46	0.01	6.95	3.2E+05	1.8E+05
5.18	30	90	11.0	1.41	2.74	1.47	0.47	0.01	6.95	7.2E+05	1.8E+05
4.00	30	30	8.0	1.41	2.74	1.25	0.50	0.01	9.56	1.1E+05	1.1E+05
4.60	30	60	8.0	1.41	2.74	1.25	0.58	0.01	9.56	4.4E+05	1.1E+05
4.97	30	90	8.0	1.41	2.74	1.25	0.62	0.01	9.56	9.9E+05	1.1E+05
5.63	30	30	14.0	1.77	2.74	1.66	0.40	0.02	5.46	6.3E+04	2.5E+05
6.77	30	60	14.0	1.77	2.74	1.66	0.48	0.02	5.46	2.5E+05	2.5E+05
8.00	30	90	14.1	1.77	2.74	1.66	0.57	0.02	5.43	5.6E+05	2.6E+05
5.27	30	30	11.0	1.77	2.74	1.47	0.48	0.02	6.95	8.0E+04	1.8E+05
6.47	30	60	11.2	1.77	2.74	1.48	0.58	0.02	6.83	3.2E+05	1.8E+05
7.50	30	90	11.1	1.77	2.74	1.48	0.68	0.02	6.89	7.2E+05	1.8E+05
5.50	30	30	8.1	1.77	2.74	1.26	0.68	0.02	9.44	1.1E+05	1.1E+05
7.17	30	60	8.0	1.77	2.74	1.25	0.90	0.02	9.56	4.4E+05	1.1E+05
7.23	30	90	8.0	1.77	2.74	1.25	0.90	0.02	9.56	9.9E+05	1.1E+05
5.73	30	30	14.0	2.14	2.74	1.66	0.41	0.02	5.46	6.3E+04	2.5E+05
6.90	30	60	14.0	2.14	2.74	1.66	0.49	0.02	5.46	2.5E+05	2.5E+05
7.93	30	90	14.0	2.14	2.74	1.66	0.57	0.02	5.46	5.7E+05	2.5E+05
6.10	30	30	11.0	2.14	2.74	1.47	0.55	0.02	6.95	8.0E+04	1.8E+05
6.90	30	60	11.1	2.14	2.74	1.48	0.62	0.02	6.89	3.2E+05	1.8E+05
8.17	30	90	11.0	2.14	2.74	1.47	0.74	0.02	6.95	7.2E+05	1.8E+05
6.10	30	30	7.7	2.14	2.74	1.23	0.79	0.02	9.93	1.1E+05	1.0E+05
6.73	30	60	8.2	2.14	2.74	1.27	0.82	0.02	9.33	4.3E+05	1.1E+05

(continued)

Table 11.39 (continued)

H_v, Cm	Q, l/min	t, s	H_0, cm	d, cm	U_1, m/s	U_2, m/s	π_1	π_2	π_3	π_4	π_5
7.17	30	90	7.9	2.14	2.74	1.24	0.91	0.02	9.68	1.0E+06	1.1E+05
1.93	20	30	14.1	1.05	1.83	1.66	0.14	0.01	2.41	6.3E+04	2.6E+05
2.03	20	60	14.0	1.05	1.83	1.66	0.15	0.01	2.43	2.5E+05	2.5E+05
2.03	20	90	14.0	1.05	1.83	1.66	0.15	0.01	2.43	5.7E+05	2.5E+05
2.17	20	30	11.0	1.05	1.83	1.47	0.20	0.01	3.09	8.0E+04	1.8E+05
2.23	20	60	11.0	1.05	1.83	1.47	0.20	0.01	3.09	3.2E+05	1.8E+05
2.17	20	90	11.0	1.05	1.83	1.47	0.20	0.01	3.09	7.2E+05	1.8E+05
2.10	20	30	7.9	1.05	1.83	1.24	0.27	0.01	4.30	1.1E+05	1.1E+05
2.10	20	60	8.0	1.05	1.83	1.25	0.26	0.01	4.25	4.4E+05	1.1E+05
2.20	20	90	8.0	1.05	1.83	1.25	0.28	0.01	4.25	9.9E+05	1.1E+05
4.68	20	30	14.0	1.41	1.83	1.66	0.33	0.01	2.43	6.3E+04	2.5E+05
4.60	20	60	14.0	1.41	1.83	1.66	0.33	0.01	2.43	2.5E+05	2.5E+05
4.98	20	90	14.1	1.41	1.83	1.66	0.35	0.01	2.41	5.6E+05	2.6E+05
4.08	20	30	11.0	1.41	1.83	1.47	0.37	0.01	3.09	8.0E+04	1.8E+05
4.23	20	60	11.0	1.41	1.83	1.47	0.38	0.01	3.09	3.2E+05	1.8E+05
4.55	20	90	11.0	1.41	1.83	1.47	0.41	0.01	3.09	7.2E+05	1.8E+05
3.47	20	30	8.1	1.41	1.83	1.26	0.43	0.01	4.20	1.1E+05	1.1E+05
3.90	20	60	8.0	1.41	1.83	1.25	0.49	0.01	4.25	4.4E+05	1.1E+05
5.07	20	90	8.0	1.41	1.83	1.25	0.63	0.01	4.25	9.9E+05	1.1E+05
5.07	20	30	14.1	1.77	1.83	1.66	0.36	0.02	2.41	6.3E+04	2.6E+05
6.20	20	60	14.1	1.77	1.83	1.66	0.44	0.02	2.41	2.5E+05	2.6E+05
6.83	20	90	14.0	1.77	1.83	1.66	0.49	0.02	2.43	5.7E+05	2.5E+05
5.70	20	30	11.0	1.77	1.83	1.47	0.52	0.02	3.09	8.0E+04	1.8E+05
6.30	20	60	11.0	1.77	1.83	1.47	0.57	0.02	3.09	3.2E+05	1.8E+05
6.93	20	90	11.0	1.77	1.83	1.47	0.63	0.02	3.09	7.2E+05	1.8E+05
5.85	20	30	8.0	1.77	1.83	1.25	0.73	0.02	4.25	1.1E+05	1.1E+05
6.30	20	60	8.0	1.77	1.83	1.25	0.79	0.02	4.25	4.4E+05	1.1E+05
7.45	20	90	8.0	1.77	1.83	1.25	0.93	0.02	4.25	9.9E+05	1.1E+05
6.37	20	30	14.0	2.14	1.83	1.66	0.45	0.02	2.43	6.3E+04	2.5E+05
7.07	20	60	14.0	2.14	1.83	1.66	0.50	0.02	2.43	2.5E+05	2.5E+05
8.90	20	90	14.0	2.14	1.83	1.66	0.64	0.02	2.43	5.7E+05	2.5E+05
6.10	20	30	11.0	2.14	1.83	1.47	0.55	0.02	3.09	8.0E+04	1.8E+05
6.73	20	60	11.0	2.14	1.83	1.47	0.61	0.02	3.09	3.2E+05	1.8E+05
8.10	20	90	11.1	2.14	1.83	1.48	0.73	0.02	3.06	7.2E+05	1.8E+05
5.93	20	30	8.0	2.14	1.83	1.25	0.74	0.02	4.25	1.1E+05	1.1E+05
6.80	20	60	8.0	2.14	1.83	1.25	0.85	0.02	4.25	4.4E+05	1.1E+05

(continued)

Table 11.39 (continued)

H_v, Cm	Q, l/min	t, s	H_0, cm	d, cm	U_1, m/s	U_2, m/s	π_1	π_2	π_3	π_4	π_5
7.40	20	90	8.0	2.14	1.83	1.25	0.93	0.02	4.25	9.9E+05	1.1E+05

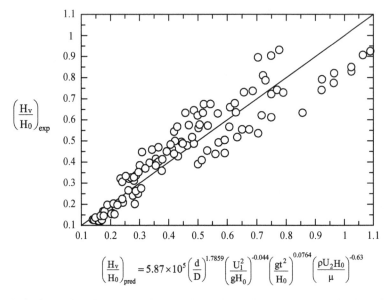

$$\left(\frac{H_v}{H_0}\right)_{pred} = 5.87 \times 10^5 \left(\frac{d}{D}\right)^{1.7859} \left(\frac{U_1^2}{gH_0}\right)^{-0.044} \left(\frac{gt^2}{H_0}\right)^{0.0764} \left(\frac{\rho U_2 H_0}{\mu}\right)^{-0.63}$$

Fig. 11.49 Comparison between predicted and experimental data to describe the dimensionless critical vortex height using dimensional analysis. Experimental data from Kamaraj et al. [66]

linearly from 1970 until the end of the 1990s, from 4 to 90%, respectively. The early concept of CC was developed by Bessemer in 1856. Siegfried Junghans (1887–1954) developed the oscillating (reciprocating) mold in 1933 and was applied in the non-ferrous industry. Irving Rossi (1889–1991) acquired patent rights to commercialize the technology developed by Junghans, founding Concast in 1954.

The oscillating mold was the first major milestone in continuous casting. Szekers has explained in the detail the variables that control the oscillation mold [72]. These variables are directly related with the rate of consumption of mold fluxes. A strand is a thread of steel leaving the model. The Stroke (S) is the total distance the mold moves in one direction; downward stroke and upward stroke. An oscillation cycle is formed by the downward and upward movements and the frequency of oscillation (f) is the number of cycles per second (Hz). The original Junghans oscillation cycle consisted of a downward velocity equal to the casting speed. Three-quarters of each cycle corresponding to the downward stroke (also called lead) following by a rapid upward stroke to the starting position. It was called 3:1 oscillation profile. In 1954, I. Halliday proposed another major milestone, the negative strip time (t), i.e., moving the mold slightly faster than the strand during the downstroke of the cycle. This operation reduces shell sticking which then allows higher casting speeds. The negative strip time was high in the 1970s, from 0.50 to 1.0 s. Now is in the range from 0.1–0.25 s.

In the late 1950s, Russian plants developed another important development, the sinusoidal oscillation profile. With this oscillation mode, the maximum instantaneous mold velocity is defined by the product πsf. The oscillation mode is responsible for the formation of oscillation marks. Its spacing (or pitch) is given by the ratio between casting speed and oscillation frequency, which is equal to the distance the strand travels during one cycle. The stroke doesn't affect the spacing of oscillation marks. The oscillation marks, in particular those from the broadface are wiped out during the rolling process. The depth of an oscillation mark should be as low as possible. It is decreased by increasing casting speed or increasing the mold flux viscosity.

Mold powders are a very important component of the casting process. A full description of mold powders is given by Mills and Däcker [73]. Mold powders are placed on top of the mold. Three layers are gradually formed: an upper solid phase (sinterized), a mushy zone and liquid slag in contact with liquid steel.

Mold powders have the following functions:

- Protect liquid steel from the surrounding atmosphere
- Form an insulation layer that decreases heat losses, preventing meniscus solidification
- Absorbs non-metallic inclusions
- Form a lubricant at the interface mold-solidified shell
- The metal/slag interfacial tension defines the shape of the meniscus.

During continuous casting, the molten powder infiltrates through the gap between the solidifying steel strand and the mold wall. It forms a lubricant. The liquid slag in contact with the mold walls crystallizes. Its thickness and physical properties controls heat extraction in the horizontal direction.

The break (or solidification) temperature (T_{br}) is the temperature where the viscosity has a large increase in viscosity during cooling due to the precipitation of solids. This temperature depends on the cooling rate. It decreases when the cooling rate increases. For comparison purposes it is usually reported at a cooling rate of 10 K/min. the breaking temperature separates the solid from the liquid fraction of the infiltrated mold powder, as shown in Fig. 11.50. Increasing the thickness of the solid decreases heat transfer in the horizontal direction. Decreasing the thickness of the liquid fraction affects lubrication of the strand.

Mold powder adheres to the mold walls and the strand steel surface and it is consumed; therefore, it should be replenished by power feeding. To achieve a good lubrication, Andrezejewski et al. [74] suggest a consumption rate of approximately 0.3 kg/m². Both Kamaraj et al. [75] and Saraswat et al. [76] have reviewed the relationships describing the consumption rate of mold fluxes from 1978 to 2000. Shridar et al. [77] indicate that powder consumption is a measure of the amount of liquid slag infiltrating into the mold/strand channel.

There is no full agreement about the variables that affect powder consumption. Saraswat et al. [76] reported the following variables in decreasing order: (1) casting speed, (2) Powder viscosity, (3) Stroke length, (4) Oscillation frequency and (5)

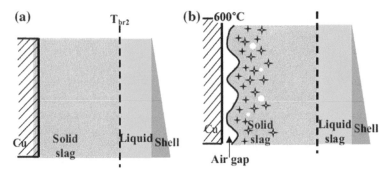

Fig. 11.50 Lubrication due to mold powders. After Mills and Däcker [73]

Break temperature. The consumption rate decreases by increasing the casting speed, increasing viscosity and increasing the break temperature. Mold powder chemical and physical composition is designed for different types of steel and casting speeds. For thin strip casting, with casting velocities up to 6 m/min, a low powder consumption is not critical as long as shows a good melting rate and an undisturbed slag infiltration which requires a lower viscosity [78].

The melting rate of mold powders is affected by several variables: carbon content, particle size, vertical heat flux, exothermic components of the powder, melting heat of mold powder. The melting rate increases by [73]:

- Decreasing free carbon. Carbon promotes non-wetting conditions among powder particles, decreasing its agglomeration. The decrease should be controlled because free carbon is an exothermic component and too low can affect steel meniscus solidification.
- Increase carbonate content to improve the permeability of the bed.
- Increase the vertical heat flux by increasing the casting speed and turbulence up to certain limit.
- Decrease the bulk density of mold power by increasing its particle size.
- Decreasing the thickness of the mold flux [79].

The thickness of the liquid slag is in the range from 15 to 20 mm [80]. A shallow slag pool (<10 mm) leads to poor power consumption.

There are several methods to measure the melting rate of mold powders: (1) Crucible test. Mold powder is heated unidirectionally and melted in a crucible at 1400 °C. After melting the crucible is cut thorough the center and the fraction of liquid is measured. The melting time is recorded. The melting time is expressed as mass of the liquid/time. In a variation of this method, the crucible has a hole through which the molten flux can drain (drain test). Experiments carried out by CSM placed the powder in a quartz tube with a diameter of 40 mm, the bottom was an aluminum foil. Heating was carried out in an induction furnace. The powder was placed on a plate of SiC at 1400 °C. They reported the melting rate as a function of free carbon [81]

$$Q_m(kg/s) = \frac{0.13857}{\left(\%C_{free}\right)^{1.0669}}$$

(2) Using a simulator. Kawamoto et al. used a simulator with a capacity of 1 ton of liquid steel. The powder was placed on top of the steel and then covered with a magnesia board. The melting rate was measured using wire dips. Their results indicate that the melting rate is a function of the content of carbonates:

$$Q_m(kg/s) = \rho_m A \left(16.8 \sum r_i(\%carbonate) - 0.00336C_v + 0.0477\right)$$

where: r_i is a rate constant using different amounts of Na_2CO_3 in the mold flux, C_V is the carbon content per unit volume (kg/m^3), ρ_m is the density of liquid flux, A is the cross-sectional area of the mold or crucible. (3) Theoretically estimated from heat transfer calculations. Mills and Däcker [73] proposed the calculation of the melting rate based on the vertical heat flux and melting enthalpy, as follows:

$$Q_m(kg/s) = \frac{q_{vert}}{\Delta H} = \frac{kA\frac{\Delta T}{d}}{\Delta H}$$

where: q_{vert} is the vertical heat flux from the liquid steel to the top mold powder in kJ/s, ΔH is the melting enthalpy of the mold powder in kJ/kg, k is the mean thermal conductivity of the mold powder in kW/mK, A is the cross-sectional area of the mold in m^2, ΔT is the temperature gradient from the liquid bath to the atmosphere and d is the thickness of the mold powder in m. The air temperature above the mold powder is in the range from 200 to 400 °C. Hanao et al. [82] reported the apparent thermal conductivity for a mold powder as a function of the slag binary basicity ratio, which can be defined through the following relationship [75],

$$k \ (W/m \ K) = -0.14484 \times B^3 + 5.8955 \times B^2 - 8.2831 \times B + 6.5702$$

The value of k ranges from 2.3 to 3.0 W/m K in the basicity range from 0.8 to 1.8.
The enthalpy of melting includes heating to the melting temperature and latent heat of melting. Kamaraj et al. [75] assumes a value for the latent heat of 500 kJ/kg.

Dimensional analysis:
The following dimensional analysis is based on the variables suggested by Kamaraj et al. [75]. The mold flux consumption rate (Q_C) depends on the following 6 variables: mold flux melting rate (Q_m), casting speed (U), mold cross section (A), negative strip time (t_N), frequency of mold oscillation (f) and stroke (S).
$Q_C = f (Q_m, U, A, S, t_N, f)$.

Variable	Symbol	Units	Dimensions
Consumption rate	Q_C	kg/m^2	$[ML^{-2}]$
Melting rate	Q_m	kg/s	$[MT^{-1}]$
Casting speed	U	m/s	$[LT^{-1}]$
Area	A	m^2	$[L^2]$
Stroke	S	m	$[L]$
Negative strip time	t	N/m	$[T]$
Frequency	f	1/s	$[T^{-1}]$

There are 7 variables and 3 dimensions. According with the π-theorem this yields 4 dimensionless groups. As non-repeating variables we choose; Q_m, t and U. Next, we define the values for the four dimensionless groups as follows:

$$\pi_1 = Q_C(Q_m)^{a_1}(t)^{b_1}(U)^{c_1}$$
$$\pi_2 = A\,(Q_m)^{a_2}(t)^{b_2}(U)^{c_2}$$
$$\pi_3 = S\,(Q_m)^{a_3}(t)^{b_3}(U)^{c_3}$$
$$\pi_4 = f\,(Q_m)^{a_4}(t)^{b_4}(U)^{c_4}$$
$$\pi_1 = f\,(\pi_2)(\pi_3)(\pi_4)$$

First π-group:

$$\pi_1 = Q_C(Q_m)^{a_1}(t)^{b_1}(U)^{c_1}$$
$$\pi_1 = \left(ML^{-2}\right)\left(MT^{-1}\right)^{a_1}(T)^{b_1}\left(LT^{-1}\right)^{c_1} = M^0 T^0 L^0$$

Balance:

M: $1 + a_1 = 0$.
  L: $-2 + c_1 = 0$
  T: $-a_1 + b_1 - c_1 = 0$

Solution:

$a_1 = -1$.
  $c_1 = 2$
  $b_1 = 1$

Substitution of these values on the π_1-group yields:

$$\pi_1 = \frac{Q_C U^2 t}{Q_m}$$

Second π-group:

$$\pi_1 = Q_C(Q_m)^{a_1}(t)^{b_1}(U)^{c_1}$$
$$\pi_1 = (ML^{-2})(MT^{-1})^{a_1}(T)^{b_1}(LT^{-1})^{c_1} = M^0T^0L^0$$

Balance:

M:$a_2 = 0$.
  L: $2 + c_2 = 0$
  T: $-a_2 + b_2 - c_2 = 0$

Solution:

$a_2 = 0$.
  $c_2 = -2$
  $b_2 = -2$

Substitution of these values on the π_2-group yields:

$$\pi_2 = \frac{A}{U^2 t^2}$$

Third π-group:

$$\pi_3 = S(Q_m)^{a_3}(t)^{b_3}(U)^{c_3}$$
$$\pi_3 = (L)(MT^{-1})^{a_3}(T)^{b_3}(LT^{-1})^{c_3} = M^0T^0L^0$$

Balance:

M: $a_3 = 0$.
  L: $1 + c_3 = 0$
  T: $-a_3 + b_3 - c_3 = 0$

Solution:

$a_3 = 0$.
  $c_3 = -1$
  $b_3 = -1$

Substitution of these values on the π_3-group yields:

$$\pi_3 = \frac{S}{Ut}$$

Fourth π-group:

$$\pi_4 = f(Q_m)^{\alpha_4}(t)^{b_4}(U)^{c_4}$$
$$\pi_2 = \left(T^{-1}\right)\left(MT^{-1}\right)^{\alpha_4}(T)^{b_4}\left(LT^{-1}\right)^{c_4} = M^0T^0L^0$$

Balance:

M: $a_4 = 0$.
   L: $c_4 = 0$
   T: $-1 - a_4 + b_4 - c_4 = 0$

Solution:

$a_4 = 0$.
   $c_4 = 0$
   $b_4 = 1$
   Substitution of these values on the π_2-group yields:

$$\pi_4 = f \cdot t$$

Then,

$$\pi_1 = f(\pi_2)(\pi_3)(\pi_4)$$
$$\frac{Q_C U^2 t}{Q_m} = f\left(\frac{A}{U^2 t^2}\right)\left(\frac{S}{Ut}\right)(f \cdot t)$$

Therefore, the form of the general function is as follows:

$$\frac{Q_C U^2 t}{Q_m} = C\left(\frac{A}{U^2 t^2}\right)^{x_1}\left(\frac{S}{Ut}\right)^{x_2}(f \cdot t)^{x_3}$$

Experimental data from Kamaraj et al. [75, 83]

Previous analysis of mold flux consumption rate (Q_C) have neglected the effect of the mold flux melting rate (Q_m). These authors measured the melting rate of mold powders in contact with interstitial free and peritectic grade steels. The chemical composition of mold powders and steels is shown in Table 11.40.

A quartz crucible with a diameter of 26 mm was placed inside an induction furnace. Steel scrap, 40 gr, was melted under an inert atmosphere up to 1550 °C, then mold powders were gradually added to reach a bed height of 5 or 10 cm. The powder was left for a certain time, in the range from 30 to 540 s to allow its melting process using a current of 275 A and then the quartz tube was air cooled. The mass of molten flux was measured. The experimental set up is shown in Fig. 11.51. The average melting rate for the IF-mold flux was higher than for P-mold flux, 0.0045 and 0.0058 g/s, respectively. The melting rate increased by decreasing the bed height.

Table 11.40   Chemical composition of mold powders and liquid steel

	CaO	SiO$_2$	MgO	Al$_2$O$_3$	Fe$_2$O$_3$	(Na,K)O	F	C$_{free}$	C$_t$	CO$_2$	kg/m^3
IF	32.1	37.6	4.5	5.2	2.1	4.7	5.2	1.5	3.3	6.6	900
P	36.5	30.5	4.0	5.0	<1.5	6.5	5.0	4.25	5.5	5.75	1005
	C	Mn	S	P	Si	Al	Cr	Nb	Ti	B	
IF	0.002	0.39	0.007	0.046	0.004	0.04	0.013	0.001	0.04	0.0005	
P	0.073	1.11	0.005	0.016	0.014	0.04	0.023	0.01	–	0.0001	

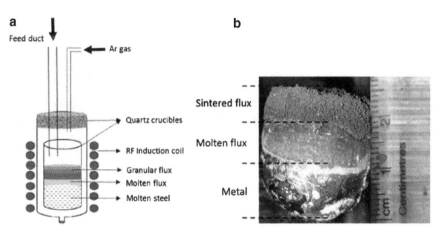

Fig. 11.51   a Experimental set up, **b** layers of mold powder. After [75, 83]. Used with permission of TMS/ASM

The experimental data were compared with the predictions using the previous relationships to estimate the melting rate. It was found a better agreement using the equation suggested by Mills and Däcker [73]. Consumption rates of mold powders reported by Pinheiro [84], Normanton et al. [85], Shin et al. [86] and Kania and Gawor [87] were employed to define the dimensionless groups. The chemical composition and thermal properties of these mold powders are shown in Tables 11.41 and 11.42, respectively. Thickness of the industrial trials was estimated as 100 and 75 mm.

Dimensional analysis will be carried out using the following relationships.

Table 11.41   Chemical composition of model powders

References	CaO	SiO$_2$	MgO	Al$_2$O$_3$	Fe$_2$O$_3$	(Na, K)O	F	C$_{free}$	C$_t$	CO$_2$	kg/m^3
[84]	30.1	32	1.4	6.7	–	14.6	8.2	3.5	–	–	700
[85]	26.9	32.7	1.1	4.8	1.5	12	9.7	–	6.3	8.45	–
[86]	39.8	36.3	3.4	0.8	0.3	3.5	6	–	3	3.5	–
[87]	26.5	31.6	1.5	4.5	0.93	9.33	8.62	8.31	–	11.9	1230

Table 11.42 Thermal properties of model powders

References	ΔH, kJ/kg	A, m^2	B	k, W/mK	T$_L$, °C	d, m
[84]	1608	0.043	0.94	2.78	1540	0.1
[85]	1784	0.0038	0.82	2.93	1550	0.075
[86]	2008	0.299	1.09	2.66	1534	0.1
[87]	1664	0.226	0.83	2.91	1533	0.1

$$\pi_1 = f(\pi_2)(\pi_3)(\pi_4)$$
$$\frac{Q_C U^2 t}{Q_m} = C\left(\frac{A}{U^2 t^2}\right)^{x_1}\left(\frac{S}{Ut}\right)^{x_2}(f \cdot t)^{x_3}$$

Table 11.43 indicates the experimental database and the values of the experimental dimensionless groups.

Applying the statistical tools of excel the following values are obtained: log C = 1.411, $x_1 = -1.09$, $x_2 = 0.83$, $x_3 = -0.21$. The regression coefficient (R^2) was 0.986. Replacing these values in the general equation:

$$\frac{Q_C U^2 t}{Q_m} = 25.8\left(\frac{A}{U^2 t^2}\right)^{-1.09}\left(\frac{S}{Ut}\right)^{0.83}(f \cdot t)^{-0.21}$$

The previous expression is illustrated in Fig. 11.52.

11.9   Penetration of a High-Velocity Gas Jet Through a Liquid Surface

Gas injection in metallurgical systems is widespread. One major application is the injection of oxygen for decarburization in steelmaking, both in the BOF and EAF processes. Figure 11.53 is a schematic of top gas injection into a liquid. It is observed that the gas jet creates a depression or cavity in the liquid.

The reaction kinetics of processes, for example the decarburization rate and slag formation, depend on the interaction between the gas jet and the liquid. As a result of this interaction the reaction interface and motion of the liquid is largely influenced. In the BOF oxygen is injected using a vertical top lance, the jet of oxygen impacts the liquid steel bath, oxygen is transferred to the liquid surface and then diffuses to the bulk liquid. Many research works have been conducted in the last 70 years on the subject [85–98]. Collins and Lubanska [89] investigated the effect of inclination angle (15–90°), momentum of the jet, axial distance from lance to liquid surface and lance diameter. The water model had a diameter of 60 cm. This ensured the disturbed cross-sectional area was small, in the order of 1%. The water tank was illuminated from two sides with mercury arc lamps to get good pictures. The lance diameter, in the range from 0.6 to 1.12 cm didn't have a significant effect on the penetration

Table 11.43 Experimental data and experimental dimensionless groups

Q_C, kg/m^2	Q_m, kg/s	U, m/s	S, m	f, 1/s	t_n, s	A, m^2	π_1	π_2	π_3	π_4
0.247	0.0268	0.0241	0.0064	2.65	0.107	0.299	0.001	44964	2.48	0.28
0.232	0.0268	0.0241	0.005	2.43	0.116	0.299	0.001	38258	1.79	0.28
0.225	0.0268	0.0241	0.005	2.94	0.101	0.299	0.000	50465	2.05	0.30
0.253	0.0268	0.0241	0.005	3.49	0.083	0.299	0.000	74728	2.50	0.29
0.223	0.0268	0.0241	0.006	2.08	0.131	0.299	0.001	29998	1.90	0.27
0.229	0.0268	0.0241	0.006	2.45	0.111	0.299	0.001	41782	2.24	0.27
0.23	0.0268	0.0241	0.006	2.9	0.124	0.299	0.001	33481	2.01	0.36
0.248	0.0268	0.0241	0.007	1.77	0.142	0.299	0.001	25531	2.05	0.25
0.208	0.0268	0.0241	0.007	2.09	0.155	0.299	0.001	21428	1.87	0.32
0.211	0.0268	0.0241	0.007	2.49	0.129	0.299	0.001	30936	2.25	0.32
0.271	0.0268	0.0225	0.00625	2.46	0.114	0.299	0.001	45446	2.44	0.28
0.247	0.0268	0.023	0.0063	2.52	0.112	0.299	0.001	45059	2.45	0.28
0.256	0.0268	0.0237	0.00637	2.58	0.109	0.299	0.001	44804	2.47	0.28
0.238	0.0268	0.0247	0.00647	2.69	0.106	0.299	0.001	43618	2.47	0.29
0.237	0.0268	0.0246	0.00646	2.67	0.106	0.299	0.001	43973	2.48	0.28
0.215	0.0268	0.0253	0.00653	2.74	0.104	0.299	0.001	43188	2.48	0.28
0.212	0.0268	0.0258	0.00658	2.79	0.102	0.299	0.001	43175	2.50	0.28
0.21	0.0268	0.0275	0.00675	2.96	0.097	0.299	0.001	42021	2.53	0.29
0.194	0.0268	0.0277	0.00677	2.97	0.097	0.299	0.001	41416	2.52	0.29
0.247	0.0268	0.0244	0.00644	2.66	0.107	0.299	0.001	43866	2.47	0.28
0.234	0.0268	0.025	0.0065	2.71	0.105	0.299	0.001	43392	2.48	0.28
0.45	0.00458	0.021	0.006	2.5	0.14	0.043	0.006	4975	2.04	0.35
0.53	0.0042	0.021	0.006	2.5	0.14	0.043	0.008	4975	2.04	0.35
0.45	0.00458	0.022	0.006	2.83	0.12	0.043	0.006	6170	2.27	0.34
0.53	0.0042	0.022	0.006	2.83	0.12	0.043	0.007	6170	2.27	0.34
0.45	0.00458	0.021	0.006	2.28	0.14	0.043	0.006	4975	2.04	0.32
0.53	0.0042	0.021	0.006	2.28	0.14	0.043	0.008	4975	2.04	0.32
0.392	0.021	0.0173	0.006	1.62	0.19	0.226	0.001	20917	1.83	0.31
0.54	0.0195	0.0173	0.006	1.62	0.19	0.226	0.002	20917	1.83	0.31
0.42	0.019	0.0173	0.006	1.62	0.19	0.226	0.001	20917	1.83	0.31
0.45	0.00035	0.02	0.007	2	0.17	0.0038	0.087	329	2.06	0.34
1.17	0.00035	0.01	0.007	2	0.21	0.0038	0.070	862	3.33	0.42
0.45	0.00035	0.02	0.007	2	0.17	0.0038	0.087	329	2.06	0.34
0.7	0.00035	0.02	0.007	1	0.14	0.0038	0.112	485	2.50	0.14
0.45	0.00035	0.02	0.007	2	0.17	0.0038	0.087	329	2.06	0.34
0.34	0.00035	0.02	0.007	4	0.106	0.0038	0.041	845	3.30	0.42

(continued)

Table 11.43 (continued)

Q_C, kg/m²	Q_m, kg/s	U, m/s	S, m	f, 1/s	t_n, s	A, m²	π_1	π_2	π_3	π_4
0.28	0.00035	0.04	0.007	2	0.07	0.0038	0.090	485	2.50	0.14
0.33	0.00045	0.02	0.004	2	0.103	0.0038	0.030	895	1.94	0.21
0.28	0.039	0.02	0.0068	1.75	0.18	0.369	0.001	28472	1.89	0.32
0.28	0.039	0.02	0.004	2	0.1	0.369	0.000	92250	2.00	0.20
0.28	0.039	0.02	0.0088	1.75	0.21	0.369	0.001	20918	2.10	0.37
0.28	0.039	0.02	0.005	2	0.14	0.369	0.000	47066	1.79	0.28

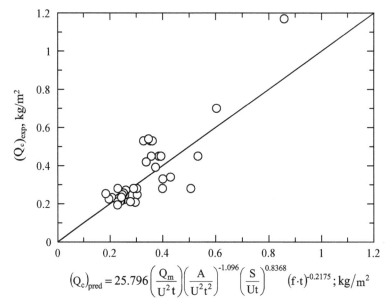

$$(Q_c)_{pred} = 25.796 \left(\frac{Q_m}{U^2 t} \right) \left(\frac{A}{U^2 t^2} \right)^{-1.096} \left(\frac{S}{Ut} \right)^{0.8368} (f \cdot t)^{-0.2175} ; kg/m^2$$

Fig. 11.52 Comparison between predicted and experimental data to describe the consumption rate of mold fluxes using dimensional analysis. Experimental data from Kamaraj et al. [75, 83]

depth. The axial distance was increased up to 16 cm. Shimada et al. [90] related the penetration depth with the jet velocity, axial distance and nozzle diameter.

Banks and Chandrasekhara [91] conducted experiments in a water model using circular and plane nozzles. The cavity depth was measured using a hook gauge. They applied dimensional analysis to this problem in addition to two analytical analysis to define the penetration depth. From turbulence theory, it is known that the centerline velocity decays with distance from the nozzle, as follows:

$$\frac{U_c}{U_j} = K_2 \frac{d_j}{h}$$

Fig. 11.53 Schematic indicating the variables during top gas injection. After Wakelin [88]

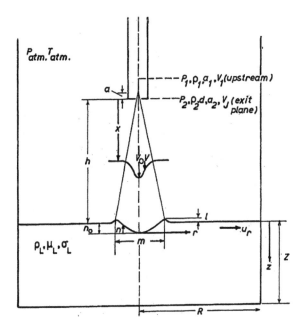

where: U_c is the centerline velocity, U_j is the velocity of the jet, d is the diameter of the nozzle and h is the nozzle distance to the liquid. Applying an energy balance; the displaced liquid model assumes the force due to the jet momentum on the liquid is equal to the weight of the displaced liquid. Using the previous equation in the energy balance, the authors arrived to the following relationship:

$$\frac{\dot{M}}{\gamma_l h^3} = \frac{\pi}{2K_2^2}\left(\frac{n_0}{h}\right)\left(1 + \frac{n_0}{h}\right)^2$$

where: n_0 is the penetration depth, h is the distance from tip of the lance to the surface of the liquid, $\dot{M}$ is the momentum of the jet. K_2 is a constant. The left hand side is a dimensionless jet momentum.

Wakelin [88] in addition to the penetration depth, also investigated the effect of liquid density (water and mercury), mass transfer using CO_2 as injected gas and the velocity fields in the liquid. He also conducted high temperature experiments related to mass transfer.

Cheslak et al. [92] employed a water-cement mixture to study the penetration depth. The mixture was allowed to dry and the dimensions of the cavity can be defined more precisely in comparison to pure water. They reported that the change in the vertical momentum is equal to the weight of displaced fluid and that liquid and gas viscosity as well as liquid surface tension have no effect on the penetration depth.

Ishikawa et al. [93] studied the penetration depth using a lance with one and three holes. They carried out dimensional analysis employing the following variables: penetration depth, gas flow rate, axial distance, nozzle diameter, gas density, liquid density and gravity force.

Koria and Lange [94] conducted high temperature experiments using hot metal with 3.5–4% C. The hot metal was placed in a preheated transparent quartz crucible with a diameter of 10 cm. Oxygen was blown through a water-cooled lance with a throat diameter (vena contracta) of 1.2 mm. A high-speed camera recorded the depression, which oscillated about 20%. The effect of oxygen supply pressure (3–9 bars), lance distance, number and inclination angle of nozzles as well as carbon content on the penetration depth was investigated. Carbon was found to have no effect. Similar result has been previous reported by Sharma et al. [95]. They indicated that chemical reaction at high temperatures do not affect cavity dimensions.

Koria and Lange [94] collected all previous work at high temperature and were able to describe the experimental results with two dimensionless numbers, as follows:

$$\frac{n_0}{h} = 4.469\left(\frac{\dot{M}\cos\theta}{\iota g h^3}\right)^{0.66}$$

where: θ is the nozzle inclination angle, ρ_l is the density of the liquid and g is the gravitational constant. The numerical modeling results obtained by Doh et al. [96] are in good agreement with the values predicted by the previous expression.

These authors found that the important variables are the gas pressure, lance distance and number of nozzles. The momentum of the jet can be calculated using the pressure drop. Koria and Lange [97] as well as Wakelin [88] proposed, respectively, the following expressions:

$$\dot{M} = 0.7854 \times 10^5 d^2\left(1.27\frac{p_0}{p_a} - 1\right)$$

where: $\dot{M}$ is the momentum of the jet in N, p_0 is the supply pressure in bars, p_a is the atmospheric pressure in bars, d is the lance distance in m.

$$\dot{M} = C_d\rho_{ga}A_aU^2 = C_dA_a\left(\frac{\frac{2\varphi}{\varphi}p_a\left\{\left(\frac{p_0}{p_a}\right)^{\frac{\varphi-1}{\varphi}} - 1\right\}}{1 - \left\{\left(\frac{p_a}{p_0}\right)^{\frac{1}{\varphi}}\frac{A_a}{A_0}\right\}^2}\right)$$

where: $\dot{M}$ is the momentum of the jet in dynes, C_d is a discharge coefficient, A_a and A_0 are the cross-sectional areas of the outlet and inlet positions in the nozzle in cm^2, p_0 is the supply pressure in atm, p_a is the atmospheric pressure in atm, φ is a ratio of heat capacities.

More than 10 relationships have been proposed to describe the penetration depth [94, 98]. Nordquist et al. [98] used their experimental data to compare those relationships and found four of them reproduce their results satisfactorily. The momentum of the jet was calculated from the gas flow rate, gas density and lance cross-sectional area. One reason of an inadequate prediction with some correlations was attributed to different ways to report the penetration depth. They used an average value but, in some reports, it is measured the maximum value. Their experimental data clearly indicate that decreasing the lance diameter, the penetration depth decreases. One of the limitations found in all the correlations was that fail to accurately reproduce experimental data using low lance diameters, below 2 mm.

Meidani et al. [99] conducted experiments in a large water model with a diameter of 1150 mm, filled up to a height of 900 mm. A supersonic air jet with a Mach number of 2.15 was employed, the gas flow rate was 4.25 Nm^3/min and two axial distances (76 and 250 mm) were considered. They reported a dimensional analysis considering the following variables: $n_0 = f (\dot{M}, \rho_l, g, h, \mu_l, \sigma_l)$. Their experimental data included the gas flow rate.

Wang et al. [100] extended the effect of temperature of the gas on the penetration depth through mathematical modeling of an industrial BOF. The velocity of the jet increases by increasing the temperature, in turn increasing the penetration depth.

The phenomena of cavity formation due to gas impingement has also been studied by mathematical modelling. Solorzano et al. [101] made experimental measurements in a water model, with PIV and validated a mathematical model with their experimental data. Their results suggest that an inclined lance promotes higher velocities of the liquid.

Qian et al. [102] used a very accurate system to measure the geometry of the cavity; an interface tracking electro resistivity probe. The probe signal is recorded by an oscilloscope. Additionally, this was the first systematic work using a top oil layer. A cylinder, 29.8 cm in diameter and 20.3 cm in height, made of plexiglass was placed in a rectangular vessel to avoid optical distortion. The range of the variables without a top oil layer was; nozzle diameter from 0.58 to 2.3 cm, nozzle distance from 1.5 to 21 cm, gas flow rate from 0.5 to 12.5 Nm^3/h. Kerosene and corn oil were used as top layers. As shown below, kerosene has a viscosity similar to water but corn oil has a much higher viscosity than water. The viscosity ratio corn oil/water is 460, much higher than that of slag/steel which is 33. The oil thickness was 0.8 cm. The experiments with an oil layer were made with a lance diameter of 1.2 cm, a nozzle distance from 3 to 7 cm and gas flow rate from 1.8 to 9.2 Nm^3/h.

	Kerosene	Corn oil	Water	Slag	Steel
μ, cP	1.5	46	1.0	200	6
ρ, g/cm^3	0.81	0.86	0.99	2–4	7.0

Ratio of momentum diffusivities: kerosene/water = 1.9, corn oil/water = 55, slag/steel = 900

In the experiments with an oil layer it was observed that in a lower viscosity oil, the gas flow rate required to decrease the oil thickness to zero is larger in comparison with an oil of increased viscosity; 3.1 Nm^3/h with kerosene in comparison with 2.0 Nm^3/h with corn oil. The higher turbulence with kerosene oil promoted higher entrainment of water droplets and air bubbles into the upper phase. The previous expression developed by Banks and Chandrasekhara [91] was applied to describe their experimental data. They needed to modify the original expression to describe all of their experimental data. This was done by dividing the values of the dimensionless penetration depth into three regions:

$$\frac{\dot{M}}{\gamma_l h^3} = 1.04; 0 < \left(\frac{h + n_0}{d}\right) < 3.5$$

$$\frac{\dot{M}}{\gamma_l h^3} = 0.84 + 0.0072\left(\frac{h + n_0}{d}\right) + 0.024\left(\frac{h + n_0}{d}\right)^2; 3.5 < \left(\frac{h + n_0}{d}\right) < 10$$

$$\frac{\dot{M}}{\gamma_l h^3} = \frac{\pi}{2K_2^2}\left(\frac{h + n_0}{d}\right)^2; \left(\frac{h + n_0}{d}\right) > 10$$

Dimensional analysis:

The variables that affect the penetration depth have been investigated by different researchers [89, 91, 93, 99]. In this example, several dimensional analyses will be carried out based on the suggested variables in these previous works.

- Dimensional analysis based on Collins and Lubanska [89]: They described the penetration depth as a function of 7 independent variables: momentum of the jet ($\dot{M}$), specific weight of the liquid (γ), distance from tip of the lance to the surface of the liquid (h), lance diameter (d), gas density (ρ_g), liquid viscosity (μ_l) and lance inclination angle (θ).

The momentum of the jet, more specifically thrust or rate of momentum involves three variables: lance diameter, gas density and gas velocity.

$$\dot{M} = A\rho_g U^2 = \frac{\pi}{4}d^2\rho_g U^2$$

The specific weight of the liquid is also a composite variable that includes the density of the liquid and the gravity force:

$$= \rho_l g$$

The total number of variables involved is 8:

$$n_0 = f\left(\dot{M}, \gamma, h, d, \rho_g, \mu_l, \theta\right)$$

The variables suggested by Collins and Lubanska [89] seem inadequate. The density of the gas is already included in the rate of momentum of the jet, in addition to this, the viscosity and surface tension of the liquid play a minor role. In addition to this, it should be noticed that they suggest two composite variables in the same problem.

Variable	Symbol	Units	Dimensions
Penetration depth	n_0	cm	[L]
Momentum of the jet	$\dot{M}$	dynes	$[MLT^{-2}]$
Specific weight of the liquid	γ	g/cm²s²	$[ML^{-2}\,T^{-2}]$
Distance lance-liquid surface	h	cm	[L]
Lance diameter	d	cm	[L]
Gas density	ρ_g	g/cm³	$[ML^{-3}]$
Viscosity of the liquid	μ_l	kg/(m × s)	$[ML^{-1}T^{-1}]$
Lance inclination angle	θ	–	

There are 8 variables and 3 dimensions. According with the π-theorem this yields 5 dimensionless groups. As non-repeating variables we choose; $\dot{M}$, γ and d.

Next, we define the values for the five dimensionless groups as follows:

$$\pi_1 = n_0\left(\dot{M}\right)^{a_1}(\gamma)^{b_1}(d)^{c_1}$$
$$\pi_2 = h\left(\dot{M}\right)^{a_2}(\gamma)^{b_2}(d)^{c_2}$$
$$\pi_3 = \rho_g\left(\dot{M}\right)^{a_3}(\gamma)^{b_3}(d)^{c_3}$$
$$\pi_4 = \mu\left(\dot{M}\right)^{a_4}(\gamma)^{b_4}(d)^{c_4}$$

The fifth dimensionless group is not required because is formed only by θ.

First π-group:

$$\pi_1 = n_0(\dot{M})^{a_1}(\gamma)^{b_1}(d)^{c_1}$$
$$\pi_1 = (L)\left(MLT^{-2}\right)^{a_1}\left(ML^{-2}T^{-2}\right)^{b_1}(L)^{c_1} = M^0T^0L^0$$

Balance:

M: $a_1 + b_1 = 0$.
  L: $1 + a_1 - 2b_1 + c_1 = 0$
  T: $-2a_1 - 2b_1 = 0$

Solution:

From M balance:
  $a_1 = -b_1$
  From T balance:
  $2b_1 = 2b_1 \therefore b_1$ has not a unique solution.
  From L balance:
  $c_1 = -1 - a_1 + 2b_1 = -1 + b_1 + 2b_1 = -1 + 3b_1$

Important note: In all of the previous examples in this book, the π -theorem has resulted in unique solutions. This example shows that the system of equations can result sometimes with a variable with no single solution. This is not a problem because, as long as the results are non-dimensional, all of them are equivalent.

Now, in order to reach the same results as those reported by Collins and Lubanska [89] the value of $b_1 = 1/3$.

then $c_1 = 0$ and $a_1 = -1/3$.

Substitution of these values on the π_1-group yields:

$$\pi_1 = \frac{n_0}{\left(\dot{M}/\gamma\right)^{\frac{1}{3}}}$$

Second π-group:

$$\pi_2 = h(\dot{M})^{a_2}(\gamma)^{b_2}(d)^{c_2}$$
$$\pi_2 = (L)\left(MLT^{-2}\right)^{a_2}\left(ML^{-2}T^{-2}\right)^{b_2}(L)^{c_2} = M^0T^0L^0$$

Balance:

M: $a_2 + b_2 = 0$.
  L: $1 + a_2 - 2b_2 + c_2 = 0$
  T: $-2a_2 - 2b_2 = 0$

Solution:

From M balance:
   $a_2 = - b_2$
   From T balance:
   $2b_2 = 2b_2 \therefore b_2$ has not a unique solution.
   From L balance:
   $c_2 = - 1 - a_2 + 2b_2 = - 1 + b_2 + 2b_2 = - 1 + 3b_2$
   Defining $b_1 = 1/3$
   then $c_1 = 0$ and $a_1 = -1/3$.
   Substitution of these values on the π_2-group yields:

$$\pi_2 = \frac{h}{\left(\dot{M}/\gamma\right)^{\frac{1}{3}}}$$

Third π-group:

$$\pi_3 = \rho_g(\dot{M})^{a_3}(\gamma)^{b_3}(d)^{c_3}$$
$$\pi_3 = \left(ML^{-3}\right)\left(MLT^{-2}\right)^{a_3}\left(ML^{-2}T^{-2}\right)^{b_3}(L)^{c_3} = M^0T^0L^0$$

Balance:
   M: $1 + a_3 + b_3 = 0$
   L: $-3 + a_3 - 2b_3 + c_3 = 0$
   T: $-2a_3 - 2b_3 = 0$

Solution:

This system of equations has no solution. Equations M and T yield a violation of units: $0 = -2$

Important note: The example seems to arise additional problems, in this case, the set of variables cannot form a dimensionless group.

Fourth π-group:

$$\pi_4 = \mu (\dot{M})^{a_4} (\gamma)^{b_4} (d)^{c_4}$$
$$\pi_4 = \left(ML^{-1}T^{-1}\right)\left(MLT^{-2}\right)^{a_4}\left(ML^{-2}T^{-2}\right)^{b_4}(L)^{c_4} = M^0 T^0 L^0$$

Balance:
   M: $1 + a_4 + b_4 = 0$
   L: $-1 + a_4 - 2b_4 + c_4 = 0$
   T: $-1 - 2a_4 - 2b_4 = 0$

Solution:

This system of equations has no solution. Equations M and T yield $0 = 0$

In one an attempt to solve the problems encountered, $\dot{M}$ is set as the dependent variable. Two of the π-groups are:

$$\pi_1 = \dot{M}(n_0)^{a_1} (\gamma)^{b_1} (d)^{c_1}$$
$$\pi_3 = \rho_g (n_0)^{a_3} (\gamma)^{b_3} (d)^{c_3}$$

Third π-group:

$$\pi_3 = \rho_g (n_0)^{a_3} (\gamma)^{b_3} (d)^{c_3}$$
$$\pi_3 = \left(ML^{-3}\right)(L)^{a_3}\left(ML^{-2}T^{-2}\right)^{b_3}(L)^{c_3} = M^0 T^0 L^0$$

Balance:
   M: $1 + b_3 = 0$
   L: $-3 + a_3 - 2b_3 + c_3 = 0$
   T: $-2b_3 = 0$
   Equations M and T yield two different values for b_3, clearly not a feasible solution.

It can be noticed that two of the variables are composite variables; the momentum of the jet includes the density of the gas, lance cross sectional area and the velocity of the gas at the exit plane of the lance. Since, the gas velocity is already included, it will be removed as a separate variable in the following analysis. The problem reduces to 7 variables and 3 dimensions. According with the π-theorem this yields 4 dimensionless groups.

$$\pi_1 = n_0(\dot{M})^{a_1}(\gamma)^{b_1}(d)^{c_1}$$
$$\pi_2 = h(\dot{M})^{a_2}(\gamma)^{b_2}(d)^{c_2}$$
$$\pi_3 = \mu(\dot{M})^{a_3}(\gamma)^{b_3}(d)^{c_3}$$

The fourth dimensionless group is θ.
The two first groups are solved as before.

Third π-group:

$$\pi_3 = \mu(\dot{M})^{a_3}(\gamma)^{b_3}(d)^{c_3}$$
$$\pi_3 = \left(ML^{-1}T^{-1}\right)\left(MLT^{-2}\right)^{a_3}\left(ML^{-2}T^{-2}\right)^{b_3}(L)^{c_3} = M^0T^0L^0$$

Balance:
M: $1 + a_3 + b_3 = 0$
L: $-1 + a_3 - 2b_3 + c_3 = 0$
T: $-1 - 2a_3 - 2b_3 = 0$

Solution:

This system of equations has no solution. Equations M and T yield $0 = 0$

No more attempts were explored considering that from the beginning, some variables suggested by Collins and Lubanska [89] were considered redundant. In fact, considering the experimental results from these authors who proved that the lance diameter (d) has a minor effect on the penetration depth because at points more than 8 diameters from the exit plane, the flow does not depend on the lance diameter, only on the jet momentum. It is clear that a more appropriate set of variables from their experimental work, for gas injection perpendicular to the bath, is:

$$n_0 = f(\dot{M}, \gamma, h)$$

This simplifies to only 4 variables and 3 dimensions, therefore only one π-group. The authors reported the following 5 dimensionless groups:

$$\frac{n_0}{(\dot{M}/\gamma)^{\frac{1}{3}}} = C\left(\frac{h}{(\dot{M}/\gamma)^{\frac{1}{3}}}\right)^{x_1}(\theta)^{x_2}\left(\frac{d}{h}\right)^{x_3}\left(\frac{\sqrt{\dot{M}\rho_g}}{\mu}\right)^{x_4}$$

Which reduces to two groups if in addition to vertical gas injection, the effects of lance diameter and liquid viscosity are negligible on the penetration depth.

$$\frac{n_0}{(\dot{M}/\gamma)^{\frac{1}{3}}} = C' \left(\frac{h}{(\dot{M}/\gamma)^{\frac{1}{3}}}\right)^{x_1}$$

- Dimensional analysis based on Banks and Chandrasekhara [91]. They defined the penetration depth (n_0) as a function of 5 variables: momentum of the jet (M), specific weight of the liquid (γ), distance from tip of the lance to the surface of the liquid (h) and lance diameter (d).

$n_0 = f (\dot{M}, \gamma, h, d)$.

Variable	Symbol	Units	Dimensions
Penetration depth	n_0	cm	[L]
Momentum of the jet	$\dot{M}$	dynes	$[MLT^{-2}]$
Specific weight of the liquid	γ	gr/cm^{2}sec^2	$[ML^{-2}\,T^{-2}]$
Distance lance-liquid surface	H	cm	[L]
Lance diameter	d	cm	[L]

There are 5 variables and 3 dimensions. According with the π-theorem this yields 2 dimensionless groups.

$$\pi_1 = f(\pi_2)$$

As non-repeating variables we choose; d, γ and h. Next, we define the values for the two dimensionless groups as follows:

$$\pi_1 = n_0(d)^{a_1}(\gamma)^{b_1}(h)^{c_1}$$
$$\pi_2 = \dot{M}(d)^{a_2}(\gamma)^{b_2}(h)^{c_2}$$

First π-group:

$$\pi_1 = n_0(d)^{a_1}(\gamma)^{b_1}(h)^{c_1}$$
$$\pi_1 = (L)(L)^{a_1}\left(ML^{-2}T^{-2}\right)^{b_1}(L)^{c_1} = M^0T^0L^0$$

Balance:

M: $b_1 = 0$.
  L: $1 + a_1 - 2b_1 + c_1 = 0$
  T: $-2b_1 = 0$

Solution:

$b_1 = 0$.
  $1 + a_1 + c_1 = 0$

The system of equations doesn't have a unique solution.
One solution: setting $a_1 = 0$, then $c_1 = -1$.
Substitution of these values on the π_1-group yields:

$$\pi_1 = \frac{n_0}{h}$$

Second π-group:

$$\pi_2 = M(d)^{a_2}(\gamma)^{b_2}(h)^{c_2}$$
$$\pi_2 = \left(MLT^{-2}\right)(L)^{a_2}\left(ML^{-2}T^{-2}\right)^{b_2}(L)^{c_2} = M^0T^0L^0$$

Balance:

M: $1 + b_2 = 0$.
  L: $1 + a_2 - 2b_2 + c_2 = 0$
  T: $-2 - 2b_2 = 0$

Solution:

$b_2 = -1$.
  $a_2 = -3 - c_2$
  if $c_2 = 0$.
  This group can have multiple variations depending on the values defined for c_2.
  Substitution of these values on the π_2-group yields:

$$\pi_2 = \frac{\dot{M}}{d^3}$$

The general function developed is:

$$\pi_1 = f(\pi_2)$$
$$\frac{n_0}{h} = C\left(\frac{\dot{M}}{\gamma d^3}\right)^{x_1}$$

Alternatively;

$$\frac{n_0}{h} = C\left(\frac{\dot{M}}{\rho_l g d^3}\right)^{x_1}$$

The equation indicates that by decreasing the density of the liquid, the penetration depth increases.

Instead of two groups, Banks and Chandrasekhara [91] reported 3 dimensionless groups. This is possible if one of the compound variables is separated, for example, if the specific weight of the liquid is replaced by the density of the liquid and the gravitational constant.

$$n_0 = f\left(\dot{M}, \rho_1, g, h, d\right)$$

Variable	Symbol	Units	Dimensions
Penetration depth	n_0	cm	[L]
Momentum of the jet	$\dot{M}$	dynes	$[MLT^{-2}]$
Density of the liquid	ρ_1	gr/cm³	$[ML^{-3}]$
Gravitational constant	g	cm/s	$[LT^{-2}]$
Distance lance-liquid surface	h	cm	[L]
Lance diameter	d	cm	[L]

There are 6 variables and 3 dimensions. According with the π-theorem this yields 3 dimensionless groups.

$$\pi_1 = f(\pi_2)(\pi_3)$$

As non-repeating variables we choose; ρ_1, g and h. Next, we define the values for the three dimensionless groups as follows:

$$\pi_1 = n_0(\rho_1)^{a_1}(g)^{b_1}(h)^{c_1}$$
$$\pi_2 = \dot{M}(\rho_1)^{a_2}(g)^{b_2}(h)^{c_2}$$
$$\pi_3 = d(\rho_1)^{a_3}(g)^{b_3}(h)^{c_3}$$

First π-group:

$$\pi_1 = n_0(\rho_1)^{a_1}(g)^{b_1}(h)^{c_1}$$
$$\pi_1 = (L)\left(ML^{-3}\right)^{a_1}\left(LT^{-2}\right)^{b_1}(L)^{c_1} = M^0T^0L^0$$

Balance:

M: $a_1 = 0$.
  L: $1 - 3a_1 + b_1 + c_1 = 0$
  T: $-2b_1 = 0$

Solution:

$a_1 = 0$.
  $b_1 = 0$.
  $c_1 = -1$

Substitution of these values on the π_1-group yields:

$$\pi_1 = \frac{n_0}{h}$$

Second π-group:

$$\pi_1 = \dot{M}(\rho_1)^{a_1}(g)^{b_1}(h)^{c_1}$$
$$\pi_2 = \left(MLT^{-2}\right)\left(ML^{-3}\right)^{a_2}\left(LT^{-2}\right)^{b_2}(L)^{c_2} = M^0T^0L^0$$

Balance:

M: $1 + a_2 = 0$.
   L: $1 - 3a_2 + b_2 + c_2 = 0$
   T: $-2 - 2b_2 = 0$

Solution:

$a_2 = -1$.
   $b_2 = -1$
   $c_2 = -1 + 3a_2 - b_2 = -3$.

Substitution of these values on the π_2-group yields:

$$\pi_2 = \frac{\dot{M}}{\iota g h^3}$$

This dimensional group and equivalents, which represent a dimensionless jet momentum, has been called "jet number".

Third π-group:

$$\pi_3 = d(\rho_1)^{a_3}(g)^{b_3}(h)^{c_3}$$
$$\pi_3 = (L)\left(ML^{-3}\right)^{a_3}\left(LT^{-2}\right)^{b_3}(L)^{c_3} = M^0T^0L^0$$

Balance:

M: $a_3 = 0$.
   L: $1 - 3a_3 + b_3 + c_3 = 0$
   T: $-2b_3 = 0$

Solution:

$a_3 = 0$.
   $b_3 = 0$.
   $c_3 = -1$
Substitution of these values on the π_3-group yields:

$$\pi_3 = \frac{d}{h}$$

The general functional relationship becomes:

$$\pi_1 = f(\pi_2)(\pi_3)$$
$$\frac{n_0}{h} = C\left(\frac{\dot{M}}{\rho_1 g h^3}\right)^{x_1}\left(\frac{d}{h}\right)^{x_2}$$

This result is slightly different to the dimensionless groups reported by Banks and Chandrasekhara [91]. One reason is that they considered $\dot{M}$ as the dependent variable and not the penetration depth. Using this scheme, the obtained results are:

$$\dot{M} = f(n_0, \rho_1, g, h, d)$$
$$\frac{\dot{M}}{\rho_1 g n_0^3} = C\left(\frac{d}{n_0}\right)^{x_1}\left(\frac{h}{n_0}\right)^{x_2}$$

Which is almost identical to the dimensionless groups reported by Banks and Chandrasekhara [91], except for the third group, which they reported as h/d. If needed, it is possible to obtain that value dividing π_3/π_2.

They also arranged the terms of the first group in terms of the Froude number, defined as a product of a density ratio, area ratio and the Froude number.

$$\frac{\dot{M}}{\rho_1 g n_0^3} = \frac{\pi}{4}\frac{\rho_g}{\rho_1}\frac{d^2}{n_0^2}\frac{U^2}{g n_0} = \frac{\pi}{4}\frac{\rho_g}{\rho_1}\frac{d^2}{n_0^2}Fr$$

- Dimensional analysis based on Ishikawa et al. [93]: These authors defined the following variables; penetration depth (n_0), gas flow rate (Q), axial distance (h), nozzle diameter (d), gas density (ρ_g), liquid density (ρ_1) and gravity force (g).

$$n_0 = f(Q, h, d, \rho_g, \rho_1, g)$$

Variable	Symbol	Units	Dimensions
Penetration depth	n_0	cm	[L]
Gas flow rate	Q	cm^3/s	$[L^3\,T^{-1}]$
Density of the liquid	ρ_1	gr/cm^3	$[ML^{-3}]$
Density of the gas	ρ_g	gr/cm^3	$[ML^{-3}]$
Distance lance-liquid surface	h	cm	[L]
Lance diameter	d	cm	[L]
Gravity constant	g	cm/s^2	$[LT^{-2}]$

There are 7 variables and 3 dimensions. According with the π-theorem this yields 4 dimensionless groups.

$$\pi_1 = f(\pi_2)(\pi_3)(\pi_4)$$

As non-repeating variables we choose; ρ_1, g and d. Next, we define the values for the three dimensionless groups as follows:

$$\pi_1 = n_0 \,(_1)^{a_1} (Q)^{b_1} (d)^{c_1}$$
$$\pi_2 = h \,(_1)^{a_2} (Q)^{b_2} (d)^{c_2}$$
$$\pi_3 = \rho_g \,(_1)^{a_3} (Q)^{b_3} (d)^{c_3}$$
$$\pi_4 = g \,(_1)^{a_4} (Q)^{b_4} (d)^{c_4}$$

First π-group:

$$\pi_1 = n_0 (\rho_1)^{a_1} (Q)^{b_1} (d)^{c_1}$$
$$\pi_1 = (L)(ML^{-3})^{a_1} (L^3 T^{-1})^{b_1} (L)^{c_1} = M^0 T^0 L^0$$

Balance:

M: $a_1 = 0$.
   L: $1 - 3a_1 - b_1 + c_1 = 0$
   T: $-b_1 = 0$

Solution:

$a_1 = 0$.
   $b_1 = 0$.
   $c_1 = -1$
   Substitution of these values on the π_1-group yields:

$$\pi_1 = \frac{n_0}{d}$$

Second π-group:

$$\pi_2 = h(\rho_1)^{a_2} (Q)^{b_2} (d)^{c_2}$$
$$\pi_2 = (L)(ML^{-3})^{a_2} (L^3 T^{-1})^{b_2} (L)^{c_2} = M^0 T^0 L^0$$

Balance:

M: $1 + a_2 = 0$.
   L: $1 - 3a_2 - b_2 + c_2 = 0$
   T: $-b_2 = 0$

Solution:

$a_2 = 0$.

$$b_2 = 0$$
$$c_2 = -1.$$

Substitution of these values on the π_2-group yields:

$$\pi_2 = \frac{h}{d}$$

Third π-group:

$$\pi_3 = \rho_g (\rho_1)^{a_3} (Q)^{b_3} (d)^{c_3}$$
$$\pi_3 = (ML^{-3})(ML^{-3})^{a_3} (L^3 T^{-1})^{b_3} (L)^{c_3} = M^0 T^0 L^0$$

Balance:

M: $1 + a_3 = 0.$
  L: $-3 - 3a_3 + 3b_3 + c_3 = 0$
  T: $-b_3 = 0$

Solution:

$a_3 = -1.$
  $b_3 = 0.$
  $c_3 = 0$

Substitution of these values on the π_3-group yields:

$$\pi_3 = \frac{\rho_g}{l}$$

Fourth π-group:

$$\pi_4 = g(\rho_1)^{a_4} (Q)^{b_4} (d)^{c_4}$$
$$\pi_4 = (LT^{-2})(ML^{-3})^{a_4} (L^3 T^{-1})^{b_4} (L)^{c_4} = M^0 T^0 L^0$$

Balance:

M: $a_4 = 0.$
  L: $1 - 3a_4 + 3b_4 + c_4 = 0$
  T: $-2 - b_4 = 0$

Solution:

$a_4 = 0.$
  $b_4 = -2$
  $c_4 = 5$

Substitution of these values on the π_4-group yields:

$$\pi_4 = \frac{gd^5}{Q^2}$$

The general function then becomes:

$$\pi_1 = f(\pi_2)(\pi_3)(\pi_4)$$

$$\frac{n_0}{d} = C\left(\frac{h}{d}\right)^{x_1}\left(\frac{\rho_g}{\rho_1}\right)^{x_2}\left(\frac{gd^5}{Q^2}\right)^{x_3}$$

Ishikawa et al. [93] obtained a different version for the π_4-group. They used a different method, other than the π -method. This method usually provides unique solution for the system of equations.

- Dimensional analysis based on Meidani et al. [99]: These authors defined the following variables; penetration depth (n_0), momentum of the jet ($\dot{M}$), axial distance (h), liquid's density (ρ_1), liquid's viscosity (μ_1), surface tension of the liquid (σ_1) and gravity force (g). The authors didn't explain why the nozzle diameter was excluded in the analysis.

$$n_0 = f\left(\dot{M}, \rho_1, g, h, \mu_1, \sigma_1\right)$$

Variable	Symbol	Units	Dimensions
Penetration depth	n_0	cm	[L]
Momentum of the jet	$\dot{M}$	dynes	$[MLT^{-2}]$
Density of the liquid	ρ_1	g/cm^3	$[ML^{-3}]$
Gravitational constant	g	cm/s	$[LT^{-2}]$
Distance lance-liquid surface	h	cm	[L]
Liquid's viscosity	μ_1	kg/(m × s)	$[ML^{-1}T^{-1}]$
Liquid's surface tension	σ_1	N/m	$[MT^{-2}]$

There are 7 variables and 3 dimensions. According with the π-theorem this yields 4 dimensionless groups.

$$\pi_1 = f(\pi_2)(\pi_3)(\pi_4)$$

As non-repeating variables we choose; $\dot{M}$, h and g. Next, we define the values for the four dimensionless groups as follows:

$$\pi_1 = n_0(\dot{M})^{a_1}(h)^{b_1}(g)^{c_1}$$

$$\pi_2 = n_1 \left(\dot{M}\right)^{a_2} (h)^{b_2} (g)^{c_2}$$
$$\pi_3 = n_1 \left(\dot{M}\right)^{a_3} (h)^{b_3} (g)^{c_3}$$
$$\pi_4 = n_1 \left(\dot{M}\right)^{a_4} (h)^{b_4} (g)^{c_4}$$

First π-group:

$$\pi_1 = n_0 (\dot{M})^{a_1} (h)^{b_1} (g)^{c_1}$$
$$\pi_1 = (L)\left(MLT^{-2}\right)^{a_1} (L)^{b_1} \left(LT^{-2}\right)^{c_1} = M^0 T^0 L^0$$

Balance:

M: $a_1 = 0.$
   L: $1 + a_1 + b_1 + c_1 = 0$
   T: $-2a_1 - 2c_1 = 0$

Solution:

$a_1 = 0.$
   $c_1 = 0.$
   $b_1 = -1$

Substitution of these values on the π_1-group yields:

$$\pi_1 = \frac{n_0}{h}$$

Second π-group:

$$\pi_2 = \rho_1 (\dot{M})^{a_2} (h)^{b_2} (g)^{c_2}$$
$$\pi_2 = \left(ML^{-3}\right)\left(MLT^{-2}\right)^{a_2} (L)^{b_2} \left(LT^{-2}\right)^{c_2} = M^0 T^0 L^0$$

Balance:

M: $1 + a_2 = 0.$
   L: $-3 + a_2 + b_2 + c_2 = 0$
   T: $-2a_2 - 2c_2 = 0$

Solution:
   $a_2 = -1.$
   $c_2 = 1$
   $b_2 = 3.$

Substitution of these values on the π_2-group yields:

$$\pi_2 = \frac{\rho_1 g h^3}{\dot{M}} \equiv \frac{\dot{M}}{\rho_1 g h^3}$$

Third π-group:

$$\pi_3 = \dot{\mu}_1(M)^{a_3}(h)^{b_3}(g)^{c_3}$$
$$\pi_3 = \left(ML^{-1}T^{-1}\right)\left(MLT^{-2}\right)^{a_3}(L)^{b_3}\left(LT^{-2}\right)^{c_3} = M^0T^0L^0$$

Balance:

M: $1 + a_3 = 0.$
  L: $-1 + a_3 + b_3 + c_3 = 0$
  T: $-1 - 2a_3 - 2c_3 = 0$

Solution:

$a_3 = -1.$
  $c_3 = \frac{1}{2}$
  $b_3 = 3/2$

Substitution of these values on the π_3-group yields:

$$\pi_3 = \frac{\mu_1 g^{\frac{1}{2}} h^{\frac{3}{2}}}{\dot{M}} \equiv \frac{\dot{M}^2}{\mu_1^2 g h^3}$$

Fourth π-group:

$$\pi_4 = \sigma_1(M)^{a_4}(h)^{b_4}(g)^{c_4}$$
$$\pi_4 = \left(MT^{-2}\right)\left(MLT^{-2}\right)^{a_4}(L)^{b_4}\left(LT^{-2}\right)^{c_4} = M^0T^0L^0$$

Balance:

M: $1 + a_4 = 0.$
  L: $a_4 + b_4 + c_4 = 0$
  T: $-2 - 2a_4 - 2c_4 = 0$

Solution:

$a_4 = -1.$
  $c_4 = 0$
  $b_4 = 1$

Substitution of these values on the π_4-group yields:

$$\pi_4 = \frac{\sigma_1 h}{\dot{M}} \equiv \frac{\dot{M}}{\sigma_1 h}$$

The general function then becomes:

$$\pi_1 = f(\pi_2)(\pi_3)(\pi_4)$$

$$\frac{n_0}{h} = C\left(\frac{\dot{M}}{\rho_1 gh^3}\right)^{x_1}\left(\frac{\dot{M}^2}{\mu_1^2 gh^3}\right)^{x_2}\left(\frac{\dot{M}}{\sigma_1 h}\right)^{x_3}$$

It has been shown before that π_2 represents a modified Froude number:

$$\frac{\dot{M}}{\rho_1 gh^3} = \frac{\pi}{4}\frac{\rho_g}{\rho_1}\frac{d^2}{h^2}\frac{U^2}{gh} = \frac{\pi}{4}\frac{\rho_g}{\rho_1}\frac{d^2}{h^2}\text{Fr} = \frac{\pi}{4}\text{Fr}_m$$

In a similar way, π_3 and π_4 can be rearranged as follows:

$$\pi_3' = \sqrt{\frac{\pi_3}{\pi_2}} = \sqrt{\frac{\dot{M}^2\rho_1 gh^3}{\dot{M}\mu_1^2 gh^3}} = \sqrt{\frac{\dot{M}\rho_1}{\mu_1^2}} \cong \sqrt{\frac{\rho_1\rho_g d^2 U^2}{\mu_1^2}} = \sqrt{\frac{\rho^2 U^2 d^2}{\mu_1^2}} = \frac{\rho Ud}{\mu_1} = \text{Re}$$

$$\pi_4 = \frac{\dot{M}}{\sigma_1 h} = \frac{\pi}{4}\frac{\rho_g d^2 U^2}{\sigma_1 h} = \frac{\pi}{4}\text{We}_m$$

Then, the final general solution can be described as follows:

$$\frac{n_0}{h} = C(Fr_m)^{x_1}(\text{Re})^{x_2}(We_m)^{x_3}$$

Table 11.44 summarizes the previous results.

The only problem to define the dimensionless groups, with the π-method, was the case with two composite variables and a redundant number of variables.

Experimental data from Wakelin [88]

Wakelin [88] conducted experimental work using a water model, a square Perspex cylinder with a diameter of 73 cm. Air gas injected from a lance in a vertical position. A converging nozzle with the divergent section was employed to ensure subsonic

Table 11.44 Results from dimensional analysis applying the π-method

1954	Collins and Lubanska [89]	$n_0 = f(\dot{M}, \gamma, h, d, \rho_g, \mu_1, \theta)$	Incomplete solution
1963	Banks and Chandrasekhara [91]	$n_0 = f(\dot{M}, \gamma, h, d)$	$\frac{n_0}{h} = C\left(\frac{\dot{M}}{\gamma d^3}\right)^{x_1}$
		$n_0 = f(\dot{M}, \rho_1, g, h, d)$	$\frac{n_0}{h} = C\left(\frac{\dot{M}}{\rho_1 gh^3}\right)^{x_1}\left(\frac{d}{h}\right)^{x_2}$
		$\dot{M} = f(n_0, \rho_1, g, h, d)$	$\frac{\dot{M}}{\rho_1 gn_0^3} = C\left(\frac{d}{n_0}\right)^{x_1}\left(\frac{h}{n_0}\right)^{x_2}$
1972	Ishikawa et al. [93]	$n_0 = f(Q, h, d, \rho_g, \rho_1, g)$	$\frac{n_0}{d} =$ $C\left(\frac{h}{d}\right)^{x_1}\left(\frac{\rho_g}{\rho_1}\right)^{x_2}\left(\frac{gd^5}{Q^2}\right)^{x_3}$
2014	Meidani et al. [99]	$n_0 = f(\dot{M}, \rho_1, g, h, \mu_1, \sigma_1)$	$\frac{n_0}{h} =$ $C(Fr_m)^{x_1}(\text{Re})^{x_2}(We_m)^{x_3}$

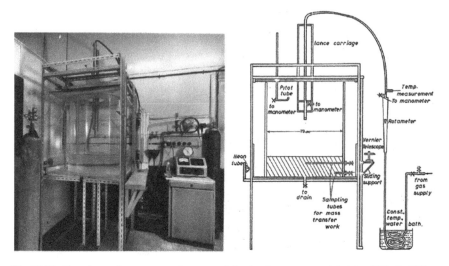

Fig. 11.54 Experimental set up to study top gas injection in a water model. After Wakelin [88]

speed. Air at constant temperature was injected. The temperature was adjusted by passing air through a copper spiral immersed in a thermostatically controlled water bath. A rotameter was used to control the gas flow rate. The expected impurities in tap water were: 60 ppm $CaCO_3$, 30 ppm $MgCO_3$, 170 ppm Na_2CO_3, 200 ppm NaCl. The experimental set up is shown in Fig. 11.54.

The rate of momentum was computed from an expression developed, assuming that flow through the nozzle was isentropic, expanded adiabatically, the pressure at the exit plane was atmospheric and the velocity was uniform across the exit plane. The experimental value was defined from measurements of the jet velocity using a pitot tube.

The penetration depth or depression was found to be dynamic with surface oscillations both in lateral and vertical directions. The oscillation of the bottom of the depression about its mean position increased by increasing the lance height, from about 7% at h = 4 cm to 20% at h = 40 cm. A fixed meter rule and a vernier telescope were used to measure the penetration depth. The vernier had a sensitivity of ± 0.01 cm. Figure 11.55 is an illustration of the depression due to top gas injection.

The experimental results reported by Wakelin are indicated in Table 11.45. The analysis was carried out initially with the conventional approach using n_0 as the dependent variable:

$$\log \pi_1 = \log C + x_1 \log \pi_2 + x_2 \log \pi_3$$
$$\log \tfrac{n_0}{h} = \log C + x_1 \log\left(\tfrac{\dot{M}}{\rho_1 g h^3}\right) + x_2 \log\left(\tfrac{d}{h}\right)$$

However, the coefficient of linear regression, R^2 was low, about 0.8, compared with the regression coefficient of 0.95 using the dimensionless groups reported by Banks

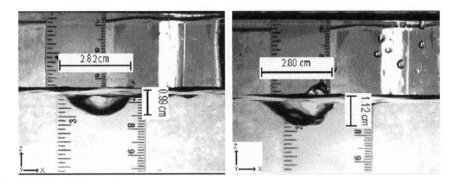

Fig. 11.55 Depression due to top gas injection. After Solorzano et al. [101]

and Chandrasekhara [91]. As indicated before, they applied dimensional analysis using the jet momentum as the dependent variable.

In this example the variable of interest is the penetration depth (n_0). One of the reasons to exclude the dependent variable as a non-repeating variable is to make it appear only in one term. In this way it is simpler to define the effect of the other forces on the dependent variable. If the variable of interest is a repeating variable it will appear in several dimensionless groups. However, it is not a limitation to do the dimensional analysis with the variable of interest as a repeating variable, in the end by algebraic operations it can be possible to separate variables.

In order to focus on a more accurate analysis of the experimental data, the alternative was to include the equation using n_0 as a repeating variable, which leads to the following relationship:

$$\log \pi_1 = \log C + x_1 \log \pi_2 + x_2 \log \pi_3$$

$$\log \frac{\dot{M}}{\rho_1 g n_0^3} = \log C + x_1 \log \left(\frac{d}{n_0}\right) + x_2 \log \left(\frac{h}{n_0}\right)$$

This expression reports a coefficient of linear regression, $R^2 = 0.95$. The π groups in Table 11.45 correspond to the previous equation.

Applying the statistical tools of excel the following values are obtained: $\log C = -8.6$, $x_1 = 0.698$, $x_2 = 1.52$. The regression coefficient (R^2) was 0.95. Replacing these values in the general equation:

$$\frac{\dot{M}}{\rho_1 g n_0^3} = 2.48 \times 10^{-9} \left(\frac{d}{n_0}\right)^{0.7} \left(\frac{h}{n_0}\right)^{1.5}$$

The previous expression is illustrated in Fig. 11.56.

Table 11.45 Experimental data and dimensionless groups using $\dot{M}$ part the dependent dimensionless group

n_0, cm	h, cm	$\dot{M}$, dynes	π_1	π_2	π_3
1.42	4	1.5E−05	0.355	2.31E−10	0.0633
1.29	4.02	1.5E−05	0.321	2.35E−10	0.0629
0.95	5.95	1.5E−05	0.160	7.26E−11	0.0425
1.29	5.95	2.5E−05	0.217	1.21E−10	0.0425
0.98	6.01	1.5E−05	0.163	6.81E−11	0.0421
0.62	7.81	1.5E−05	0.079	3.21E−11	0.0324
0.89	7.81	2.5E−05	0.114	5.35E−11	0.0324
1.22	7.81	3.9E−05	0.156	8.35E−11	0.0324
1.93	8.07	5.6E−05	0.239	1.09E−10	0.0314
2.46	8.07	7.8E−05	0.305	1.51E−10	0.0314
2.97	8.07	1.0E−04	0.368	2.00E−10	0.0314
0.61	8.09	1.5E−05	0.075	2.89E−11	0.0313
1.3	8.09	1.5E−05	0.161	2.89E−11	0.0313
2.26	8.09	1.5E−05	0.279	2.89E−11	0.0313
0.48	9.39	1.5E−05	0.051	1.85E−11	0.0269
0.62	9.39	2.5E−05	0.066	3.08E−11	0.0269
1.02	9.39	4.0E−05	0.109	4.86E−11	0.0269
1.36	9.39	5.8E−05	0.145	7.14E−11	0.0269
0.76	11.45	4.0E−05	0.066	2.68E−11	0.0221
1.17	11.45	5.8E−05	0.102	3.94E−11	0.0221
1.4	11.45	8.0E−05	0.122	5.43E−11	0.0221
0.85	11.5	3.9E−05	0.074	2.58E−11	0.0220
1.49	11.5	8.0E−05	0.130	5.33E−11	0.0220
2.25	11.5	1.4E−04	0.196	9.25E−11	0.0220
0.964	11.53	1.5E−05	0.084	9.64E−12	0.0219
1.63	11.53	2.5E−05	0.141	1.63E−11	0.0219
6.85	11.53	1.0E−04	0.594	6.85E−11	0.0219
8.95	11.53	1.3E−04	0.776	8.94E−11	0.0219
11.5	11.53	1.7E−04	0.997	1.15E−10	0.0219
2.28	13.77	5.8E−05	0.166	2.26E−11	0.0184
3.13	13.77	8.0E−05	0.227	3.12E−11	0.0184
4.11	13.77	1.1E−04	0.298	4.10E−11	0.0184
0.555	13.88	1.5E−05	0.040	5.53E−12	0.0182
0.94	13.88	2.5E−05	0.068	9.34E−12	0.0182
1.46	13.88	3.8E−05	0.105	1.45E−11	0.0182

(continued)

Table 11.45 (continued)

n_0, cm	h, cm	$\dot{M}$, dynes	π_1	π_2	π_3
1.62	17.12	8.0E−05	0.095	1.62E−11	0.0148
2.81	17.12	1.1E−04	0.164	2.31E−11	0.0148
4.51	17.12	2.2E−04	0.263	4.51E−11	0.0148
0.294	17.15	1.5E−05	0.017	2.93E−12	0.0148
0.496	17.15	2.5E−05	0.029	4.95E−12	0.0148
0.759	17.15	3.8E−05	0.044	7.58E−12	0.0148
1.11	17.15	5.5E−05	0.065	1.11E−11	0.0148
1.58	17.15	7.8E−05	0.092	1.58E−11	0.0148
3.5	17.15	1.7E−04	0.204	3.49E−11	0.0148
4.41	17.15	2.2E−04	0.257	4.41E−11	0.0148
5.5	17.15	2.7E−04	0.321	5.50E−11	0.0148
2.1	17.2	1.1E−04	0.122	2.10E−11	0.0147
2.75	17.2	1.4E−04	0.160	2.74E−11	0.0147
1.59	20.69	1.4E−04	0.077	1.58E−11	0.0122
2.05	20.69	1.8E−04	0.099	2.05E−11	0.0122
0.276	20.86	2.5E−05	0.013	2.75E−12	0.0121
0.421	20.86	3.8E−05	0.020	4.21E−12	0.0121
0.55	20.86	5.5E−05	0.026	6.18E−12	0.0121
0.74	20.86	7.8E−05	0.035	8.76E−12	0.0121
0.91	20.86	1.0E−04	0.044	1.16E−11	0.0121
1.07	20.86	1.4E−04	0.051	1.53E−11	0.0121
0.85	23.97	2.2E−04	0.035	1.62E−11	0.0106
1.28	23.97	3.3E−04	0.053	2.46E−11	0.0106
1.8	23.97	1.3E−04	0.075	9.96E−12	0.0106
0.35	24.11	3.8E−05	0.015	2.73E−12	0.0105
0.46	24.11	5.5E−05	0.019	4.00E−12	0.0105
0.61	24.11	7.8E−05	0.025	5.67E−12	0.0105
0.73	24.11	1.0E−04	0.030	7.53E−12	0.0105
0.93	24.11	1.4E−04	0.039	9.96E−12	0.0105
1.07	24.11	1.8E−04	0.044	1.27E−11	0.0105
1.28	24.11	2.2E−04	0.053	1.61E−11	0.0105
1.59	24.11	2.7E−04	0.066	2.00E−11	0.0105
1.83	24.11	3.4E−04	0.076	2.45E−11	0.0105
2.18	24.11	4.1E−04	0.090	3.01E−11	0.0105
1.08	27.17	2.2E−04	0.040	1.13E−11	0.0093

(continued)

Table 11.45 (continued)

n_0, cm	h, cm	$\dot{M}$, dynes	π_1	π_2	π_3
1.28	27.17	2.8E−04	0.047	1.43E−11	0.0093
0.38	27.24	5.6E−05	0.014	2.82E−12	0.0093
0.46	27.24	7.9E−05	0.017	3.96E−12	0.0093
0.58	27.24	1.0E−04	0.021	5.24E−12	0.0093
0.71	27.24	1.4E−04	0.026	6.91E−12	0.0093
0.93	27.24	1.8E−04	0.034	8.83E−12	0.0093

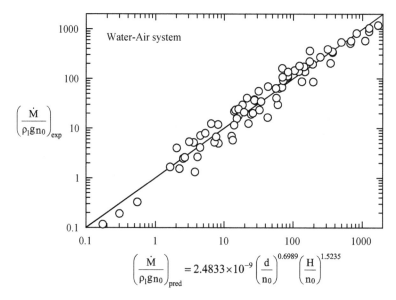

$$\left(\frac{\dot{M}}{\rho_l g n_0}\right)_{pred} = 2.4833 \times 10^{-9} \left(\frac{d}{n_0}\right)^{0.6989} \left(\frac{H}{n_0}\right)^{1.5235}$$

Fig. 11.56 Comparison between predicted and experimental data to describe the dimensionless momentum of the jet due to gas injection into a liquid using dimensional analysis. Experimental data from Wakelin [88]

References

1. K. Ito, R.J. Fruehan, Study on the foaming of CaO-SiO2-FeO slags: part I. Foaming parameters and experimental results. Metall. Trans. B **20**, 509–514 (1989)
2. K. Ito, R.J. Fruehan, Study on the foaming of CaO-SiO2-FeO slags: part II. Dimensional analysis and foaming in iron and steelmaking processes. Metall. Trans. B **20**, 515–521 (1989)
3. R. Jiang, R.J. Fruehan, Slag foaming in bath smelting. Metall. Trans. B **22**, 481–489 (1991)
4. Y. Zhang, R.J. Fruehan, Effect of the bubble size and chemical reactions on slag foaming. Metall. Mater. Trans. B **26**, 803–812 (1995)
5. Y. Zhang, R.J. Fruehan, Effect of carbonaceous particles on slag foaming. Metall. Mater. Trans. B **26**, 813–819 (1995)
6. Y. Zhang, R.J. Fruehan, Effect of gas type and pressure on slag foaming. Metall. Mater. Trans. B **26**, 1088–1091 (1995)

7. B. Ozturk, R.J. Fruehan, Effect of temperature on slag foaming. Metall. Mater. Trans. B **26**, 1086–1088 (1995)
8. S.M. Jung, R.J. Fruehan, Foaming characteristics of BOF slags. ISIJ Int. **40**, 348–355 (2000)
9. H. Matsuura, R.J. Fruehan, Slag foaming in an electric arc furnace. ISIJ Int. **49**, 1530–1535 (2009)
10. Q.F. Shu, K.C. Chou, Calculation for density of molten slags using optical basicity. Ironmak. Steelmak. **40**, 571–577 (2013)
11. K.C. Mills, B.J. Keene, Physical properties of BOS slags. Int. Mater. Rev. **32**, 1–120 (1987)
12. G. Urbain, Viscosity estimation of slags. Steel Res. **58**, 111–116 (1987)
13. D. Skupien, D.R. Gaskell, The surface tensions and foaming behavior of melts in the system CaO-FeO-SiO2. Metall. Mater. Trans. B Process Metall. Mater. Process. Sci. **31**, 921–925 (2000)
14. Y.-E. Lee, D.R. Gaskell, The densities and structures of silicate melts. Metall. Trans. B **5**, 853–860 (1974)
15. K. Ruff, Bildung von Gasblasen an Düsen bei konstantem Volumendurchsatz. Chemie Ing. Tech. **44**, 1360–1366 (1972)
16. S. Ghag, P.C. Hayes, H.G. Lee, Physical model studies on slag foaming. ISIJ Int. **38**, 1201–1207 (1998)
17. S.S. Ghag, P.C. Hayes, H.G. Lee, Model development of slag foaming. ISIJ Int. **38**, 1208–1215 (1998)
18. L. Pilon, A.G. Fedorov, R. Viskanta, Steady-state thickness of liquid-gas foams. J. Colloid Interface Sci. **242**, 425–436 (2001)
19. L. Pilon, R. Viskanta, Minimum superficial gas velocity for onset of foaming. Chem. Eng. Process. Process Intensif. **43**, 149–160 (2004)
20. D. Lotun, L. Pilon, Physical modeling of slag foaming for various operating conditions and slag compositions. ISIJ Int. **45**, 835–840 (2005)
21. L. Pilon, Foams in glass manufacturing. Chapter 16. *Foam Engineering: Fundamentals and Applications* (2012), pp. 355–409
22. J.A. Attia, S. Kholi, L. Pilon, Scaling laws in steady-state aqueous foams including Ostwald ripening. Colloids Surfaces A **436**, 1000–1006 (2013)
23. L. Pilon, Personal communication (2020)
24. K. Yonezawa, K. Schwerdtfeger, Spout eyes formed by an emerging gas plume at the surface of a slag-covered metal melt. Metall. Mater. Trans. A Phys. Metall. Mater. Sci. **30**, 411–418 (1999)
25. M. Peranandhanthan, D. Mazumdar, Modeling of slag eye area in argon stirred ladles. ISIJ Int. **50**, 1622–1631 (2010)
26. M.A.S.C. Castello-Branco, K. Schwerdtfeger, Large-scale measurements of the physical characteristics of round vertical bubble plumes in liquids. Metall. Mater. Trans. B **25**, 359–371 (1994)
27. G. Ebneth, W. Pluschkell, Dimensional analysis of the vertical heterogeneous buoyant plume. Steel Res. **56**, 513–518 (1985)
28. V.T. Mantripragada, S. Sarkar, Wall stresses in dual bottom purged steel making ladles. Chem. Eng. Res. Des. **139**, 335–345 (2018)
29. V.T. Mantripragada, S. Sarkar, Slag eye formation in single and dual bottom purged industrial steelmaking ladles. Can. Metall. Q. **59**, 159–168 (2020)
30. A. Tripathi, J.K. Saha, J.B. Singh, S.K. Ajmani, Numerical simulation of heat transfer phenomenon in steel making ladle. ISIJ Int. **52**, 1591–1600 (2012)
31. L.I. Linmin, L.I. Baokuan, Z. Liu, Modeling of gas-steel-slag three-phase flow in ladle metallurgy: part II. Multi-scale mathematical model. ISIJ Int. **57**, 1980–1989 (2017)
32. Y. Pan, D. Guo, J. Ma, W. Wang, F. Tang, C. Li, Mixing time and fluid flow pattern of composition adjustment by sealed argon bubbling with ladles of large height/diameter ratio. ISIJ Int. **34**, 794–801 (1994)
33. K. Mandal, D. Mazumdar, Dimensional analysis and mixlng phenomena in bubble stirred ladles. ISIJ Int. **38**, 1150–1152 (1998)

34. K. Mandal, Modelling of mixing and fluid flow phenomena in steelmaking. M.Sc. Thesis, Indian Institute of Technology Kanpur, 1998
35. G. Reiter, K. Schwerdtfeger, Characteristics of entrainment at liquid/liquid interfaces due to rising bubbles. ISIJ Int. **32**, 57–65 (1992)
36. G. Reiter, K. Schwerdtfeger, Observations of physical phenomena occurring during passage of bubbles through liquid/liquid interfaces. ISIJ Int. **32**, 50–56 (1992)
37. G.J. Hills, H. Høiland, Pressure dependence of the surface tension of mercury. J. Colloid Interface Sci. **99**, 463–467 (1984)
38. G. Wiegand, E.U. Franck, Interfacial tension between water and non-polar fluids up to 473 K and 2800 bar. Berichte der Bunsengesellschaft/Physical Chem. Chem. Phys. **98**, 809–817 (1994)
39. J. Zhang, L.I.U. Qing, S. Yang, Z. Chen, L.I. Jingshe, Z. Jiang, Advances in ladle shroud as a functional device in tundish metallurgy: a review. ISIJ Int. **59**, 1167–1177 (2019)
40. S. Chang, X. Cao, C.H. Hsin, Z. Zou, M. Isac, R.I.L. Guthrie, Removal of inclusions using micro-bubble swarms in a four-strand, full-scale, water model tundish. ISIJ Int. **56**, 1188–1197 (2016)
41. S. Chang, S. Ge, Z. Zou, M.M. Isac, R.I.L. Guthrie, Modeling slag behavior when using micro-bubble swarms for the deep cleaning of liquid steel in tundishes. Steel Res. Int. **88**, 1–11 (2017)
42. R. Guthrie, M. Isac, Towards forming micro-bubbles in liquid steel, in *Proceedings of the Extraction 2018. P. C. Hayes Symposium on Pyrometallurgical Processing*, TMS, Ottawa, Ontario, Canada, 26–29 Aug 2018, pp. 729–740
43. K. Chattopadhyay, H. Mainul, M. Isac, R.I.L. Guthrie, Physical and mathematical modeling of Inert gas-shrouded ladle nozzles and their role on slag behavior and fluid flow patterns in a delta-shaped, four-strand tundish. Metall. Mater. Trans. B Process Metall. Mater. Process. Sci. **41**, 225–233 (2010)
44. K. Chattopadhyay, M. Isac, R.I.L. Guthrie, Effect of submergence depth of the ladle shroud on liquid steel quality output from a delta shaped four strand tundish. Ironmak. Steelmak. **38**, 398–400 (2011)
45. K. Chattopadhyay, M. Isac, R.I.L. Guthrie, Physical and mathematical modelling of inert gas shrouding in a tundish. ISIJ Int. **51**, 573–580 (2011)
46. S. Chatterjee, K. Chattopadhyay, Formation of slag "eye" in an inert gas shrouded tundish. ISIJ Int. **55**, 1416–1424 (2015)
47. S. Chatterjee, K. Chattopadhyay, Physical modeling of slag 'Eye' in an inert gas-shrouded tundish using dimensional analysis. Metall. Mater. Trans. B Process Metall. Mater. Process. Sci. **47**, 508–521 (2016)
48. S. Chatterjee, K. Chattopadhyay, Tundish open eye formation in inert gas-shrouded tundishes: a macroscopic model from first principles. Metall. Mater. Trans. B Process Metall. Mater. Process. Sci. **47**, 3099–3114 (2016)
49. S. Chatterjee, D. Li, J. Leung, J. Sengupta, K. Chattopadhyay, Tundish open eye formation: what is the effect on liquid steel cleanliness? AISTech Iron Steel Technol. Conf. Proc. **3**, 2537–2550 (2016)
50. S. Chatterjee, D. Li, J. Leung, J. Sengupta, J. Young, K. Chattopadhyay, Open eye formation in a slab caster tundish: cause and effects, in *Proceedings of 9th European Continuous Casting Conference* (2017), pp. 506–517
51. S. Chatterjee, D. Li, J. Leung, J. Sengupta, K. Chattopadhyay, Investigation of eccentric open eye formation in a slab caster tundish. Metall. Mater. Trans. B Process Metall. Mater. Process. Sci. **48**, 1035–1044 (2017)
52. S. Chatterjee, D. Li, K. Chattopadhyay, Tundish open eye formation: a trivial event with dire consequences. Steel Res. Int. **88**, 1–12 (2017)
53. S. Chatterjee, D. Li, K. Chattopadhyay, Modeling of liquid steel/slag/argon gas multiphase flow during tundish open eye formation in a two-strand tundish. Metall. Mater. Trans. B Process Metall. Mater. Process. Sci. **49**, 756–766 (2018)

54. A. Ma, S. Chatterjee, K. Chattopadhyay, Ensemble prediction of tundish open eyes using artificial neural networks. ISIJ Int. **59**, 1287–1294 (2019)
55. S. Chatterjee, D. Li, K. Chattopadhyay, Criticality of sampler locations in inert-gas shrouded tundishes during open eye events and unbalanced throughputs. ISIJ Int. **56**, 1889–1892 (2016)
56. K. Chattopadhyay, Modelling of transport phenomena for improved steel quality in a delta shaped four strand tundish. Ph.D. Thesis, Mining and Materials Engineering, McGill University (2011)
57. G. Wen, Y. Huang, P. Tang, M. Zhu, Influence of ladle shroud's shapes on characteristics of fluid flow in tundish. J. Chongqing Univ. **34**, 69–74 (2011)
58. P.K. Singh, Ph.D. Thesis, Indian Institute of Technology Kanpur, India (2019)
59. P.K. Singh, D. Mazumdar, Macroscopic modelling of argon-steel flow inside a ladle shroud, in *Proceedings of the 3rd International Conference on Science and Technology of Ironmaking and Steelmaking* (2017), pp. 399–402
60. D. Mazumdar, P.K. Singh, R.K. Tiwari, Shrouded transfer of molten steel from ladle to tundish: current understanding, mathematical modelling and new insight. ISIJ Int. **58**, 1545–1547 (2018)
61. P.K. Singh, D. Mazumdar, A physical model study of two-phase gas–liquid flows in a ladle shroud. Metall. Mater. Trans. B Process Metall. Mater. Process. Sci. **49**, 1945–1962 (2018)
62. P.K. Singh, D. Mazumdar, Mathematical modelling of gas–liquid, two-phase flows in a ladle shroud. Metall. Mater. Trans. B Process Metall. Mater. Process. Sci. **50**, 1091–1103 (2019)
63. G. Solorio, A. Ramnos, J. Barreto, R. Morales, Modeling study of turbulent flow effect on inclusion rfemoval in a tundish with swirling ladle shroud. Steel Res. Int. **80**, 223–234 (2009)
64. G. Solorio-Diaz, R. Davila-Morales, J. De Jesus Barreto-Sandoval, H.J. Vergara-Hernández, A. Ramos-Banderas, S.R. Galvan, Numerical modelling of dissipation phenomena in a new ladle shroud for fluidynamic control and its effect on inclusions removal in a slab tundish. Steel Res. Int. **85**, 863–874 (2014)
65. K. Kuwana, M.I. Hassan, P.K. Singh, K. Saito, J. Nakagawa, Scale-model experiment and numerical simulation of a steel teeming process. Mater. Manuf. Process. **23**, 407–412 (2008)
66. A. Kamaraj, G.K. Mandal, G.G. Roy, Control of slag carryover from the BOF vessel during tapping: BOF cold model studies. Metall. Mater. Trans. B Process Metall. Mater. Process. Sci. **50**, 438–458 (2019)
67. A. Kamaraj, G.G. Roy, G.K. Mandal, Modeling and simulation studies on BOF tapping process, in *Proceedings of the Third International Conference on Science and Technology of Ironmaking and Steelmaking*, ed. by D. Mazumdar, Kanpur India, 11–13 Dec 2017, pp. 403–407
68. A. Kamaraj, Personal communication (2020)
69. World Steel Association, World Steel in Figures 2019 (2019)
70. D. Ameling, J. Petry, M. Sittard, W. Ulbrich, J. Wolf, Untersuchungen zur Schaumschlackenbildung im Elektrolicht bogenofen. Stahl Eisen 625–630 (1986)
71. V. Smil, *Still the Iron Age* (BH Elsevier, 2016)
72. E.S. Szekeres, Overview of mold oscillation in continuous casting. Iron Steel Eng. **73**, 29–37 (1996)
73. K.C. Mills, C.Å. Däcker, *The Casting Powders Book* (2017). ISBN 9783319536163
74. P. Andrezejewski, A. Drastik, U. Kohler, H. Lachmund, Y. Xie, T. Buhles, W. Pluschkell, New aspects of oscillation mode operation and results in slab casting, in *Proceedings of the Process Technology Conference*, Warrendale PA, USA, 1990, pp. 173–181
75. A. Kamaraj, A. Dash, P. Murugaiyan, S. Misra, Investigation on mold flux melting and consumption during continuous casting of liquid steel. Metall. Mater. Trans. B (2020)
76. R. Saraswat, A.B. Fox, K.C. Mills, P.D. Lee, B. Deo, The factors affecting powder consumption of mould fluxes. Scand. J. Metall. **33**, 85–91 (2004)
77. S. Sridhar, K.C. Mills, S.T. Mallaband, Powder consumption and melting rates of continuous casting fluxes. Ironmak. Steelmak. **29**, 194–198 (2002)
78. J.A. Kromhout, R.C. Schimmel, Understanding mould powders for high-speed casting. Ironmak. Steelmak. **45**, 249–256 (2018)

79. J.A. Kromhout, D.W. Van der Plas, Melting speed of mould powders: determination and application in casting practice. Ironmak. Steelmak. **29**, 303–307 (2002)
80. M. Görnerup, M. Hayashi, C. Däcker, S. Seetharaman, Mould fluxes in continuous casting of steel—characterization and performance tuning, in *VII International Conference on Molten Slags, Fluxes & Salts* (2004), pp. 745–752
81. V. Ludlow, S. McKay, B. Harris, J. Kromhout, T. Cimarelli, F. Ors, M. Thalhammer, Mould powder development for faster casting speeds and thin slab casting, Luxembourg. ECSC Report 19490 (1996)
82. M. Hanao, M. Kawamoto, A. YamanaKa, Influence of mold flux on initial solidification of hypo-peritectic steel in a continuous casting mold. Tetsu-To-Hagane **100**, 581–590 (2014)
83. A. Kamaraj, N. Haldar, P. Murugaiyan, S. Misra, High-temperature simulation of continuous casting mould phenomena. Trans. Indian Inst. Met. (2020)
84. C. Pinheiro, Mould thermal response, billet surface quality and mould-flux behaviour in the continuous casting of steel billets with powder lubrication. PhD Thesis, University of British Columbia, CA (1997)
85. A. Normanton, V. Ludlow, B. Harris, M. Hecht, C. Däcker, A. DiDonato, T. Sohlgren, Mould powder consumption, melting and lubrication and their effects on mould heat transfer and subsequent surface quality of continuously cast slab, Luxembourg. ECSC Report 21907 (2005)
86. H.J. Shin, S.H. Kim, B.G. Thomas, G.G. Lee, J.M. Park, J. Sengupta, Measurement and prediction of lubrication, powder consumption, and oscillation mark profiles in ultra-low carbon steel slabs. ISIJ Int. **46**, 1635–1644 (2006)
87. H. Kania, J. Gawor, Impact of mould powder density on surface quality and near-surface zone microstructure of cast slab. Arch. Metall. Mater. **57**, 339–345 (2012)
88. D.H. Wakelin, The interaction between gas jets and the surfaces of liquids, inclduing molten metals. Ph.D. Thesis, University of London (1966)
89. R.D. Collins, H. Lubanska, The depression of liquid surfaces by gas jets. J. Appl. Phys. **5**, 22–26 (1954)
90. M. Shimada, K. Sagawa, S. Maehara, M. Ishibachi, Model experiments on the profile and operation of the L.D. converter. Tetsu to Hagane **44**, 1056–1058 (1958)
91. R.B. Banks, D.V. Chandrasekhara, Experimental investigation of the penetration of a high-velocity gas jet through a liquid surface. J. Fluid Mech. **15**, 13–34 (1963)
92. F.R. Cheslak, J.A. Nicholls, M. Sichel, Cavities formed on liquid surfaces by impinging gaseous jets. J. Fluid Mech. **36**, 55–63 (1969)
93. H. Ishikawa, S. Mizoguchi, K. Segawa, Model study on jet penetration and slopping in the LD converter. Tetsu-To-Hagane **58**, 76–84 (1972)
94. S.C. Koria, K.W. Lange, Penetrability of impinging gas jets in molten steel bath. Steel Res. **58**, 421–426 (1987)
95. S.K. Sharma, J.W. Hlinka, D.W. Kern, The bath circulation, jet penetration and high-temperature reaction zone in BOF steelmaking. Iron Steelmak. **4**, 7–18 (1977)
96. Y. Doh, P. Chapelle, A. Jardy, G. Djambazov, K. Pericleous, G. Ghazal, P. Gardin, Toward a full simulation of the basic oxygen furnace: deformation of the bath free surface and coupled transfer processes associated with the post-combustion in the gas region. Metall. Mater. Trans. B Process Metall. Mater. Process. Sci. **44**, 653–670 (2013)
97. S.C. Koria, K.W. Lange, Experimental study on the behaviour of an underexpanded supersonic gas jet. Arch. Fur Das Eisenhuttenwes. **55**, 427–432 (1984)
98. A. Nordquist, N. Kumbhat, L. Jonsson, P. Jönsson, The effect of nozzle diameter, lance height and flow rate on penetration depth in a top-blown water model. Steel Res. Int. **77**, 82–90 (2006)
99. A.R.N. Meidani, M. Isac, A. Richardson, A. Cameron, R.I.L. Guthrie, Modelling shrouded supersonic jets in metallurgical reactor vessels. ISIJ Int. **44**, 1639–1645 (2004)
100. H. Wang, R. Zhu, Y.L. Gu, C.J. Wang, Behaviours of supersonic oxygen jet injected from four-hole lance during top-blown converter steelmaking process. Can. Metall. Q. **53**, 367–380 (2014)

101. J. Solórzano-López, R. Zenit, M.A. Ramírez-Argáez, Mathematical and physical simulation of the interaction between a gas jet and a liquid free surface. Appl. Math. Model. **35**, 4991–5005 (2011)
102. F. Qian, R. Mutharasan, B. Farouk, Studies of interface deformations in single- and multi-layered liquid baths due to an impinging gas jet. Metall. Mater. Trans. B **27**, 911–920 (1996)
103. J. Mietz, S. Schneider, F. Oeters, Model experiments on mass transfer in ladle metallurgy. Steel Res. **62**, 1–9 (1991)
104. Y. Liu, H. Bai, H. Liu, M. Ersson, P.G. Jönsson, Y. Gan, Physical and numerical modelling on the mixing condition in a 50 t ladle. Metals (Basel). **9**, 1–19 (2019)

Chapter 12
Scaling and Similarity

12.1 Introduction

A physical model is the representation of a full-scale system or prototype, built at a smaller scale, operating under similar or equivalent forces, designed to study heat, mass and momentum transport and based on the model results improve the performance of the full-scale system. Due to a smaller scale, use of common fluids and in some cases the use of transparent vessels, it is cheaper to operate and flow visualization is possible. Physical models are used to describe a particular phenomenon occurring in a larger system, called prototype. For example, the lift and drag forces of an airplane model can be measured in a wind tunnel. The model is scaled down and the results can be scaled up.

In order to scale down a prototype and built a physical model, several similarity criteria should be satisfied. Fulfillment of this criteria is the key to scale up the results obtained in a physical model. This chapter describes in detail the criteria on similarity.

12.2 Similarity Criteria

Metallurgical operations can involve simultaneous heat, mass and momentum transfer. Chemical reactions, in particular those at high temperatures, cannot be described by dimensionless numbers, therefore, all transport phenomena occurring in a full metallurgical process cannot be studied completely at the same time in a scale model. A physical model usually focuses on the description of one particular phenomena, for example momentum transfer.

In fluid mechanics two flows are similar if one flow can be made identical to the other by the application of constant scale factors to each variable, in other words if dimensions, velocities, pressures, density and viscosity, are multiplied by factors that make the other characteristic parameters the same, then the two flows will be

A. N. Conejo, *Fundamentals of Dimensional Analysis*,
https://doi.org/10.1007/978-981-16-1602-0_12

identical [1]. The relationship between scale and the property values for the two systems is defined as follows,

$$\lambda_i = \frac{i_1}{i_2}$$

where i represents a property value like dimensions, velocities or pressures.

Two processes can be considered completely similar if they take place in a similar geometrical system and if all the dimensionless numbers necessary to describe them have the same numerical value. To satisfy these conditions the following similarity criteria should be achieved:

1.   Geometric similarity
2.   Material similarity
3.   Mechanical similarity

    3.1   Static similarity
    3.2   Kinematic similarity
    3.3   Dynamic similarity

4.   Thermal similarity
5.   Chemical similarity
6.   Electromagnetic similarity

Most physical models are operated under isothermal conditions, neglecting all chemical reactions and electromagnetic forces. In these conditions it is only needed to satisfy geometric, material and mechanical similarity.

There is an order to achieve similarity, first the model should be designed to achieve geometric similarity, then material similarity and subsequently dynamic similarity. In each case full similarity, depending on the problem, probably will not be fully satisfied, however, depending on the phenomena of interest the achievable conditions can be satisfactory. This is discussed in more detail in this chapter.

12.2.1   Geometric Similarity

The first step in scaling a prototype is to achieve geometric similarity. Geometric similarity refers to the condition of having the same geometric proportions and shape. Two geometric figures are similar if the ratios of all the corresponding lengths are identical. The ratio of the corresponding lengths is called geometric scale ratio. If geometric similarity is fulfilled in a vessel, the coordinates at one point in the prototype are equivalent to the coordinates in the physical model, as shown below in cylindrical coordinates (r, θ, z) [2].

$$\lambda = \frac{r_m}{r_{fs}} = \frac{z_m}{z_{fs}}; \theta_m = \theta_{fs}$$

where; λ is the geometric scale factor.

Homologous points are points that have the same relative location in both model and prototype.

A familiar example of geometric similarity is the case of similar triangles. For similar triangles the following ratio of lengths is valid:

$$\frac{L_1}{l_1} = \frac{L_2}{l_2} = \frac{L_3}{l_3}$$

Are there any limits on how small the geometric scale should be?: Limits on how small the model can be, have not been defined yet. Zlokarnik [3] asks the question; how small can a model be?, but lacks of an answer. His general comment is that the smaller the geometric scale, much smaller the experimental error, for example for a geometric scale of 10%, the experimental error should be much lower than 10%.

Cengel [4] points out the case of big prototypes because of the need to build very small scales, several hundred times smaller when dealing with big prototypes, for example a river model to study flooding. A model with a small geometric scale, for example 1:100, brings two problems; first, if the depth of the model is built using that scale it will be so small that surface tension forces will be involved and second, the flow will be laminar instead of the turbulent flow in the actual river. This is called *scale effects*. One solution is to build a distorted model with a larger depth, however complete geometric similarity will not be satisfied. Another solution is to increase the geometric scale as much as possible. The distorted model results will need empirical corrections before scaling up.

It is important to clarify that for the same geometric scale factor, the final dimensions of the physical model can be completely different because, the same prototype, for example a steelmaking reactor, can have a wide range of sizes, almost by one order of magnitude, therefore a geometric scale 1:10 could mean a large physical model if the original prototype is big but it could mean a very small physical model if the original prototype is small. Probably there is not a unique response to the lower limit of a geometric scale but depends on each case; how accurate and reliable is the information obtained from a physical model with a low geometric scale. Are there any sensors that introduce perturbations to the system as the model decreases in size?, a property value in the model multiplied by the geometric scale no longer reproduces the property value of the prototype?

The answer to an almost perfect geometric similarity is to design a system with a geometric scale 1:1. It has additional benefits because in some cases, it is the only way to achieve full dynamic similarity, however with a big scale the resources needed drastically increases not just materials but time to do every experiment. Then, as a second choice would be to make a physical model as large as possible depending on the availability of space and resources. Perfect similarity would include surface roughness therefore, even perfect geometric similarity is not possible.

12.2.2 Materials Similarity

The second step in scaling a prototype is to achieve material similarity.

The physical properties of the materials in the model and prototype should be similar; density, viscosity and surface tension.

In the case of steelmaking operations, liquid steel is properly represented by water since water at room temperature has almost the same kinematic viscosity as liquid steel at 1600 °C. The viscosity of water at 20 °C is 1×10^{-3} N·s·m^{-2}. Liquid water and liquid steel are Newtonian fluids; therefore, its physical properties (density and viscosity) remain constant as a function of time. Examples of non-Newtonian fluids; whole blood, paints and milk.

Metallurgical slags on the other are not properly represented by conventional oils. Steel slags have a density about half of liquid steel but its viscosity is much higher than liquid steel, at least 100 times higher. Commercial oils with high viscosity also have a high density. Then only one property can be similar, either density or viscosity. It is better to choose an oil based on its density because if its density is similar to that of water there is a large oil emulsification which affects fluid flow phenomena. Oils with lower density can have viscosities even lower than water.

A proper selection of model materials has also an effect on the experimental conditions, for example, in mechanical stirring the rotational speed is directly proportional to the viscosity of the fluid. Differences on fluid's viscosities can represent from one to two orders of magnitude in the rotational speed.

12.2.3 Mechanical Similarity

The third step in scaling a prototype is to achieve mechanical similarity.

Gupta et al. [5] divides mechanical similarity into three groups: (1) static which represents the similarity of deformation in both model and prototype, (2) kinematic which represents the similarity of motion and (3) dynamic which represents the similarity of forces.

Static similarity

Static similarity applies to solid bodies subject to loads. It is applied in mechanical engineering. It can be torque or static loads which should be similar in model and prototype.

Kinematic similarity

The motion of fluid elements should follow similar trajectories in corresponding intervals of time.

Dynamic similarity

In the construction of a physical model, usually a low low-temperature analogue, dynamic similarity is one of the main parameters to scale down a prototype. Dynamic similarity refers to scaling the main forces present in the prototype. The corresponding forces acting at corresponding times at corresponding locations in the model should also correspond [2]. Two systems are said to be dynamically similar if corresponding homologous points of the systems have similar values of the relevant dimensionless groups. As a consequence, the ratios of different forces in the model and prototype should be similar.

$$\frac{F_{i,m}}{F_{i,fs}} = \frac{F_{j,m}}{F_{j,fs}} = \text{constant}$$

which is equivalent to,

$$\left(\frac{F_i}{F_j}\right)_m = \left(\frac{F_i}{F_j}\right)_{fs} = \text{constant}$$

where; i and j represent different types of forces (inertial, gravitational, etc.)

Each force at homologous points should be similar in the model and the prototype and also the total forces should be similar. According with Newton's second law the sum of all forces should be equal to the acceleration term or inertia force.

$$\sum F = F_p + F_g + F_f = F_i$$

Similarity in general involve governing equations applicable under laminar flow conditions. Problems dealing with turbulent flow conditions employ effective viscosity, turbulent thermal conductivity, etc. that are problem specific [6].

The definition of a criterion for dynamic similarity can be simple and similarity can be complete if the problem to be investigated defines the independent variable as a function of only one dimensionless number. For example, the steady motion of a body in a fluid that fills a very large vessel can be described with the Reynolds number, and the motion of a streamlined surface ship at high speeds can be described with the Froude number [7]. The pressure drop of a homogeneous fluid in a straight, smooth pipe can also be represented as a function of the Reynolds number. There is only one relationship between pressure drop and the Re number independently if the scale of the pipe changes. In this example, the model characteristics can be defined if its Re number is equal to the prototype.

$$Re_m = Re_p$$

$$\left(\frac{UL}{\nu}\right)_m = \left(\frac{UL}{\nu}\right)_p$$

However, even if the process achieves complete dynamic similarity it should not be expected that the model and the prototype will behave identically; only on the particular phenomena that is investigated [7]. It can also happen that large changes in the scale, for example for an extremely small diameter of the pipe, surface tension forces become important [8].

Frequently complete dynamic similarity is not possible. This is because there are several criteria that cannot be satisfied simultaneously. Each dimensionless group defines one criterion. In those cases, it is still possible to scale down a prototype applying the following alternatives:

- Each criterion should be analyzed carefully, defining the forces involved and quantify those forces for the conditions of the problem. Depending on its magnitude some of forces can be ignored and consequently its dimensionless number can be excluded.
- The relevance of some forces changes depending on the process conditions, for example, the magnitude of the inertial forces decreases when the gas flow rate decreases.
- Dimensionless groups can be combined. It reduces the number of criteria.
- The number of dimensionless groups can be reduced combining variables. One example is density differences under natural convection, the buoyance force parameter incorporates two variables; $\Delta \rho g$.

The following is an example of analysis to select the dominant dimensionless group.

Motion of a ship: The general problem studied by Froude to define the drag coefficient of a streamlined surface ship, was described by Zlokarnik [6]. Complete dynamic similarity is not possible in this case. The problem involves 7 variables; drag resistance (F), ship's length (l), speed of the ship (U), displaced volume (V), density (ρ) and kinematic viscosity (ν) of water, and gravity (g), and 4 dimensions. The problem can be described with 4 dimensionless groups,

$$\pi_1 = f(\pi_2)(\pi_3)(\pi_4)$$

$$\frac{F}{\rho U^2 l^2} = f\left(\frac{\nu}{Ul}\right)\left(\frac{gl}{U^2}\right)\left(\frac{V}{l^3}\right)$$

$$Ne = f\left(Re^{-1}\right)\left(Fr^{-1}\right)\left(\frac{V}{l^3}\right)$$

where; Ne is the Newton number and V/l^3 represent the dimensionless displaced volume.

Fixing the dimensions of the model, fixes the value of π_4. The drag coefficient is calculated from changes in Fr and Re numbers. The Froude number can only be changed with changes in the ship's velocity. Equating the Fr number of model and prototype, assuming a geometric scale of 10%, the value of the ship's velocity in the model is calculated. Equating the Re numbers for the model and the prototype, inserting the velocity values, it is obtained the ratio of viscosities of the fluid and the prototype. Since the fluid's prototype is water, complete dynamic similarity requires a liquid with a viscosity just 3% with respect to water. Not considering the difficulty to find such a fluid, the result indicates that water cannot be employed. Therefore, only the Fr number can be satisfied but not the Re number. Therefore, complete dynamic similarity is not possible.

Partial dynamic similarity implies that the model obeys only the most dominant forces but not all of them. This type of physical model is called a *distorted model*. Extrapolation of model results to the full-scale system might require a proper interpretation. One example can be described as follows;

$$(\pi_1)_m = (\pi_1)_p$$

$$(\pi_2)_m \neq (\pi_2)_p$$

To increase the velocity of any vehicle in motion (ships, automobiles, airplanes, etc.) and reduce fuel consumption it is important to decrease the drag force. The drag force depends on the flow at the boundary layer and therefore is affected by the roughness of the surface. In these applications, geometric similarity requires the surface roughness to be similar in both model and prototype.

Guthrie points out that usually only one dimensionless group can be satisfied in practice and is the physical modeler's experience who decide which is the most important criteria and which factors can be ignored in setting up the model [2].

Examples of dynamic similarity

Bottom gas stirring in the ladle furnace

If the process is isothermal, without chemical reactions and only the interest is to study fluid flow phenomena, then satisfying geometric, material and dynamic similarity is equivalent to complete similarity. The Navier–Stokes equation is the governing equation describing all forces that affect fluid flow. This equation in dimensionless form contains three dimensionless groups, however under steady state it reduces to two; Re and Fr numbers, therefore, complete dynamic similarity would require to satisfy equality of their values on model and prototype.

To satisfy the Fr number, the velocity or its equivalent value, a corresponding gas flow rate in the model is calculated, as will be shown in a section below. Using as an example an industrial size ladle of 220 ton of nominal capacity and the following dimensions; height of liquid steel 2.83 m, average radius 3.59 m, argon flow rate 40 Nm^3/h through one porous plug. The physical model has a geometric scale 1:10.

The plume velocities correspond to 0.95 and 0.30 m/s, for the full-scale system and model, respectively.

$$\left(\frac{U_p^2}{gH}\right)_m = \left(\frac{U_p^2}{gH}\right)_{fs}$$

$$\left(\frac{0.30^2}{9.81 \times 0.283}\right)_m = \left(\frac{0.95^2}{9.81 \times 2.83}\right)_{fs}$$

$$(0.03304)_m = (0.03381)_{fs}$$

The fluid in the physical model is water and its dimensions have been fixed by the previous scale. Since the Re number depends on three parameters that have already been fixed (plume velocity to satisfy Fr number, dimensions fixed by the scale and kinematic viscosity of water) therefore, it has a unique value in the water model.

$$Re = \frac{U_pH}{\nu}$$

The relationships between velocity (or gas flow rate) and geometric scale based on Re number is given below,

$$(U_p)_m = \sqrt{\frac{H_m}{H_{fs}}} (U_p)_{fs} = \lambda^{\frac{1}{2}} (U_p)_{fs}$$

Since satisfying material similarity requires both model and prototype to have the same kinematic viscosity, the velocity to satisfy the Re number is as follows,

$$(U_p)_m = \frac{H_{fs}}{H_m} (U_p)_{fs} = \lambda^{-1} (U_p)_{fs}$$

It is clearly seen that with a scaled down model, it is impossible to satisfy both criteria. The only possibility using a water model would be to use a geometric scale 1:1 because $\lambda^{-1} = 1$ and $\lambda^{\frac{1}{2}} = 1$, in this condition the velocity of the liquid in the model and the prototype is the same. Another option is to choose a different fluid. The two dimensionless groups are equated and solved for the geometric scale, to define the ratio of viscosities between model and prototype:

Froude number similarity:

$$\left(\frac{U^2}{gL}\right)_m = \left(\frac{U^2}{gL}\right)_p$$

$$\therefore \frac{L_m}{L_p} = \frac{U_m^2}{U_p^2}$$

Reynolds number similarity

$$\left(\frac{UL}{\nu}\right)_m = \left(\frac{UL}{\nu}\right)_p$$

$$\therefore \frac{L_m}{L_p} = \frac{\nu_m}{\nu_p} \frac{U_p}{U_m}$$

Since the geometric scale ratio is the same to satisfy both Fr and Re similarity:

$$\frac{L_m}{L_p} = \frac{\nu_m}{\nu_p} \frac{U_p}{U_m} = \frac{U_m^2}{U_p^2}$$

This is equivalent to:

$$\frac{\nu_m}{\nu_p} = \frac{U_m^2}{U_p^2} \frac{U_m}{U_p} = \left(\frac{U_m}{U_p}\right)^3$$

From Fr similarity, the following relationship can be used:

$$\frac{U_m}{U_p} = \left(\frac{L_m}{L_p}\right)^{\frac{1}{2}}$$

$$\frac{\nu_m}{\nu_p} = \left(\left(\frac{L_m}{L_p}\right)^{\frac{1}{2}}\right)^3$$

The final result provides the ratio of kinematic viscosities between model and prototype

$$\frac{\nu_m}{\nu_p} = \lambda^{\frac{3}{2}}$$

The kinematic viscosity of liquid steel at 1600 °C is about 1×10^{-6} m²/s. In order to satisfy both Re and Fr numbers, the fluid to represent liquid steel in a model with a geometric scale 1:10, should have the following kinematic viscosity:

$$\nu_m = \lambda^{\frac{3}{2}} \nu_p = \left(\frac{1}{10}\right)^{\frac{3}{2}} \times 10^{-6} = 3.16 \times 10^{-8} \text{m}^2/\text{s}.$$

The fluid should have a kinematic viscosity two orders of magnitude lower than water. The only fluid closer to that value is mercury with a kinematic viscosity in the order of 10^{-7} m²/s. On practical grounds the model can use water and carry out the experiments under conditions of incomplete similarity.

Having the conflict to satisfy only one criterion, one of the two dimensionless groups should be chosen to scale down the prototype. At this point it is convenient to recall the nature of the forces involved. The Fr number is a ratio of inertial/gravitational forces and the Re number is a ratio of inertial/molecular viscous forces. To define which forces are higher, the values of Fr and Re are computed. In the example, U_p for the prototype was 0.95 m/s, density of liquid steel 7000 kg/m^3, absolute viscosity 7×10^{-3} kg/m s.

$$(Re)_p = \left(\frac{\rho U_p H}{\mu} \right)_p = \frac{(7000)(0.95)(2.83)}{7 \times 10^{-3}} = 6.45 \times 10^6.$$

This value clearly indicates that the viscous forces are negligible in comparison with the inertial forces. The Fr number was already calculated before and the value reported was 3.38×10^{-2} indicating that the gravitational forces are higher that the inertial forces, therefore the gravitational and inertial forces dominate fluid flow in this case. Since the Fr number contains those forces, it would be convenient to choose the Fr number to scale down the prototype.

An incompressible fluid forms a free surface or an interface between two immiscible fluids due to the effect of gravity forces, therefore, gravity forces are important in any flow with a free surface. Practical examples involving a free surface or interfaces are wave formation caused by a ship through water and gas injection in a ladle furnace.

Stirring due to bottom gas injection involves the formation of bubbles. Bubble and droplet formation is controlled by the Weber number (We). This number is a ratio of inertial to surface tension forces. To assess the importance of surface tension forces, the We number is calculated. The surface tension of liquid steel is taken as 1600 dynes/cm, equivalent to 1.63 kg/s [2, 5].

$$(We)_p = \left(\frac{\rho_l U_p^2}{\sigma} \right)_p = \frac{(7000)(0.95)^2}{1.63} = 3.8 \times 10^4$$

The result indicates that surface tension in this problem are 4 orders of magnitude lower than inertial forces.

Once the relevant forces have been identified and the dimensionless number that better describe them is chosen, the next step is to define the gas flow rate in the physical model (Q_m). A relationship between Q_m and Q_{fs} can be derived from the equality of Froude numbers between model and full-scale systems. Since the problem involves gas injection, the motion of the liquid is due to the motion of the bubbles, therefore is more convenient to use the modified Froude number (Fr') rather than the conventional one.

$$\left(Fr' \right)_m = \left(Fr' \right)_{fs}$$

$$\left(\frac{\rho_g U_0^2}{\rho_l g L} \right)_m = \left(\frac{\rho_g U_0^2}{\rho_l g L} \right)_{fs}$$

Two problems arise here; to decide the values of the characteristic velocity (U_0) and the characteristic length (L). Since the gas bubble rise from bottom to top of the liquid, the height of the liquid is usually taken as the characteristic length. Several

Table 12.1 Relationships to predict plume velocity due to bottom gas injection in ladles

Year	Authors	Characteristic velocity	n	Units
1982	Y. Sahai and R. Guthrie	$U_p = 4.4Q^{\frac{1}{3}}H^{\frac{1}{4}}R^{-\frac{1}{3}}$	–	U in m/s, Q in m³/s, R and H in m
1985	D. Mazumdar and R. Guthrie	$U_0 = \overline{U}$	2.75	
1985	G. Ebneth and W. Plushkell	$U_p = 15.1Q^{0.244}H^{-0.08}$	–	
1987	S.-H. Kim and R. Fruehan	$U_T = Q_n/(\pi d_n^2/4)$	2.5	U in cm/s, Q in lt/min, d_n in cm
1990	D. Mazumdar	$U_p = 4.5Q^{\frac{1}{3}}H^{\frac{1}{4}}R^{-\frac{1}{4}}$	1.5	U in m/s, Q in m³/s, R and H in m
1993	D. Mazumdar and R. Guthrie	$\overline{U} = 0.86Q^{\frac{1}{3}}H^{\frac{1}{4}}R^{-0.58}$	–	U in m/s, Q in m³/s, R and H in m
1994	M. Castello-Branco and K. Schwerdtfeger	$U_p =$ $17.4Q^{0.244}d^{-0.0288}\left(\frac{\rho_g}{\rho_l}\right)^{0.0218}H^{-0.08}$	–	U in cm/s, Q in cm³/s, d and H in cm
2002	D. Mazumdar	$U_p = 3.1Q^{\frac{1}{3}}H^{\frac{1}{4}}R^{-0.58}$	2.5	U in m/s, Q in m³/s, R and H in m

researchers [9–15] have applied different approaches to define the characteristic velocity which includes; plume velocity (U_p), average velocity of the liquid ($\overline{U}$) and superficial gas velocity at the tuyere exit (U_T). Table 12.1 shows a summary of the proposed relationships. Applying these expressions, a simple relationship of the form $Q_m = \lambda^n Q_{fs}$ is developed. The value of the exponent "n" is shown in this table.

Mazumdar et al. [16] investigated the values for the exponent "n" reported up to 2000, corresponding to 1.5, 2.5 and 2.75. The objective was to clarify which exponent in the geometric scale factor fully satisfied the Fr number. They reported mixing time from 4 ladles, without a top layer, central gas injection and using vessel A as the prototype.

To satisfy Fr similarity based on measurements on mixing time two conditions should be met:

– Similarity for the time scale in Froude number dominated flows is defined by the following relationship:

$$\tau_m = \lambda^{\frac{1}{2}}\tau_p$$

– A tracer is employed to measure mixing time. The tracer concentration gradients in the model and the prototype should be similar. The mass conservation equation for incompressible fluids involving Fick's law in dimensionless form includes the inverse of the product of Re·Sc numbers, therefore this parameter should be the same.

	A	B	C	D
L, m	0.93	0.49	0.41	0.25
D, m	1.12	0.60	0.495	0.3
λ	1	0.535	0.44	0.27
Q, $\times 10^{-4}$ m^3/s	6–10			

For each gas flow rate in the prototype, three values were defined for each water model. The gas flow rate in the model was calculated for models B, C and D, using the following expression:

$$Q_m = \lambda^n \cdot Q_p$$

Calculation of Re and Sc numbers require the plume velocity, kinematic viscosity of water and diffusion coefficient. The plume velocity was calculated from Sahai and Guthrie's expression [9]. The diffusion coefficient was approx. 8.5×10^{-4} m^2/s. The calculated 1/Re·Sc values ranged from $2–3 \times 10^{-3}$, a fairly constant value, indicating chemical similarity. Then, mixing times for vessels B, C and D were compared with the prototype (vessel A), for the three expressions used to calculate the gas flow rate in the water models, using different exponents. Similarity was complete using the exponent 2.5 with the larger vessel B, satisfactory with vessel C and unsatisfactory for vessel D. The first conclusion was that Froude number similarity should be calculated with the expression $Q_m = \lambda^{2.5} Q_p$.

In regard with the discrepancies with the smallest vessel, the values of Re and Fr numbers were compared for all the experiments and they found that, the order of magnitude is the one expected for Froude number dominated flows (Re $> 4 \times 10^5$ and gas flow rates in the range 0.001–0.01 Nm^3/min ton). Decreasing the geometric scale, the Re number decreases. For the smallest vessel the value was the lowest, about 0.4×10^5 but still a high value. Then, a reasonable explanation by the authors indicated that by decreasing the scale, the dead zones also reduced its size but the sensor position to measure the electric conductivity was not adjusted accordingly. Large differences, up to 1.7 times, can be found if the sensor is located or not in a dead zone. Applying a correction factor of about 1.7, vessel D also matched the expected mixing times.

Derivation of a dynamic similarity criteria:

The most common value currently employed for the exponent is 2.5. Kim and Fruehan [12] derived a relationship using the superficial gas velocity and Mazumdar [15] using the plume velocity. The derivation of their relationships is shown below as option 1 and option 2.

Option 1: Kim and Fruehan

The modified Froude number is given by the following equation:

$$Fr' = \frac{\rho_g U_T^2}{\rho_l g L}$$

where: Fr' is the modified Froude number, U_T is the linear velocity of gas at the tuyere outlet (cm/s), ρ_g is the density of the gas (gr/cm^3), ρ_l is the density of the liquid (gr/cm^3), g is the gravity constant ($981 \ cm \cdot s^{-2}$), L is a characteristic length (cm).

$$\left(\frac{\rho_g U_T^2}{\rho_l g L}\right)_{fs} = \left(\frac{\rho_g U_T^2}{\rho_l g L}\right)_m$$

Since we are interested in obtained an expression for the gas flow rate, the following relationship is applied,

$$Q = U_T A = U_T \cdot \pi d_T^2 / 4$$

$$U_T = \frac{Q}{\pi d_T^2 / 4}$$

where: Q is gas flow rate (l/min), d_T is the diameter of tuyere (cm), A is the tuyere cross sectional area (cm^2).

The density of the gas can be defined in terms of the molecular mass of a gas, considering that the molar volume, $\overline{V}$, of any gas is equal to 22.4 L.

$$\rho_g = \frac{w}{V} = \frac{nM}{V} = \frac{M}{V/n} = \frac{M}{\overline{V}} = \frac{M}{22.4}$$

The product $\rho_g U_T$ is:

$$\rho_g U_T = \frac{Q}{22.4} \cdot \frac{M}{60} \cdot \frac{1}{\pi d_T^2 / 4}$$

M is the molecular mass of the gas and 60 is a conversion factor from seconds to minutes.

$$\frac{\rho_g U_T^2}{\rho_l g L} = \frac{\rho_g U_T^2}{\rho_l g L} \cdot \frac{\rho_g}{\rho_g} = \frac{\left(\rho_g U_T\right)^2}{\rho_l \rho_g g L}$$

Replacing the value for $\rho_g U_T$ into the previous equation:

$$\frac{(\rho_g U_T)^2}{\rho_1 \rho_g g L} = \frac{\left(\frac{Q}{22.4} \cdot \frac{M}{60} \cdot \frac{1}{\pi d_T^2/4}\right)^2}{\rho_1 \rho_g g L} = \frac{\frac{Q^2 M^2}{22.4^2 60^2 \pi^2 d_T^4/4^2}}{\rho_1 \rho_g g L}$$

$$\frac{(\rho_g U_T)^2}{\rho_1 \rho_g g L} = \left(\frac{\frac{M^2}{22.4^2 60^2 \pi^2/4^2}}{\rho_1 \rho_g g}\right) \cdot \frac{Q^2}{L d_T^4} = C \cdot \frac{Q^2}{L d_T^4}$$

where C is a constant,

$$C = \frac{\frac{M^2}{22.4^2 60^2 \pi^2/4^2}}{\rho_1 \rho_g g} = \frac{M^2}{\rho_g \rho_1(981)\left(22.4^2 60^2 \pi^2/4^2\right)} = \frac{9.159 \times 10^{-10} M^2}{\rho_g \rho_1}$$

Substituting the result $\frac{(\rho_g U_T)^2}{\rho_1 \rho_g g L}$ into the expression the defines equal Froude number on both model and prototype:

$$\left(\frac{CQ^2}{L d_T^4}\right)_m = \left(\frac{CQ^2}{L d_T^4}\right)_{fs}$$

$$Q_m^2 = \left(\frac{CQ^2}{L d_T^4}\right)_{fs} \cdot \left(\frac{L d_T^4}{C}\right)_m$$

$$Q_m^2 = \left(\frac{C_{fs}}{C_m}\right) \cdot \frac{\left(L d_T^4\right)_m}{\left(L d_T^4\right)_{fs}} \cdot Q_{fs}^2$$

$$Q_{fs} = \left(\frac{C_{fs}}{C_m}\right)^{\frac{1}{2}} \cdot \left[\frac{\left(L d_T^4\right)_m}{\left(L d_T^4\right)_{fs}}\right]^{\frac{1}{2}} \cdot Q_{fs}$$

Defining the geometric scale coefficient as λ:

$$\lambda = \frac{L_m}{L_{fs}} = \frac{(d_T)_m}{(d_T)_{fs}}$$

$$\lambda^5 = \frac{\left(L d_T^4\right)_m}{\left(L d_T^4\right)_{fs}}$$

Replacing this expression, the previous relationship Q_m–Q_{fs} reduce to,

$$Q_m = \left(\frac{C_{fs}}{C_m}\right)^{\frac{1}{2}} \cdot \lambda^{\frac{5}{2}} \cdot Q_{fs}$$

The value of the constants is now calculated,

$$C_{fs} = \frac{9.159 \times 10^{-10} M_{Ar}^2}{\rho_{Ar} \cdot \rho_{steel}}$$

$$C_m = \frac{9.159 \times 10^{-10} M_{air}^2}{\rho_{air} \cdot \rho_{water}}$$

$$\frac{C_{fs}}{C_m} = \frac{M_{Ar}^2 \cdot \rho_{air} \cdot \rho_{water}}{M_{air}^2 \cdot \rho_{Ar} \cdot \rho_{steel}}$$

The density of argon at 1000 °C obtained from the density of argon at 0 °C

$$\rho_{Ar}^{STP} = 1.761 \text{ kg/m}^3$$

$$\rho_{Ar}^{1000} = \rho_{Ar}^{STP} \cdot \frac{273}{1273} = 0.377$$

$$\rho_{Ar}^{1000\ °C} = 0.377 \text{ kg/m}^3$$

also,

$$M_{air} = 29 \text{ g/mol}$$

$$M_{Ar} = 39.95 \text{ g/mol}$$

$$\rho_{air} = 1.293 \text{ kg/m}^3$$

$$\rho_{water} = 1000 \text{ kg/m}^3$$

$$\rho_{steel} = 7000 \text{ kg/m}^3$$

Then,

$$\frac{C_{fs}}{C_m} = \frac{M_{Ar}^2 \cdot \rho_{air} \cdot \rho_{water}}{M_{air}^2 \cdot \rho_{Ar} \cdot \rho_{steel}} = 0.93$$

$$\left(\frac{C_{fs}}{C_m}\right)^{\frac{1}{2}} = 0.97$$

Replacing the value of the constants,

$$Q_m = \left(\frac{C_{fs}}{C_m}\right)^{\frac{1}{2}} \cdot \lambda^{\frac{5}{2}} \cdot Q_{fs}$$

So,

$$Q_m = 0.97 \cdot \lambda^{\frac{5}{2}} \cdot Q_{fs}$$

Since the constant is close to 1, it is taken as one to simplify the expression:

$$Q_m = \lambda^{\frac{5}{2}} \cdot Q_{fs}$$

This is the final expression that defines the gas flow rate in the physical model that allows similarity of Froude numbers in both model and full-scale systems however, the gas flow rate in the physical model correspond to STP conditions and for the full-scale system should be computed for the pressure and temperature of liquid steel.

$$Q_m^{STP} = \lambda^{\frac{5}{2}} \cdot Q_{fs}^{TP}$$

The gas flow rates reported in the steel plant correspond to STP conditions but the real values correspond to the pressure and temperature of the gas in the ladle, therefore corrections for pressure and temperature should be taken into consideration [17, 18]. Since argon heats rapidly as soon as it enters the liquid steel it is justified to assume that it has the temperature of the liquid, this temperature is taken as 1600 °C. the pressure of the gas changes with height. To simplify the calculation an average value it is taken at half height of the ladle. The industrial ladle in this example has a height of liquid steel of 2.833 m.

$$Q_{fs}^{TP} = \left(\frac{T}{T_0}\right)\left(\frac{P_0}{\overline{P}}\right) \cdot Q_{fs}^{STP} = \left(\frac{1873}{273}\right)\left(\frac{101325}{101325 + \frac{1}{2}(7000 \times 9.81 \times 2.833)}\right) \cdot Q_{fs}^{STP} = 3.5 \cdot Q_{fs}^{STP}$$

The final equation to compute the gas flow rate for the physical model is the following,

$$Q_m^{STP} = \lambda^{2.5} \cdot 3.5 \cdot Q_{fs}^{STP}$$

The constant should be adjusted for ladles with different heights and temperatures of liquid steel.

Option 2: Mazumdar

The analysis is simplified using the conventional Froude number and replacing the value of plume velocity with the formula developed; plume velocity as a function of gas flow rate and ladle dimensions.

$$Fr = \frac{U_p^2}{gH}$$

$$U_p = 3.1 Q^{\frac{1}{3}} H^{\frac{1}{4}} R^{-0.58}$$

$$\left[\frac{\left(3.1 Q^{\frac{1}{3}} H^{\frac{1}{4}} R^{-0.58}\right)^2}{gH}\right]_m = \left[\frac{\left(3.1 Q^{\frac{1}{3}} H^{\frac{1}{4}} R^{-0.58}\right)^2}{gH}\right]_{fs}$$

$$Q_m^{\frac{2}{3}} = \left(\frac{R_m}{R_{fs}}\right)^{1.16} \left(\frac{H_m}{H_{fs}}\right)^{\frac{1}{2}} \cdot Q_{fs}^{\frac{2}{3}}$$

$$Q_m = \left(\frac{R_m}{R_{fs}}\right)^{1.74} \left(\frac{H_m}{H_{fs}}\right)^{\frac{3}{4}} \cdot Q_{fs}$$

$$Q_m = \lambda^{2.5} \cdot Q_{fs}$$

Summarizing the previous results; the scaling equation for dynamic similarity has been based on the Fr number. Geometric and dynamic similar systems with Froude number dominated flows are summarized through the following relationships:

$$L_m = \lambda L_p$$

$$V_m = \lambda^3 V_p$$

$$A_m = \lambda^2 V_p$$

$$U_m = \lambda^{\frac{1}{2}} U_p$$

$$Q_m = \lambda^{\frac{5}{2}} \cdot Q_{fs}$$

$$t_m = \lambda^{\frac{1}{2}} t_p$$

The volume (L^3), area (L^2) and time scales are defined from geometric similarity and velocity similarity.

The next stage is to apply the scaling equations to define the parameters of the physical model. The prototype is an industrial size ladle furnace that has two porous plugs for bottom gas injection. The gas flow rate (STP) in each porous plug range from 0 to 40 Nm^3/h. For this ladle, the gas flow rate for the model and plume velocities are calculated using the following equations.

$$Q_m^{STP} = \lambda^{2.5} \cdot 3.2569 \cdot Q_{fs}^{STP}$$

$$U_p = 3.1 \, Q^{\frac{1}{3}} H^{\frac{1}{4}} R^{-0.58}$$

Prototype			Model	Plume velocities		Froude similarity	
Q_p, STP		Q_p, TP	Q_m				
Nm^3/h	Nl/min	l/min	Nl/min	U_m	U_p	Fr_m	Fr_p
2	33.3	108.5	0.343	0.114	0.356	0.00466	0.00457
10	166.6	542.8	1.716	0.194	0.606	0.01354	0.01323
20	333.3	1085.6	3.433	0.243	0.762	0.02140	0.02091
40	666.7	2171.3	6.866	0.306	0.958	0.03381	0.03304

Prototype dimensions: Average diameter (D) = 3.588 m, height of liquid steel (H) = 2.833 m, λ = 0.1

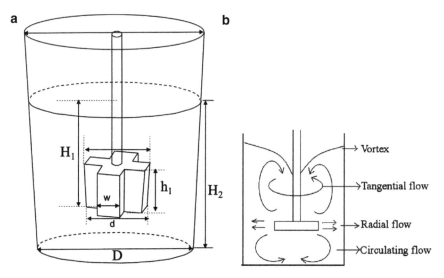

Fig. 12.1 a KR reactor, **b** flow regimes in KR reactor

Mechanically stirred reactors

The source of mechanical stirring is the rotation of an impeller. The power to drive
the impeller (P) depends on a large number of variables: rotation speed (N), impeller
geometric parameters such as diameter (d), width (w), height (h_1); ladle diameter (D),
immersion depth (H_1), water height (H_2), distance impeller-ladle bottom (C); fluid's
properties, density (ρ) and viscosity (μ); blades characteristics like blade number
(b_N), blade position (b_θ) and blade thickness (b_t); gravity force, g. Gas flow rate (Q)
if there is gas injection through the impeller. Figure 12.1a shows schematically the
system. Figure 12.1b shows different flow regimes in the reactor, according to Liu
[19].

The relationship between impeller power and the previous variables is expressed
as follows:

$$P = f(N, d, w, h_1, D, H_1, H_2, C, \rho, \mu, Q, g, \text{etc})$$

The result from dimensional analysis includes the following dimensionless
groups:

$$\frac{P}{\rho N^3 d^5} = f\left(\frac{\rho N d^2}{\mu}, \frac{N^2 d}{g}, \frac{D}{d}, \frac{H_2}{d}, \ldots\right)$$

The first group, π_1, is called power number, the second group, π_2, is the Re
number and thee third group, π_3, is the Fr number. Dynamic similarity should be
completed satisfying Re and Fr numbers:

$$\left(\frac{\rho N d^2}{\mu}\right)_m = \left(\frac{\rho N d^2}{\mu}\right)_p$$

$$\left(\frac{N^2 d}{g}\right)_m = \left(\frac{N^2 d}{g}\right)_p$$

The rotation speed can be defined by the Fr number,

$$\frac{N_m}{N_p} = \left(\frac{d_p}{d_m}\right)^{\frac{1}{2}}$$

To satisfy both criteria, this equation is substituted into Re number similarity,

$$\frac{N_m}{N_p} = \left(\frac{\rho}{\mu}\right)_p \left(\frac{\mu}{\rho}\right)_m \left(\frac{d_p}{d_m}\right)^2 = \left(\frac{d_p}{d_m}\right)^{\frac{1}{2}}$$

Which defines the kinematic viscosity of the fluid in the physical model.

$$\left(\frac{\mu}{\rho}\right)_m = \left(\frac{\mu}{\rho}\right)_p \left(\frac{d_m}{d_p}\right)^{\frac{3}{2}}$$

$$\nu_m = \nu_p \left(\frac{d_m}{d_p}\right)^{\frac{3}{2}}$$

Since the model has usually a scale lower than 1:1, the kinematic viscosity of the fluid in the physical model should be lower than the prototype.

Pangarkar suggests the Fr number is unimportant when the vortex is suppressed. This can be accomplished with four baffles of width equal to 10% of the stirred tank reactor [20].

12.2.4 Thermal Similarity

Most physical modeling work is carried out under isothermal conditions. This is a convenient assumption but not necessarily realistic.

Heat transfer can be dissipated by radiation, conduction or convection. The rate at which this is happening in the prototype should be proportional with the rate at which is happening in the model,

$$\frac{\dot{Q}_{r,m}}{\dot{Q}_{r,fs}} = \frac{\dot{Q}_{k,m}}{\dot{Q}_{k,fs}} = \frac{\dot{Q}_{c,m}}{\dot{Q}_{c,fs}} = \text{constant}$$

Two systems are thermally similar if the temperature gradients in the model correspond to those in the prototype at corresponding positions and time.

The temperature gradients (ΔT) of a fluid depend on the following variables: fluid's velocity (U), fluid's density (ρ), fluid's viscosity (μ), fluid's heat capacity (C_p), fluid's thermal conductivity (k), dimension of characteristic length (L), gravity (g) and the thermal expansion coefficient (β).

Damle and Sahai [21] as well as Pan and Björkman [22] derived three dimensionless groups based on the analysis of two governing equations; momentum and energy conservation.

$$\left(\frac{\mu_{eff}}{\rho_{ref}UL}\right)_m = \left(\frac{\mu_{eff}}{\rho_{ref}UL}\right)_p$$

$$\left(\frac{k_{eff}}{\rho_{ref}C_pUL}\right)_m = \left(\frac{k_{eff}}{\rho_{ref}C_pUL}\right)_p$$

$$\left(\frac{gL}{U^2}\beta\Delta T\right)_m = \left(\frac{gL}{U^2}\beta\Delta T\right)_p$$

The first group is the turbulent Re^{-1} number, the second one is the turbulent Pe^{-1} number and the third one was defined by Damle and Sahai [21] as Tundish Richardson number (Tu).

In addition to those, one more dimensionless group was derived based on heat losses as boundary conditions:

$$\frac{q_m}{q_p} = \frac{C_{p,m}\rho_{ref,m}\beta_m}{C_{p,p}\rho_{ref,p}\beta_p}\sqrt{\lambda} = 0.4\sqrt{\lambda}$$

Table 12.2 summarizes the properties for water and steel.

In general, it is not possible to achieve completed similarity when several dimensionless numbers are involved. In this case there are 4 dimensionless groups. The first one is the Re^{-1} number and the second is the Pe^{-1} number. Three reasons have

Table 12.2 Thermophysical properties of water and liquid steel

Property	Symbol	Units	Steel, 1600 °C	Water, 60 °C
Density	ρ	kg/m^3	6962.8	983.2
Viscosity	μ	Pa·s	6×10^{-3}	8.9×10^{-4}
Thermal conductivity	k	W/m·K	27.9	0.637
Heat capacity	C_p	J/kg·K	787	4182
Coeff. thermal expansion	β	K^{-1}	1.5×10^{-4}	5×10^{-4}

Original source; Liquid steel from J. Chen. Handbook of diagrams and data for steelmaking, Metallurgical Industry Press, Beijing China (1984) 391, and for water from R.C. Weast. Handbook of chemistry and physics, 57th ed., CRC Press, USA (1977) F-5

been suggested to ignore their influence [21–23]. First, under isothermal conditions in the Tundish, the turbulent Re number remains almost constant and is expected to happen also under non-isothermal conditions. Then, as the Pe number is the product of RePr numbers, then the Pe number is assumed to remain constant. Second, the turbulent Re and turbulent Pe numbers involve the effective viscosity and the effective thermal conductivity, parameters difficult to assess and control experimentally, and third, both viscous forces and thermal conductivity have a small effect on the rate of momentum in the bulk of the fluid. On this basis, the groups involving Re and Pe numbers are sacrificed as main criterion to define thermal similarity. Therefore, the Tu number is chosen as the main criterion on thermal similarity. A more detailed analysis of this number indicates that Tu is a ratio of Grashof and Reynolds numbers,

$$\text{Tu} = \frac{gL}{U^2}\beta\Delta T = \frac{Gr}{Re^2}$$

Furthermore, it can also be defined in terms of the Fr number, as follows,

$$\text{Tu} = \frac{gL}{U^2}\beta\Delta T = \frac{\beta\Delta T}{Fr}$$

Since the numerator clearly defines the thermal gradients and the denominator, the Fr number clearly defines a ratio between inertial and gravitational forces, the equation is divided into two conditions to satisfy thermal similarity:

$$(\beta\Delta T)_m = (\beta\Delta T)_p$$

$$\left(\frac{gL}{U^2}\right)_m = \left(\frac{gL}{U^2}\right)_p$$

Mazumdar and Evans [23] indicate that thermal similarity is difficult to achieve with water models, due to the following reasons:

- Heat loss in an industrial reactor is different in different areas, making difficult to replicate.
- The Prandtl number of liquid steel is much lower than water about 60 times. The low Prandtl number in liquid steel leads to laminar flow in the vicinity of the wall but is not the case using a water model.

12.2.5 Chemical Similarity

A precondition to reach chemical similarity is to have achieved geometric, material, dynamic and thermal similarity. Chemical reactions involve mass transfer but also formation of new chemical species depending on reaction kinetics. Kinetics is largely influenced by temperature.

In chemically similar systems the corresponding concentration differences have a constant ratio to one another at corresponding points within the geometrical system of interest [2].

Mazumdar and Evans [23] state that chemical similarity is impossible to achieve in cold models due to the strong influence of temperature on reaction rates. It is important to clarify that sometimes we refer to chemical similarity but refers only to the evaluation of concentration gradients but not to the generation or consumption of chemical species due to chemical reactions.

References

1. D.C. Ipsen, *Units, Dimensions, and Dimensionless Numbers* (McGraw-Hill, USA, 1960), p. 134
2. R.I.L. Guthrie, Dimensional analysis and reactor design. Chapter 3 in *Engineering in Process Metallurgy* (Oxford University Press, Oxford, 1993), pp. 151–178
3. M. Zlokarnik, *Scale-up in Chemical Engineering*, 2nd edn. (Wiley, Weinham , 2006), pp. 43–47
4. Y.A. Çengel, J.M. Cimbala, Dimensional analysis and modeling. Chapter 7 in *Fluid Mechanics: Fundamentals and Applications* (2006), p. 301
5. G. Gupta, S. Sarkar, A. Chychko, L. Teng, M. Nzotta, A. Seetharaman, Process concepts for scaling-up and plant studies. Chapter 3.1 in *Treatise on Process Metallurgy*, vol. 3 (Elsevier, 2014), pp. 1100–1144
6. M. Zlokarnik, *Scale-up in Chemical Engineering*, vol. 53, 2nd edn. (Wiley, Weinham Germany, 2006), p.110
7. G.I. Barenblatt, *Scaling, Self-Similarity and Intermediate Asymptotics* (Cambridge University Press, New York USA, 2009), p. 56
8. G.I. Barenblatt, *Scaling, Self-Similarity and Intermediate Asymptotics* (Cambridge University Press, New York USA, 2009), pp. 26–27
9. Y. Sahai, R.I.L. Guthrie, Hydrodynamics of gas stirred melts: part I. Gas/liquid coupling. Metall. Trans. B 13, 193–202 (1982)
10. D. Mazumdar, R.I.L. Guthrie, Hydrodynamic modeling of some gas injection procedures in ladle metallurgy operations. Metall. Trans. B **16**, 83–90 (1985)
11. G. Ebneth, W. Plushkell, Dimensional analysis of the vertical heterogeneous buoyant plume . Steel Res. Int. **56**(10), 513–518 (1985)
12. S. Kim, R.J. Fruehan, Physical modeling of liquid/liquid mass transfer in gas stirred ladles. Metall. Trans. B **18**, 381–390 (1987)
13. D. Mazumdar, R.I.L. Guthrie, Y. Sahai, On mathematical models and numerical solutions of gas stirred ladle systems. Appl. Math. Model. **17**, 255–262 (1993)
14. M. Castello-Branco, K. Schwerdtfeger, Large-scale measurements of the physical characteristics of round vertical bubble plumes in liquids. Metall. Mater. Trans. B **25**, 359–371 (1994)
15. D. Mazumdar, On the estimation of plume rise velocity in gas-stirred ladles. Metall. Mater. Trans. B **33**, 937–941 (2002)
16. D. Mazumdar, H.B. Kim, R.I.L. Guthrie, Modelling criteria for flow simulation in gas stirred ladles: experimental study. Ironmak. Steelmak. **27**(4), 302–309 (2000)
17. J. Mietz, S. Schneider, F. Oeters, Model experiments on mass transfer in ladle metallurgy. Steel Res. **62**, 1–9 (1991)
18. Y. Liu, H. Bai, H. Liu, M. Ersson, P.G. Jönsson, Y. Gan, Physical and numerical modelling on the mixing condition in a 50 t ladle. Metals (Basel). **9**, 1–19 (2019)
19. Y. Liu, M. Sano, T. Zhang, Q. Wang, J. He, Intensification of bubble disintegration and dispersion by mechanical stirring in gas injection refining. ISIJ Int. **49**(1), 17–23 (2009)

20. V.G. Pangarkar, *Design of Multiphase Reactors* (Wiley & Sons, New Jersey USA, 2015), p. 147
21. C. Damle, Y. Sahai, A criterion for water modeling of non-isothermal melt flows in continuous casting Tundishes ISIJ Int. 36(6) 681 689 (1996)
22. Y. Pan, B. Bjorkman, Numerical analysis on the similarity between steel ladles and hot-water models regarding natural convection phenomena. ISIJ Int. **42**(1), 53–62 (2002)
23. D. Mazumdar, J. Evans. Elements of physical modeling. Chapter 3 in *Modelling of Steelmaking Processes* (CRC Press, Boca Raton FL USA, 2010), pp. 99–138

Chapter 13
Non-dimensionalization of Differential Equations

13.1 Introduction

Dimensional analysis expresses a natural phenomenon in terms of dimensionless groups, however in order to reach reliable results, its main limitation is the need of an accurate definition of the variables that control those phenomena. This is a task that requires a solid knowledge of the problem, experimental work to define those variables, etc.

Most of the natural phenomena have been studied and expressed in terms of differential equations with its initial and boundary conditions specific for every problem. In those differential equations, each term represents a property of the system, involving heat, mass and momentum transfer. Units on both sides of the equations, for each term, are the same.

Scaling, normalizing or non-dimensionalization of differential equations including their initial and boundary conditions is another way to define the dimensionless groups in a given problem. This method does not require the definition of the relevant variables because they are already defined by the governing equations, which can be, for example, the NavierStokes equations for momentum transfer, Fourier equation for heat transfer by conduction, Fick's law for mass transfer, etc. The next main step is how to scale those variables. The choice of scales requires knowledge of each problem, its selection aims at normalizing the variables between zero and unity, whenever this is possible. The main mathematical operation to normalize a differential equation involves the application of the chain rule.

Nondimensionalization of differential equations provides information on the dimensionless groups that describe a given problem. This information is useful, especially for analysis on similarity but a relationship between the dependent and independent variables is not the primary goal. In this sense, conventional dimensional analysis is more appropriate.

Analytical solution of differential equations is easy for simple equations. When the number of differential equations increases the solution involves numerical methods, which is possible for many problems but not only requires specialized knowledge to

A. N. Conejo, *Fundamentals of Dimensional Analysis*,
https://doi.org/10.1007/978-981-16-1602-0_13

handle complex algorithms but also in many cases the complexity is such that current computers cannot handle those problems. In any case, by handling the problem using dimensionless variables has several advantages, as follows:

– The number of independent variables is decreased
– The scale of dimensionless variables can be adjusted from zero to unity.

One limitation of scaling differential equations is that different dimensionless groups can be obtained which are not easily interpreted.

Langtangen and Pedersen [1] suggest the following steps to scale differential equations:

– Identify independent and dependent variables, parameters and their dimensions. The variables are clearly identified in the derivative terms.
– Introduce reference or characteristic values for the variables and define dependent and independent dimensionless variables.
– Insert the new dimensionless variables in the differential equation. Apply the chain rule.
– Make each term dimensionless.
– Estimate the characteristic values used to scale the variables. This is the most challenging part of the scaling procedure. The goal is to achieve nondimensional groups equal to one, as many as possible. If possible, make the constants that appear in initial or boundary conditions also equal to one. Also consider the possibility to eliminate dimensionless groups with a value close to zero.

The Chain Rule:

If a variable y depends on x, $y = f(x)$, and x in turn depends on t, $x = g(t)$, then y can also be defined in terms of t, $y = f(g(t)) = (f \circ g)(t)$. The derivative of this composite function is called the chain rule.

$$\frac{dy}{dt} = \frac{dy}{dx} \cdot \frac{dx}{dt}$$

This rule can be extended to more functions, for example:

$$\frac{du}{dt} = \frac{du}{dy} \cdot \frac{dy}{dx} \cdot \frac{dx}{dt}$$

Partial derivatives apply when a function depends on several variables, for example; $u = f(x, y)$. If in turn $x = g(r, t)$ and also $y = \phi(r, t)$, the chain rules can be expressed as follows:

$$\frac{\partial u}{\partial r} = \frac{\partial u}{\partial x} \cdot \frac{\partial x}{\partial r} + \frac{\partial u}{\partial y} \cdot \frac{\partial y}{\partial r}$$

$$\frac{\partial u}{\partial t} = \frac{\partial u}{\partial x} \cdot \frac{\partial x}{\partial t} + \frac{\partial u}{\partial y} \cdot \frac{\partial y}{\partial t}$$

Some rules for scaling:

In order to make one variable dimensionless, it is divided by a reference or characteristic maximum value, furthermore, it is desirable that the magnitude of that ratio changes from zero to unity.

– Length scales: Any length variable, for example a distance (x), a radius (r), and other length variables can be transformed in dimensionless form dividing by a reference length. A distance can be divided by a reference length, for example the maximum length, as follows.

$$\bar{x} = \frac{x}{L}$$

$$0 < \bar{x} < 1$$

$$\bar{r} = \frac{r}{R}$$

$$0 < \bar{r} < 1$$

– Time scales:

A reference time can be defined in terms of a ratio of characteristic length and characteristic velocity. This scale represents the time it takes a fluid particle to move a distance L_c at a speed U_c.

$$\bar{t} = \frac{t}{t_c} = \frac{t}{(L_c/U_c)} = \frac{U_c t}{L_c}$$

where: U_c and L_c are characteristic velocity and length, respectively.
In transport phenomena involving diffusivity (M^2/T), a choice for time scale is:

$$\bar{t} = \frac{L^2}{D}$$

where: L is a length scale.
In conduction heat transfer, a convenient dimensionless time is defined by the Fourier number itself:

$$\bar{t} = \frac{t}{L_c^2/\alpha} = \frac{\alpha t}{L_c^2} = Fo$$

– Temperature scales: In many problems involving heat transfer it is possible to define the temperature range, from initial or boundary conditions. For example, if the initial temperature is $T(x, 0) = T_0$ and the final temperature is $T(x, \infty) =$

T_1, any transient temperature will be in the range from T_0 to T_1. A convenient dimensionless temperature is as follows:

$$\theta = \frac{T - T_1}{T_0 - T_1}$$
$$\theta(x, 0) = 0$$
$$\theta(x, \infty) = 1$$

Alternatively:

$$\theta = \frac{T}{T_{ref}}$$

– Velocity scales:

A simple scale uses a characteristic or reference velocity (U_c):

$$\overline{U} = \frac{U}{U_c}$$

Problems involving gravity force can define as a reference velocity the product of g (L/T^2) with a characteristic length (L), as follows:

$$U_c = \sqrt{gL}$$

– Concentration scales: similar to heat transfer, in many problems involving mass transfer it is possible to define the concentration range, from initial or boundary conditions. For example, if the initial concentration is $C(x, 0) = C_0$ and the final concentration is $C(x, \infty) = C_1$, any transient concentration will be in the range from C_0 to C_1. A convenient dimensionless concentration is as follows:

$$\overline{C} = \frac{C - C_1}{C_0 - C_1}$$
$$\overline{C}(x, 0) = C_0$$
$$\overline{C}(x, \infty) = C_1$$

A simple scale is as follows:

$$\overline{C} = \frac{C}{C_{final}}$$

– Pressure scales: similar to heat and mass transfer, it is possible to define a dimensionless pressure, from initial or boundary conditions. For example, if the initial pressure is $P(x, 0) = P_0$ and the final pressure is $P(x, \infty) = P_1$, any transient concentration will be in the range from P_0 to P_1. A convenient dimensionless concentration is as follows

$$\overline{P} = \frac{P - P_\infty}{P_0 - P_\infty}$$

$$\overline{P}(x, 0) = 0$$

$$\overline{P}(x, \infty) = P_\infty$$

A second form is to replace pressure by its equivalent form: $\rho\, U^2$

$$\overline{P} = \frac{P - P_c}{\rho_c U_c^2}$$

A third form is to replace the reference pressure by the hydrostatic pressure

$$\overline{P} = \frac{P}{\rho_l g L}$$

13.2 Scaling Differential Equations

Scaling first order Ordinary Differential Equations (ODE's)

Newton's law of cooling: When a body at some temperature is placed in a cooling environment, its temperature falls rapidly in the beginning, and then the change in temperature levels off until the body's temperature equals that of the surroundings. Newton carried out some experiments on cooling hot iron and found that the temperature decayed exponentially. Later, this result was formulated as a differential equation: the rate of change of the temperature in a body is proportional to the temperature difference between the body and its surroundings. This statement is known as Newton's law of cooling. Assumes that the temperature inside the body is uniform. This approximation is valid if the rate of thermal energy transfer within the body is faster than the rate of thermal energy transfer at the surface. If heat is transferred to the surroundings by convection only, Newton's law of cooling can be expressed as follows:

$$\frac{dQ}{dt} = -hA(T - T_s)$$

where: Q is the amount of heat in Watts, h is the convective heat transfer coefficient in W/m^2 K, A is the cross-sectional area in m^2. The temperature gradient in this case is negative.

Replacing the equivalent value: $dQ = C_p dT$

$$\frac{dT}{dt} = -\frac{hA}{C_p}(T - T_s) = -k(T - T_s)$$

Which can also be expressed as:

$$T'(t) = -k(T - T_s)$$

The units of k are 1/s.
Initial condition:

$$T(0) = T_0$$

The independent variable is the temperature of the body (T) and the independent variable is time (t). T_s is the temperature of the surroundings, which may be time dependent. T_0 is the initial temperature.

The scales proposed to make T and t, dimensionless are the following:

$$\theta = \frac{T - T_s}{T_0 - T_s}$$

$$\bar{t} = \frac{t}{t_c}$$

where: θ and $\bar{t}$ are the dimensionless temperature and time, respectively. t_c is a constant reference time.

T in terms of θ is defined as follows:

$$T = (T_0 - T_s)\theta + T_s$$

Replacing these values in Newton's cooling law:

$$\frac{dT}{dt} = -k(T - T_s)$$

Applying the chain rule:

$$\frac{dT}{dt} = \frac{dT}{d\bar{t}} \cdot \frac{d\bar{t}}{dt} = \frac{d[\theta(T_0 - T_s) + T_s]}{d\bar{t}} \cdot \frac{d\left(\frac{t}{t_c}\right)}{dt} = (T_0 - T_s)\frac{d\theta}{d\bar{t}} \cdot \frac{1}{t_c}$$

$$\frac{dT}{dt} = \frac{(T_0 - T_s)}{t_c}\frac{d\theta}{d\bar{t}} = -k(T - T_s)$$

$$\frac{d\theta}{d\bar{t}} = -t_c k\frac{T - T_s}{T_0 - T_s} = -t_c k\theta$$

Newton's law of cooling in dimensionless form becomes:

$$\frac{d\theta}{d\bar{t}} = -t_c k\,\theta$$

The product $t_c k$ is a dimensionless group. A simple way to define the characteristic time is to make this product the unity, $t_c k = 1$, then the value of t_c should be equal to k. With this condition:

$$\frac{d\theta}{d\bar{t}} = -\theta$$

For the initial condition $t = 0$, $T = T_0$ and $\theta = 1$; for $t = \infty$, $T = T_s$, then $\theta = 0$

$$\theta(0) = 0$$

In many cases the chain rule can be applied in one simple step if the function involves the variable and constants, in this case for example:

$$\frac{dT}{dt} = \frac{d[\theta(T_0 - T_s) + T_s]}{d\left(\frac{t}{t_c}\right)} = \frac{(T_0 - T_s)}{t_c}\frac{d\theta}{d\bar{t}}$$

Scaling Navier–Stokes equations

1D, rectangular coordinates, one body force: The Navier–Stokes equations (NSE) are a fundamental set of equations describing a balance of momentum conservation. Due to its large importance in fluid flow in all metallurgical operations, a brief description of this equations is provided. The general form of the NSE is usually developed considering all forces affecting fluid flow.

Force is a rate of momentum (MU/t) and both pressure and stress are rate of momentum per unit area (flux of momentum):

$$F = m \cdot a = m\frac{dU}{dt} = \frac{m\Delta U}{\Delta t}$$

$$P = \tau = \frac{F}{A} = \frac{1}{A}\frac{m\Delta U}{\Delta t}$$

Momentum is classified into two types; convective momentum in the direction of the flow and viscous momentum in the normal direction to the flow. Flux of convective momentum can also be expressed as:

$$\rho\,UU$$

The general momentum balance includes; flux of convective momentum, viscous momentum due to normal shear stresses, viscous momentum due to tangential shear stresses, flux of momentum due to pressure forces and additional body forces that contribute to the motion of the fluid. In the general treatment usually gravitational forces are the ones included, however for specific problems others forces can be added, for example buoyancy forces due to thermal gradients.

The final result of the momentum balance, assuming constant physical properties of the fluid (density and viscosity), is the following general expression in three directions:

$$\rho \frac{DU}{Dt} = -\nabla P + \rho g + \mu \nabla^2 U$$

In one direction (x), reduces to the following expression:

$$\rho \frac{DU_x}{Dt} = -\frac{\partial P}{\partial x} + \rho g + \mu \frac{\partial^2 U_x}{\partial x^2}$$

The total derivative is:

$$\rho \frac{DU_x}{Dt} = \rho \frac{\partial U_x}{\partial t} + \rho \frac{\partial (U_x U_x)}{\partial x}$$

Therefore,

$$\rho \frac{\partial U_x}{\partial t} + \rho \frac{\partial (U_x U_x)}{\partial x} = -\frac{\partial P}{\partial x} + \rho g + \mu \frac{\partial^2 U_x}{\partial x^2}$$

Nondimensionalization: The variables are velocity (U), time(t), length (x) and Pressure (P).

Scales:

$$\overline{x} = \frac{x}{L_c}$$

$$\overline{U}_x = \frac{U_x}{U_c}$$

$$\overline{t} = \frac{L_c t}{U_c}$$

$$\overline{P} = \frac{P - P_c}{\rho_c U_c^2}$$

where: $\overline{x}$, $\overline{U}_x$, $\overline{t}$, $\overline{P}$ are dimensionless variables, L_c is a characteristic length, U_c is a characteristic velocity and ρ_c is a characteristic or reference density.

Replacing the previous scales in the differential equation:

$$\rho\frac{\partial\left(U_c\overline{U}_x\right)}{\partial\left(\frac{L_c\overline{t}}{U_c}\right)}+\rho\frac{\partial\left(U_c\overline{U}_x U_c\overline{U}_x\right)}{\partial(L_c\overline{x})}=-\frac{\partial\left(\rho_c U_c^2\overline{P}+P_c\right)}{\partial(L_c\overline{x})}+\rho g+\mu\left(\frac{\partial}{\partial(L_c\overline{x})}\frac{\partial\left(U_c\overline{U}_x\right)}{\partial(L_c\overline{x})}\right)$$

$$\frac{\rho\,U_c^2}{L_c}\frac{\partial\overline{U}_x}{\partial\overline{t}}+\frac{\rho\,U_c^2}{L_c}\frac{\partial\overline{U}_x\overline{U}_x}{\partial\overline{x}}=-\frac{\rho_c U_c^2}{L_c}\frac{\partial\overline{P}}{\partial\overline{x}}+\rho g+\mu\frac{U_c}{L_c^2}\frac{\partial^2\overline{U}_x}{\partial\overline{x}^2}$$

Both sides are multiplied by $\frac{L_c}{\rho_c U_c^2}$ and defining a dimensionless density as $\overline{\rho}=\frac{\rho}{\rho_c}$

$$\overline{\rho}\frac{\partial\overline{U}_x}{\partial\overline{t}}+\overline{\rho}\frac{\partial\overline{U}_x\overline{U}_x}{\partial\overline{x}}=-\frac{\partial\overline{P}}{\partial\overline{x}}+\frac{\overline{\rho}g L_c}{U_c^2}+\frac{\mu}{\rho_c U_c L_c}\frac{\partial^2\overline{U}_x}{\partial\overline{x}^2}$$

The terms. $\frac{\overline{\rho}g L_c}{U_c^2}$ and $\frac{\mu}{\rho_c U_c L_c}$ are the reciprocal of a modified Froude and Reynolds numbers, respectively.

$$\overline{\rho}\frac{\partial\overline{U}_x}{\partial\overline{t}}+\overline{\rho}\frac{\partial\overline{U}_x\overline{U}_x}{\partial\overline{x}}=-\frac{\partial\overline{P}}{\partial\overline{x}}+\frac{1}{Fr}+\frac{1}{Re}\frac{\partial^2\overline{U}_x}{\partial\overline{x}^2}$$

It can be observed that excluding gravitational forces in the original differential equation would have resulted in only one dimensionless group. The final result, including gravitational forces produced a result with two dimensionless groups. However, since this analysis was for only one direction and only one external body force, it should be expected more dimensionless group if more terms are added to the differential equation.

Applying a different pressure scale:

$$\overline{P}=\frac{P-P_\infty}{P_o-P_\infty}$$

$$P=(P_o-P_\infty)\overline{P}+P_\infty$$

where: P_o is initial pressure and P is the final or reference pressure

$$\rho\frac{\partial\left(U_c\overline{U}_x\right)}{\partial\left(\frac{L_c\overline{t}}{U_c}\right)}+\rho\frac{\partial\left(U_c\overline{U}_x U_c\overline{U}_x\right)}{\partial(L_c\overline{x})}=-\frac{\partial\left[(P_o-P_\infty)\overline{P}+P_\infty\right]}{\partial(L_c\overline{x})}+\rho g$$

$$+\mu\left(\frac{\partial}{\partial(L_c\overline{x})}\frac{\partial\left(U_c\overline{U}_x\right)}{\partial(L_c\overline{x})}\right)$$

$$\frac{\rho\,U_c^2}{L_c}\frac{\partial\overline{U}_x}{\partial\overline{t}}+\frac{\rho\,U_c^2}{L_c}\frac{\partial\overline{U}_x\overline{U}_x}{\partial\overline{x}}=-\frac{(P_o-P_\infty)}{L_c}\frac{\partial\overline{P}}{\partial\overline{x}}+\rho g+\mu\frac{U_c}{L_c^2}\frac{\partial^2\overline{U}_x}{\partial\overline{x}^2}$$

Both sides are multiplied by $\frac{L_c}{\rho U_c^2}$

$$\frac{\partial \overline{U}_x}{\partial \overline{t}} + \frac{\partial \overline{U}_x \overline{U}_x}{\partial \overline{x}} = -\frac{(P_0 - P_\infty)}{\rho\, U_c^2}\frac{\partial \overline{P}\partial \overline{x}}{} \frac{gL_c}{+\quad U_c^2} + \frac{\mu}{\rho\, U_c L_c}\frac{\partial^2 \overline{U}_x}{\partial \overline{x}^2}$$

The terms; $\frac{(P_0-P_\infty)}{\rho U_c^2}$, $\frac{gL_c}{U_c^2}$ and $\frac{\mu}{\rho U_c L_c}$ are the Euler number, inverse of Froude number and inverse of Reynolds number, respectively.

$$\frac{\partial \overline{U}_x}{\partial \overline{t}} + \frac{\partial \overline{U}_x \overline{U}_x}{\partial \overline{x}} = -\mathrm{Eu}\frac{\partial \overline{P}}{\partial \overline{x}} + \frac{1}{\mathrm{Fr}} + \frac{1}{\mathrm{Re}}\frac{\partial^2 \overline{U}_x}{\partial \overline{x}^2}$$

Now, assume the dimensionless time is defined as t/t_c, everything else as in the previous case:

$$\rho\frac{\partial\left(U_c\overline{U}_x\right)}{\partial\left(t_c\overline{t}\right)} + \rho\frac{\partial\left(U_c\overline{U}_x U_c\overline{U}_x\right)}{\partial\left(L_c\overline{x}\right)} = -\frac{\partial\left[(P_0 - P_\infty)\overline{P} + P_\infty\right]}{\partial\left(L_c\overline{x}\right)}$$

$$+ \rho g + \mu\left(\frac{\partial}{\partial(L_c\overline{x})}\frac{\partial\left(U_c\overline{U}_x\right)}{\partial(L_c\overline{x})}\right)$$

$$\frac{\rho U_c}{t_c}\frac{\partial\overline{U}_x}{\partial\overline{t}} + \frac{\rho U_c^2}{L_c}\frac{\partial\overline{U}_x\overline{U}_x}{\partial\overline{x}} = -\frac{(P_0 - P_\infty)}{L_c}\frac{\partial\overline{P}}{\partial\overline{x}} + \rho g + \mu\frac{U_c}{L_c^2}\frac{\partial^2\overline{U}_x}{\partial\overline{x}^2}$$

Both sides are multiplied by $\frac{L_c}{\rho U_c^2}$

$$\frac{L_c}{t_c U_c}\frac{\partial\overline{U}_x}{\partial\overline{t}} + \frac{\partial\overline{U}_x\overline{U}_x}{\partial\overline{x}} = -\frac{(P_0 - P_\infty)}{\rho\, U_c^2}\frac{\partial\overline{P}}{\partial\overline{x}} + \frac{gL_c}{U_c^2} + \frac{\mu}{\rho\, U_c L_c}\frac{\partial^2\overline{U}_x}{\partial\overline{x}^2}$$

The terms; $\frac{L_c}{t_c U_c}$, $\frac{(P_0-P_\infty)}{\rho U_c^2}$, $\frac{gL_c}{U_c^2}$ and $\frac{\mu}{\rho U_c L_c}$ are the Strouhal number (St or Sr), the Euler number (Eu), the inverse of Froude number (Fr) and the inverse of Reynolds number (Re), respectively. The Strouhal number can be associated with the oscillations of the flow, a high value would indicate that the number of oscillations dominate the flow and for low values, the oscillations are swept by a fast-moving fluid. For steady-state conditions, the St number vanishes.

$$\mathrm{St}\frac{L_c}{t_c U_c}\frac{\partial\overline{U}_x}{\partial\overline{t}} + \frac{\partial\overline{U}_x\overline{U}_x}{\partial\overline{x}} = -\mathrm{Eu}\frac{\partial\overline{P}}{\partial\overline{x}} + \frac{1}{\mathrm{Fr}} + \frac{1}{\mathrm{Re}}\frac{\partial^2\overline{U}_x}{\partial\overline{x}^2}$$

It has been shown that the original definition of scales, for the same differential equation also defines the maximum number of dimensionless groups. In this example, the last set of scales allowed to represent the Navier–Stokes equation with four dimensionless groups: St, Eu, Fr and Re. On the other hand, using the first set of scales resulted in a normalized equation with only two groups; Fr and Re. It is important to assign a set of scales that yields a nondimensionalized expression which contains the maximum number of dimensionless groups. It is a way to have a better description of all the force ratios in the problem investigated.

Increasing the number of dimensions does not increase the number of dimensionless groups. The following example increases from 1 to 3 dimensions, with the first set of scales.

3D, rectangular coordinates, one body force, direction x: The Navier–Stokes equation for 3D flow in the direction x is expressed as follows:

$$\rho \frac{\partial U_x}{\partial t} + \rho \left[\frac{\partial (U_x U_x)}{\partial x} + \frac{\partial (U_x U_x)}{\partial y} + \frac{\partial (U_x U_x)}{\partial z} \right] = -\frac{\partial P}{\partial x} + \rho g$$

$$+ \mu \left[\frac{\partial^2 U_x}{\partial x^2} + \frac{\partial^2 U_x}{\partial y^2} + \frac{\partial^2 U_x}{\partial z^2} \right]$$

In addition to the previous scales, two additional scales for y and z are required:

$$\overline{y} = \frac{y}{L_c}$$

$$\overline{z} = \frac{z}{L_c}$$

Nondimensionalization: the scales are replaced in the differential equation.

$$\rho \frac{\partial (U_c \overline{U}_x)}{\partial \left(\frac{L_c \overline{t}}{U_c} \right)} + \rho \left[\frac{\partial (U_c \overline{U}_x U_c \overline{U}_x)}{\partial (L_c \overline{x})} + \frac{\partial (U_c \overline{U}_x U_c \overline{U}_x)}{\partial (L_c \overline{y})} + \frac{\partial (U_c \overline{U}_x U_c \overline{U}_x)}{\partial (L_c \overline{z})} \right] =$$

$$- \frac{\partial (\rho_c U_c^2 \overline{P} + P_c)}{\partial (L_c \overline{x})} + \rho g$$

$$+ \mu \left\{ \left(\frac{\partial}{\partial (L_c \overline{x})} \frac{\partial (U_c \overline{U}_x)}{\partial (L_c \overline{x})} \right) + \left(\frac{\partial}{\partial (L_c \overline{y})} \frac{\partial (U_c \overline{U}_x)}{\partial (L_c \overline{y})} \right) + \left(\frac{\partial}{\partial (L_c \overline{z})} \frac{\partial (U_c \overline{U}_x)}{\partial (L_c \overline{z})} \right) \right\}$$

Applying the derivative and collecting common terms:

$$\frac{\rho U_c^2}{L_c} \frac{\partial \overline{U}_x}{\partial \overline{t}} + \frac{\rho U_c^2}{L_c} \left(\frac{\partial \overline{U}_x \overline{U}_x}{\partial \overline{x}} + \frac{\partial \overline{U}_x \overline{U}_x}{\partial \overline{y}} + \frac{\partial \overline{U}_x \overline{U}_x}{\partial \overline{z}} \right) =$$

$$- \frac{\rho_c U_c^2}{L_c} \frac{\partial \overline{P}}{\partial \overline{x}} + \rho g + \mu \frac{U_c}{L_c^2} \left(\frac{\partial^2 \overline{U}_x}{\partial \overline{x}^2} + \frac{\partial^2 \overline{U}_x}{\partial \overline{y}^2} + \frac{\partial^2 \overline{U}_x}{\partial \overline{z}^2} \right)$$

Both sides are multiplied by $\frac{L_c}{\rho_c U_c^2}$ and defining a dimensionless density as $\overline{\rho} = \frac{\rho}{\rho_c}$

$$\overline{\rho} \frac{\partial \overline{U}_x}{\partial \overline{t}} + \overline{\rho} \left(\frac{\partial \overline{U}_x \overline{U}_x}{\partial \overline{x}} + \frac{\partial \overline{U}_x \overline{U}_x}{\partial \overline{y}} + \frac{\partial \overline{U}_x \overline{U}_x}{\partial \overline{z}} \right) =$$

$$-\frac{\partial \overline{P}}{\partial \overline{x}} + \frac{\overline{\rho}\, g L_c}{U_c^2} + \frac{\mu}{\rho_c U_c L_c}\left(\frac{\partial^2 \overline{U}_x}{\partial \overline{x}^2} + \frac{\partial^2 \overline{U}_x}{\partial \overline{y}^2} + \frac{\partial^2 \overline{U}_x}{\partial \overline{z}^2}\right)$$

The terms. $\frac{\overline{\rho}\, g L_c}{U_c^2}$ and $\frac{\mu}{\rho_c U_c L_c}$ are the reciprocal of a modified Froude and Reynolds numbers, respectively.

$$\overline{\rho}\frac{\partial \overline{U}_x}{\partial \overline{t}} + \overline{\rho}\left(\frac{\partial \overline{U}_x \overline{U}_x}{\partial \overline{x}} + \frac{\partial \overline{U}_x \overline{U}_x}{\partial \overline{y}} + \frac{\partial \overline{U}_x \overline{U}_x}{\partial \overline{z}}\right) = -\frac{\partial \overline{P}\partial \overline{x}}{+} \frac{1}{Fr} + \frac{1}{Re}\left(\frac{\partial^2 \overline{U}_x}{\partial \overline{x}^2} + \frac{\partial^2 \overline{U}_x}{\partial \overline{y}^2} + \frac{\partial^2 \overline{U}_x}{\partial \overline{z}^2}\right)$$

Alternate nondimensionalization method: [2]

Since all terms in any equation should have the same dimensions and units, division by one of those terms would make the equation dimensionless. For example, the x-component of the Navier–Stokes equation in 3D.

$$\rho\frac{\partial U_x}{\partial t} + \rho\left[\frac{\partial(U_x U_x)}{\partial x} + \frac{\partial(U_x U_x)}{\partial y} + \frac{\partial(U_x U_x)}{\partial z}\right] =$$
$$-\frac{\partial P}{\partial x} + \rho g + \mu\left[\frac{\partial^2 U_x}{\partial x^2} + \frac{\partial^2 U_x}{\partial y^2} + \frac{\partial^2 U_x}{\partial z^2}\right]$$

Every term represents a force acting on a volume element.

In the alternate method, in the first step the differential equation is re-written without operators. In a second step, all terms are arbitrarily divided by one of the terms. In this step, it is possible to explore the terms used as a divisor because depending on each divisor it will result in different dimensionless groups. The new expression is automatically dimensionless. In addition to this, it is important to check the presence of groups dimensionally equivalent because they should share the same variables.

In the third step, the resulting terms are algebraically manipulated to obtain, if possible, known dimensionless groups.

First step: The previous Navier–Stokes equation is expressed without operators. Since the interest is in the dimensions, is irrelevant to differentiate which direction:

$$\frac{\rho U}{t} + \frac{\rho U^2}{L} = -\frac{P}{L} + \rho g + \frac{\mu U}{L^2}$$

Second step: There are five terms. The first two terms are dimensionally equivalent $\left(\rho\frac{U^2}{L} = \rho\frac{U}{t}\right)$. Any term can be chosen.

Choosing the term ρg as the divisor:

$$\frac{U}{gt} + \frac{U^2}{gL} = -\frac{P}{\rho g L} + 1 + \frac{\mu U}{\rho g L^2}$$

Choosing the term $\frac{\mu U}{L^2}$ as the divisor:

$$\left(\frac{L^2}{\mu U}\right)\left(\frac{\rho U}{t}\right)+\left(\frac{L^2}{\mu U}\right)\left(\frac{\rho U^2}{L}\right)=-\left(\frac{L^2}{\mu U}\right)\left(\frac{P}{L}\right)+\left(\frac{L^2}{\mu U}\right)(\rho g)+1$$

Third step: Observing the variables involved, is possible, in principle to form Reynolds $\left(\frac{\rho U L}{\mu}\right)$ and Euler numbers.

$$\left(\frac{\rho U L}{\mu}\right)\left(\frac{L}{Ut}\right)+\left(\frac{\rho U L}{\mu}\right)=-\left(\frac{PL}{\mu U}\right)\left(\frac{\rho U}{\rho U}\right)+\left(\frac{L^2}{\mu U}\right)(\rho g)\left(\frac{U}{U}\right)+1$$

$$\left(\frac{\rho U L}{\mu}\right)\left(\frac{L}{Ut}\right)+\left(\frac{\rho U L}{\mu}\right)=-\left(\frac{P}{\rho U^2}\right)\left(\frac{\rho U L}{\mu}\right)+\left(\frac{gL}{U^2}\right)\left(\frac{\rho U L}{\mu}\right)+1$$

In total, 5 dimensionless groups can be defined:

$$(\text{Re})(\text{St})+(\text{Re})=-(\text{Eu})(\text{Re})+\left(\frac{1}{\text{Fr}}\right)(\text{Re})+1$$

In this case, since it is known the Navier Stokes equations can be described with 4 dimensionless groups, then it can be concluded that using $\frac{\mu U}{L^2}$ as divisor provides the best results.

From the nondimensionalization process, the final result is the definition of the dimensionless groups that describe this differential equation.

$$f(\text{Re, Fr, Eu, St}) = 0$$

The method is simple but also requires to explore all options which would provide a large number of dimensionless groups to choose from. Kunes [3] calls this method, the integral analogue method.

Motion due to bubble stirring, 1D, rectangular coordinates, two body forces, direction x: The motion is due to bubbles rising in the fluid, therefore, bubbles properties, such as bubble size, density and surface tension are important variables. With bubbles in the system, the Navier–Stokes equation includes a force due to surface tension per unit volume at the gas/liquid interface.

$$\rho\frac{\partial U_x}{\partial t} + \rho\frac{\partial(U_x U_x)}{\partial x} = -\frac{\partial P}{\partial x}+\mu\frac{\partial^2 U_x}{\partial x^2}+\Delta\rho g + f_\sigma$$

The surface tension force per unit volume is the product of surface tension times the tangent of the curve (σ t). Since dt = kn ds,

$$f_\sigma = \sigma t = \sigma knds = \sigma kn\delta_s$$

where: k is the curvature, n is the unit normal, ds is the curvilinear coordinate and δ_s is the surface Dirac δ-function, which is non-zero only on the interface.

$$\rho\frac{\partial U_x}{\partial t} + \rho\frac{\partial(U_x U_x)}{\partial x} = -\frac{\partial P}{\partial x} + \mu\frac{\partial^2 U_x}{\partial x^2} + \Delta\rho g + \sigma\, kn\,\delta_s$$

Scales: Based on the scales proposed by Hua and Lou [4]

$$\bar{x} = \frac{x}{D}$$

$$\bar{U} = \frac{U}{\sqrt{gD}}$$

$$\bar{t} = \frac{t\sqrt{g}}{\sqrt{D}}$$

$$\bar{p} = \frac{p}{\rho_1 gD}$$

Replacing the scales:

$$\rho\frac{\partial\left(\sqrt{gD}\bar{U}_x\right)}{\partial\left(\frac{\sqrt{D}\bar{t}}{\sqrt{g}}\right)} + \rho\frac{\partial\left(gD\bar{U}_x\bar{U}_x\right)}{\partial(D\bar{x})} = -\frac{\partial(\rho_1 gD\bar{p})}{\partial(D\bar{x})} + \mu\left(\frac{\partial}{\partial(D\bar{x})}\frac{\partial\left(\sqrt{gD}\bar{U}_x\right)}{\partial(D\bar{x})}\right)$$

$$+ \Delta\rho g + \sigma\, kn\,\delta_s$$

$$\rho g\frac{\partial\bar{U}_x}{\partial\bar{t}} + \rho g\frac{\partial\left(\bar{U}_x\bar{U}_x\right)}{\partial(\bar{x})} = -\rho_1 g\frac{\partial\bar{p}}{\partial\bar{x}} + \frac{\mu\, g^{\frac{1}{2}}}{D^{\frac{3}{2}}}\left(\frac{\partial}{\partial\bar{x}}\frac{\partial\bar{U}_x}{\partial\bar{x}}\right) + (\rho - \rho_1)g + \sigma kn\delta_s$$

dividing by $\rho_1 g$:

$$\bar{\rho}\frac{\partial\bar{U}_x}{\partial t} + \bar{\rho}\frac{\partial\left(\bar{U}_x\bar{U}_x\right)}{\partial(\bar{x})} = -\frac{\partial\bar{p}}{\partial\bar{x}} + \frac{\mu}{\rho_1 g^{\frac{1}{2}}D^{\frac{3}{2}}}\left(\frac{\partial}{\partial\bar{x}}\frac{\partial\bar{U}_x}{\partial\bar{x}}\right) + (\bar{\rho} - 1) + \frac{\sigma}{\rho_1 g}kn\delta_s$$

The dimensionless groups $\frac{\mu}{\rho_1 g^{\frac{1}{2}}D^{\frac{3}{2}}}$ and $\frac{\sigma}{\rho_1 g}$ are equivalent to the inverse of the Re and Bo numbers.

Heat equations

Heat transfer, based on heat transfer rate per unit area is related to the variables $\rho C_p T$, as follows:

$$q_x = -k\frac{\partial T}{\partial x} = -\frac{k}{\rho C_p}\frac{\partial(\rho C_p T)}{\partial x} = -\alpha\frac{\partial(\rho C_p T)}{\partial x}$$

where: q_x is the heat transfer rate per unit area with dimensions; MT^3, α is the thermal diffusivity, with dimensions; L^2T^{-1}, C_p is the specific heat capacity with dimensions; $L^2T^{-2}\theta^{-1}$, k is the thermal conductivity with dimensions; $MLT^{-3}\theta^{-1}$.

Conduction heat transfer: Heat transfer by conduction is expressed by Fourier's law, which in 3 dimensions is defined as follows:

$$\frac{\partial T}{\partial t} = \frac{k}{\rho\,C_p}\nabla^2 T + \frac{\dot{q}}{\rho C_p}$$

For one dimension and without generation of heat, reduces to:

$$\frac{\partial T}{\partial t} = \alpha\frac{\partial^2 T}{\partial x^2}$$

Scales:

$$\theta = \frac{T - T_1}{T_0 - T_1}$$

$$\bar{t} = \frac{t}{L_c^2/\alpha} = \frac{\alpha t}{L_c^2} = Fo$$

$$x = L_c\bar{x}$$

Replacing scales:

$$\frac{\partial\left[(T_0 - T_s)\theta + T_0\right]}{\partial\left(\frac{L_c^2}{\alpha}Fo\right)} = \alpha\frac{\partial^2\left[(T_0-T_s)\theta + T_0\right]}{\partial(L_c\bar{x})^2}$$

Dimensionless heat conduction equation in one dimension:

$$\frac{\partial\theta}{\partial Fo} = \frac{\partial^2\theta}{\partial\bar{x}^2}$$

Heat convection equation without generation of heat

The rate of heat flow by convection per unit volume (MT^{-3}), assuming constant thermal conductivity and internal heat generation per unit volume $(\dot{q})$, is expressed by the following general differential equation:

$$\frac{D(\rho C_p T)}{Dt} = k\nabla^2 + \dot{q}$$

$$\frac{\partial(\rho C_p T)}{\partial t} + U_x\frac{\partial(\rho C_p T)}{\partial x} + U_y\frac{\partial(\rho C_p T)}{\partial y} + U_z\frac{\partial(\rho C_p T)}{\partial z} =$$

$$k\left(\frac{\partial^2 T}{\partial x^2} + \frac{\partial^2 T}{\partial y^2} + \frac{\partial^2 T}{\partial z^2}\right) + \dot{q}$$

Assuming density and heat capacity remain constant:

$$\frac{\partial\left(\rho C_p T\right)}{\partial t} + U_x \frac{\partial\left(\rho C_p T\right)}{\partial x} + U_y \frac{\partial\left(\rho C_p T\right)}{\partial y} + U_z \frac{\partial\left(\rho C_p T\right)}{\partial z} =$$
$$k\left(\frac{\partial^2 T}{\partial x^2} + \frac{\partial^2 T}{\partial y^2} + \frac{\partial^2 T}{\partial z^2}\right) + \dot{q}$$

Variables: Temperature, time, velocity and distance.
Scales:

$$T = (T_0 - T_s)\theta + T_s$$

$$t = \frac{L_c \bar{t}}{U_c}$$

$$U_j = U_c \bar{U}_j$$

$$x = L_c \bar{x}$$

$$y = L_c \bar{y}$$

$$z = L_c \bar{z}$$

The length scales assume a cube.
Replacing the scales in the differential equation.

$$\frac{\partial[(T_0 - T_s)\theta + T_0]}{\partial\left(\frac{L_c \bar{t}}{U_c}\right)} + U_c \bar{U}_j \sum_{j=x}^{z} \frac{\partial[(T_0 - T_s)\theta + T_0]}{\partial\left(L_c \bar{j}\right)} =$$
$$\alpha\left(\sum_{j=x}^{z} \frac{\partial^2[(T_0 - T_s)\theta + T_0]}{\partial\left(L_c \bar{j}\right)^2}\right) + \frac{\dot{q}}{\rho C_p}$$

$$(T_0 - T_s)\frac{U_c}{L_c}\frac{\partial\theta}{\partial\bar{t}} + (T_0 - T_s)\frac{U_c \bar{U}_j}{L_c}\sum_{j=x}^{z} \frac{\partial\theta}{\partial\bar{j}} = (T_0 - T_s)\frac{\alpha}{L_c^2}\left(\sum_{j=x}^{z} \frac{\partial^2\theta}{\partial\bar{j}^2}\right) + \frac{\dot{q}}{\rho C_p}$$

$$\frac{\partial\theta}{\partial\bar{t}} + \bar{U}_x\frac{\partial\theta}{\partial\bar{x}} + \bar{U}_y\frac{\partial\theta}{\partial\bar{y}} + \bar{U}_z\frac{\partial\theta}{\partial\bar{z}} = \frac{\alpha}{U_c L_c}\left(\frac{\partial^2\theta}{\partial\bar{x}^2} + \frac{\partial^2\theta}{\partial\bar{y}^2} + \frac{\partial^2\theta}{\partial\bar{z}^2}\right) + \frac{\dot{q}L_c}{\rho C_p U_c \Delta T}$$

The term $\frac{\alpha}{U_c L_c}$, is the reciprocal of the Peclet number. The final term, is also dimensionless, equivalent to a Peclet number per unit volume.

Assuming the heat generation is zero, the heat equation is defined only in terms of one dimensionless number.

$$\frac{\partial \theta}{\partial \bar{t}} + \bar{U}_x \frac{\partial \theta}{\partial \bar{x}} + \bar{U}_y \frac{\partial \theta}{\partial \bar{y}} + \bar{U}_z \frac{\partial \theta}{\partial \bar{z}} = Pe^{-1} \left(\frac{\partial^2 \theta}{\partial \bar{x}^2} + \frac{\partial^2 \theta}{\partial \bar{y}^2} + \frac{\partial^2 \theta}{\partial \bar{z}^2} \right)$$

Equation of continuity

The equation of continuity describes the law of conservation of matter.

$$\text{(rate of mass in)} - \text{(rate of mass out)} = \text{(rate of mass accumulation)}$$

$$\text{(rate of mass in/out direction j)} = \frac{\partial \rho U}{\partial j}$$

$$\text{(rate of mass accumulation)} = \frac{\partial \rho}{\partial t} V$$

$$-\left(\frac{\partial \rho U_x}{\partial x} + \frac{\partial \rho U_y}{\partial y} + \frac{\partial \rho U_z}{\partial z} \right) = \frac{\partial \rho}{\partial t}$$

If the density remains constant:

$$\left(\frac{\partial U_x}{\partial x} + \frac{\partial U_y}{\partial y} + \frac{\partial U_z}{\partial z} \right) = 0$$

$$\nabla \cdot U = 0$$

In dimensionless form is:

$$\nabla \cdot \bar{U} = 0$$

General equation of diffusion with convection

Transport of chemical species in a fluid with constant density and diffusion coefficient, is expressed as follows [5]:

$$\frac{\partial X_A}{\partial t} + U \nabla \cdot X_A = D_A \nabla^2 \cdot X_A$$

where: X_A is the mass fraction of component A, U is the diffusion velocity plus the flow due to external forces such as pressure and gravity gradients, D_A is the diffusion coefficient. The previous equations exclude a term due to chemical reactions.

Mass diffusivity has dimensions L^2T^{-1}, equivalent to momentum diffusivity of thermal diffusivity, is relate to the mass transfer rate per unit area as follows

$$\frac{\partial N_A}{\partial t} = D_A \frac{\partial C_A}{\partial x}$$

Dividing the previous equation by the molecular mass, X_A is converted into mole fraction:

$$\frac{\partial C_A}{\partial t} + U\nabla \cdot C_A = D_A \nabla^2 \cdot C_A$$

The diffusion equation with convection in rectangular coordinates in direction x:

$$\frac{\partial C_A}{\partial t} + U_x \frac{\partial C_A}{\partial x} = D_A \frac{\partial^2 C_A}{\partial x^2}$$

Nondimensionalization of diffusion equation:
Variables: C, t, U_x, x
Scales:

$$\overline{C} = \frac{C - C_1}{C_0 - C_1}$$

$$\overline{t} = \frac{U_c t}{L_c}$$

$$U_x = \; = U_c \overline{U}_x$$

$$x = L_c \overline{x}$$

Replacing scales:

$$\frac{\partial\left[(C_{A0} - C_{A1})\overline{C}_A + C_{A1}\right]}{\partial\left(\frac{L_c \overline{t}}{U_c}\right)} + U_c \overline{U}_x \frac{\partial\left[(C_{A0} - C_{A1})\overline{C}_A + C_{A1}\right]}{\partial(L_c \overline{x})} =$$

$$D_A \frac{\partial^2\left[(C_{A0} - C_{A1})\overline{C}_A + C_{A1}\right]}{\partial(L_c \overline{x})^2}$$

$$\frac{(C_{A0} - C_{A1})U_c}{L_c}\frac{\partial \overline{C}_A}{\partial \overline{t}} + \overline{U}_x \frac{(C_{A0} - C_{A1})U_c}{L_c}\frac{\partial \overline{C}_A}{\partial \overline{x}} = D_A \frac{(C_{A0} - C_{A1})}{L_c^2}\frac{\partial^2 \overline{C}_A}{\partial \overline{x}^2}$$

$$\frac{\partial \overline{C}_A}{\partial \overline{t}} + \overline{U}_x \frac{\partial \overline{C}_A}{\partial \overline{x}} = \frac{D_A}{U_c L_c}\frac{\partial^2 \overline{C}_A}{\partial \overline{x}^2}$$

The term $\frac{D_A}{U_c L_c}$ can be transformed into a product of two dimensionless groups, multiplying and dividing by the kinematic viscosity

$$\frac{D_A}{U_c L_c} = \frac{D_A}{U_c L_c} \frac{\nu}{\nu} = \frac{\nu}{U_c L_c} \frac{D_A}{\nu} = \frac{1}{ReSc}$$

$$\frac{\partial \overline{C}_A}{\partial \overline{t}} + \overline{U}_x \frac{\partial \overline{C}_A}{\partial \overline{x}} = \frac{1}{ReSc} \frac{\partial^2 \overline{C}_A}{\partial \overline{x}^2}$$

The diffusion equation with convection in cylindrical coordinates is:

$$\frac{\partial C_A}{\partial t} + \left(U_r \frac{\partial C_A}{\partial r} + U_\theta \frac{1}{r} \frac{\partial C_A}{\partial \theta} + U_z \frac{\partial C_A}{\partial z} \right)$$
$$= D_A \left[\frac{1}{r} \frac{\partial}{\partial r} \left(r \frac{\partial C_A}{\partial r} \right) + \frac{1}{r^2} \frac{\partial^2 C_A}{\partial \theta^2} + U_z \frac{\partial^2 C_A}{\partial z^2} \right]$$

The same dimensionless groups are formed.
The product ReSc is the Sherwood number:

$$Sh = f(ReSc)$$

Similarity of chemical species in a fluid will be achieved if the following condition is satisfied:

$$\left(\frac{1}{ReSc} \right)_m = \left(\frac{1}{ReSc} \right)_p .$$

Defining values of characteristic quantities

Scaling is not simple, especially if the objective is to scale the range of the dimensionless variables from zero to unity. The key to reach this objective is the value of the characteristic quantities. A common way to define their value is to make the dimensionless groups comprising the characteristic quantities equal to unity, except where a group comprises only physical constants because it would violate the properties of a fluid [6].

In principle, making a variable dimensionless seems trivial, for example time is divided by a reference time to define a dimensionless time. A differential equation can me be transformed into dimensionless form but there is no guarantee that the physical phenomena can be accurately reproduced, on the contrary, if the scales are not adequate, the solution of the dimensionless differential equation can be wrong. Tan [7] provided an example, the differential equation describing a projectile thrown-up, about choosing different scales for the same differential equation. The trajectory of a projectile thrown-up is affected by gravity force, if it thrown-up with an initial velocity U_0 and neglecting air resistance the trajectory is defined as follows:

$$\frac{d^2x}{dt^2} = -g \cdot \frac{R^2}{(R+x)^2}$$

$$t = 0 : x = 0, \quad \frac{dx}{dt} = U_0$$

where: x is the distance in the vertical direction, positive in the upward direction, R is the radius of the earth, t is the travelling time, g is the gravitational acceleration constant.

The scales shown in the table below will be compared:

Case I:

$$\frac{U_0^2 R}{R} \frac{d^2\bar{x}}{dt^2} = -g \cdot \frac{R^2}{(R+\bar{x}R)^2} = -g \cdot \frac{R^2}{[R(1+\bar{x})]^2}$$

Case II:

$$\frac{gR}{R} \frac{d^2\bar{x}}{d\bar{t}^2} = -g \cdot \frac{R^2}{(R+\bar{x}R)^2} = -g \cdot \frac{R^2}{[R(1+\bar{x})]^2}$$

Case III:

$$\frac{U_0^2 g^2}{g U_0^2} \frac{d^2\bar{x}}{d\bar{t}^2} = -g \cdot \frac{R^2}{\left(R\bar{x}\frac{U_0^2}{g}\right)^2} = -g \cdot \frac{R^2}{\frac{R^2}{R^2}\left(R+\bar{x}\frac{U_0^2}{g}\right)^2} = -g \cdot \frac{1}{\left(1+\frac{U_0^2}{gR}\bar{x}\right)^2}$$

Dimensionless length	Dimensionless time	Dimensionless differential Eq.	Dimensionless initial cond.
$\bar{x} = \frac{x}{R}$	$\bar{t} = \frac{t}{R/U_0}$	$\frac{U_0^2}{gR} \frac{d^2\bar{x}}{d\bar{t}^2} = -\frac{1}{(1+\bar{x})^2}$	$\bar{t} = 0 : \bar{x} = 0, \frac{d\bar{x}}{d\bar{t}} = \frac{U_0}{U_0} = 1$
$\bar{x} = \frac{x}{R}$	$\bar{t} = \frac{t}{\sqrt{R/g}}$	$\frac{d^2\bar{x}}{d\bar{t}^2} = -\frac{1}{(1+\bar{x})^2}$	$\bar{t} = 0 : \bar{x} = 0, \frac{d\bar{x}}{d\bar{t}} = \frac{U_0}{\sqrt{R/g}}$
$\bar{x} = \frac{x}{(U_0^2/g)}$	$\bar{t} = \frac{t}{(U_0/g)}$	$\frac{d^2\bar{x}}{d\bar{t}^2} = -\frac{1}{\left(1+\frac{U_0^2}{gR}\bar{x}\right)^2}$	$\bar{t} = 0 : \bar{x} = 0, \frac{d\bar{x}}{d\bar{t}} = \frac{U_0^2}{U_0^2} = 1$

Tan [7] solved the dimensionless differential equation for the three previous cases and reported the following: Case I: no solution exists for $\bar{x}$; Case II: negative value for $\bar{x}$; Case III: provides a realistic representation of the trajectory of a projectile thrown-up. This example clearly illustrates the need to study the scales of reference. The problem deals with a projectile thrown-up and due to the action of gravity the object has to reach a maximum height or maximum distance, at a point where the final velocity is zero. The maximum height is obtained from the basic concepts of uniform linear motion:

$$-g = U\frac{dU}{dx} \quad \therefore UdU = -gdx$$

upon integration from $x = 0$ to $x = x_{max}$ and $U = U_0$ to $U = U_{final}$

$$U_{final}^2 = U_0^2 - 2g(x_{max} - x_0)$$

This gives:

$$x_{max} = \frac{U_0^2}{2g} \cong \frac{U_0^2}{g}$$

The time to reach the maximum height is based on the relationship:

$$a = \frac{dU}{dt} \quad \therefore dU = -gdt$$

upon integration from $t = 0$ to $t = t$ and $U = U_0$ to $U = U_{final}$.

$$U_{final} - U_0 = -gt.$$

Therefore,

$$t = \frac{U_0}{g}$$

In this problem it should be noted that the dimensionless group obtained is the Froude number. Tan [7] also provided the meaning of this number in the context of this problem; represents a ratio of two lengths,

$$Fr = \frac{U_0^2/g}{R}$$

The ratio also represents the maximum height with respect to the radius of the earth.

In addition to the previous scales, DeVille [8] also analyzed the simplest scales:

$$\bar{x} = \frac{x}{x_c}$$

$$\bar{t} = \frac{t}{t_c}$$

Replacing the scales, the differential equation becomes:

$$\frac{x_c}{t_c^2}\frac{d^2\bar{x}}{d\bar{t}^2} = -g \cdot \frac{R^2}{(R + x_c\bar{x})^2} = -g \cdot \frac{1}{\left(1+\frac{x_c}{R}\bar{x}\right)^2}$$

$$\frac{x_c}{gt_c^2}\frac{d^2\bar{x}}{d\bar{t}^2} = -\frac{1}{\left(1 + \frac{x_c}{R}\bar{x}\right)^2}$$

$$\bar{t} = 0 : \bar{x} = 0, \quad \frac{d\bar{x}}{d\bar{t}} = \frac{t_c U_0}{x_c}$$

The nondimensionalization process results in three dimensionless groups:

$$\pi_1 = \frac{x_c}{gt_c^2}$$

$$\pi_2 = \frac{x_c}{R}$$

$$\pi_3 = \frac{t_c U_0}{x_c}$$

Scaling included two characteristic values (t_c and x_c). Two out of the available three dimensionless groups can be chosen to define their values. Considering the rules for scaling provided in the introduction, since R is the ratio of the earth's surface, the ratio x_c/R should be very small and can be neglected to define their characteristic values, then the other two remaining groups are set to one. In this way the value of the characteristic quantities is defined:

$$\frac{x_c}{gt_c^2} = 1 \text{ and } \frac{t_c U_0}{x_c} = 1$$

$$\therefore t_c = \frac{U_0}{g} \text{ and } x_c = \frac{U_0^2}{g}$$

These results would define the second dimensionless groups as the Froude number.

Melting rate of pellets in its own melt

One of the main sources of metallics for the production of steel is sponge iron. Sponge iron is produced by pre-reduction of iron ore pellets. Iron ore pellets are spherical particles of approximately 6–15 mm. The melting and dissolution rate has been investigated by many researchers, for example, Ehrich et al. [9] and Seaton et al. [10]. The heat equation without generation of heat, in spherical coordinates in the radial direction is the following:

$$\frac{\partial(\rho C_p T)}{\partial t} = k\left[\frac{1}{r^2}\frac{\partial}{\partial r}\left(r^2\frac{\partial T}{\partial r}\right)\right]$$

Assuming ρC_p remain constant and taking the derivative with respect to r:

$$\frac{\partial T}{\partial t} = \alpha\left[\frac{\partial^2 T}{\partial x^2} + \frac{2}{r}\frac{\partial T}{\partial r}\right]; 0 \leq r \leq r(t)$$

with the following initial and boundary conditions:

$$T = T_0; t = 0, r \leq R_0$$

$$T = T_m; t = 0, r = R_0$$

$$\frac{\partial T}{\partial r}\bigg|_{r=0} = 0$$

$$k\frac{\partial T}{\partial r}\bigg|_{r=r(t),t>0} - h(T_m - T_f) = \rho \cdot H\frac{dr(t)}{dt}$$

Scales:

$$\theta = \frac{T - T_m}{T_m - T_0}$$

$$\bar{t} = \frac{\alpha t}{R_0^2} = Fo$$

$$\bar{r} = \frac{r}{R_0}$$

where: T_0 is the initial temperature, T_m is the melting temperature, T_f is the fluid's temperature, R_0 is the original radius of the pellet.

Replacing the scales:

$$\frac{\alpha}{R_0^2}\frac{\partial[(T_m-T_0)\theta + T_m]}{\partial Fo} = \alpha\left[\frac{\partial^2[(T_m-T_0)\theta + T_m]}{\partial(R_0\bar{r})^2} + \frac{2}{(R_0\bar{r})}\frac{\partial[(T_m-T_0)\theta + T_m]}{\partial((R_0\bar{r}))}\right]$$

$$\frac{\partial\theta}{\partial Fo} = \frac{\partial^2\theta}{\partial\bar{r}^2} + \frac{2}{\bar{r}}\frac{\partial\theta}{\partial\bar{r}}; 0 \leq \bar{r} \leq \frac{\bar{r}(t)}{R_0}$$

Initial and boundary conditions:

$$\theta = -1; Fo = 0, \bar{r} \leq 1$$

$$\left. \frac{\partial \theta}{\partial \bar{r}} \right|_{\overline{r}=0} = 0$$

$$\frac{k(T_m - T_0)}{R_0} \left. \frac{\partial \theta}{\partial \bar{r}} \right|_{\overline{r}=\overline{r}(t), Fo > 0} - h(T_m - T_f) = \frac{\rho \Delta H R_0}{\underset{\alpha}{R_0^2}} \frac{\partial \overline{r}(t)}{\partial Fo}$$

It is simplified to the following expression:

$$\left. \frac{\partial \theta}{\partial \bar{r}} \right|_{\overline{r}=\overline{r}(t), Fo > 0} - \frac{h R_0}{k} \theta* = \frac{\Delta H}{C_p(T_m - T_0)} \frac{\partial \overline{r}(t)}{\partial Fo}$$

where: $\theta* = \frac{(T_m - T_f)}{(T_m - T_0)}$, $\frac{h R_0}{k}$ is the Biot number and $\frac{\Delta H}{C_p(T_m - T_0)}$ is also a dimensionless heat ratio, in the numerator is the latent heat of fusion and the denominator is the sensible heat.

13.3   How to Decrease the Number of Dimensionless Groups

The lower the number of dimensionless groups, the lower the amount of experimental work. Discriminated Dimensional Analysis (DDA) focuses on the idea of increasing the number of dimensions. Szirtes [11] has a large discussion on the subject to decrease the number of dimensionless groups. According with the π-theorem, the number of dimensionless groups results from the difference between the number of independent variables and the number of independent dimensions, therefore, to decrease the number of dimensionless groups:

- Decrease the number of variables
- Increase the number of dimensions
- Combination of two dimensionless group into a single one.

The first two methods decrease the number of dimensionless groups, however both have limitations, requiring additional knowledge and its general use is usually not recommended. This is perhaps one of the reasons why DDA is not a widespread method.

Decrease the number of variables: A decrease in the number of variables has been reported in many examples, beginning with the classical report from Rayleigh [12] in 1915, combining density and heat capacity in a group of variables called volumetric heat capacity. This work led to the known Rayleigh-Riabouchinsky controversy. The decision of grouping variables requires experience and a clever mind. If experience indicates that some collected variables affect a given phenomenon, then it is possible to fuse the variables. It is not made arbitrarily. Bridgman [13] in the problem of a small sphere falling under gravity in a viscous liquid includes force (F) in the variables gravity and viscosity and the dimensions MLT for the other variables (U, D and ρ).

Grouping variables can be made in the form of multiplication, division, subtraction and addition. A common case is when buoyancy forces are involved due to a density difference of two fluids:

$$\Delta \rho = \rho_l - \rho_g$$

The two variables, ρ_l and ρ_g are replaced by the variable $\Delta \rho$.

Combination of dimensionless groups: In some cases, the resulting dimensionless groups share a common numeration or denominator, or even this can be obtained by simple algebraic operations, in those cases division or multiplication reduces two dimensionless groups into a single one.

$$\pi_1 = f(\pi_2)(\pi_3)$$

where: $\pi_2 = \frac{x}{y}$ and $\pi_3 = \frac{x}{z}$, then:

$$\pi_1 = f\left(\frac{\pi_2}{\pi_3}\right) = f(\pi_2')$$

The combination can also be made with other combinations, depending on each problem, especially on the basis of experimental information or analytical solutions available.

Increase the number of dimensions:

1. Mixing systems of dimensions: the systems of dimensions have been discussed before. In fluid mechanics one system of dimensions involves FLT, however an equivalent system is MLT, because $F = MLT^{-2}$. Depending on the nature of each problem in some specific cases is possible to mix systems of dimensions and increase their number, for example FMLT. In the case of heat transfer problems, one system of dimensions is MLTθ but heat, Q, as a dimension can be included, increasing the number of dimensions to MLTθQ.

    The dimensions of density are mass per volume, however volume itself is another variable. Therefore, volume can be used both as a variable and as a dimension, assigning different symbols to make the difference.

    Szirtes [11] strongly advices against the use of hybrid systems of dimensions. Its correct use can be helpful; however, he suggests that any conclusion drawn from dimensional analysis using hybrid dimensions should be "thoroughly scrutinized and should be verified by other means as well".

2. Splitting dimensions (DDA): The dimensions of length and mass can be split. Length is split based on direction. When a force is applied in a given direction it should be indicated in the expression $F = ML_jT^{-1}$, this applies to gravity force,

shear forces in viscous flows, etc. Mass can have two roles; inertial mass and weight. As inertial mass, its direction should be included.

David and Nolle [14] as well as Szirtes [11] warns about the split of dimensions. Szirtes point out the following; "Splitting dimensions is always a perilous undertaking; it must be done with extreme care, and the results checked, and rechecked, for possible errors". In support of this advice is the lack of diffusion of this method in the literature, with few exceptions as shown in the next section. The following two examples are taken from Szirtes [11].

The height (h) of a meniscus in a capillary tube depends on the following variables: liquid's both density and surface tension, tube radius and gravity force. Their dimensions are: $h = L$, $\rho = ML^{-3}$, $\sigma = MLT^{-2}/L = MT^{-2}$, $r = L$, $g = LT^{-2}$. In order to split the dimension of length, first, a coordinate system should be defined. Defining cylindrical coordinates and noticing the interest is in forces affecting the vertical motion of the meniscus: $h = L_z$, $\rho = ML_r^2L_z$, $\sigma = ML_zT^{-2}/L_r$, $r = L_r$, $g = L_zT^{-2}$. The number of dimensions increases from three (MLT) to four (ML$_z$L$_r$T).

The distance (d) of a bullet shot in a horizontal direction depends on the following variables: ejection speed (U), ejection altitude (h) and gravity force (g). Their dimensions are: $d = L$, $U = LT^{-1}$, $h = L$, $g = LT^{-2}$. Defining rectangular coordinates; x-axis is the horizontal direction and y-axis is the vertical direction, the new dimensions become: $d = L_x$, $U = L_xT^{-1}$, $h = L_y$, $g = L_yT^{-2}$.

13.4 Discriminated Dimensional Analysis (DDA)

Variables that are characterized with a magnitude are called scalars, such as temperature and concentration. Variables that are characterized with magnitude and direction are called vectors, such as force and velocity. Williams [15] was the first one to propose in 1892 a more detailed description of the dimension of length according with its direction. Conventional Dimensional Analysis (CDA) uses only one dimension of length (L) and treats all variables as scalars. With the inclusion of direction in the dimension of length, the vector nature of those variables is defined. The dimension of length in the x-direction can be called X or L_x and so on. Once the direction is included, the number of dimensions is increased, for example, instead of ML^3 for density in CDA, the new method defines the dimensions of density as $ML_x^{-1}L_y^{-1}L_z^{-1}$. Other examples, using a rectangular coordinate system, are the following:

length: L_x, L_y or L_z

Area: L_xL_y, L_yL_z, or L_zL_x

Volume: $L_xL_yL_z$

Density: $ML_x^{-1}L_y^{-1}L_z^{-1}$

Velocity: L_xT^{-1}, L_yT^{-1}, L_zT^{-1}

Force: ML_xT^{-2}, ML_yT^{-2}, ML_zT^{-2}

k. viscosity: $L_x^2 T^{-1}$

The dimensions of some variables need to be defined based on the coordinates of the specific problem.

Dynamic viscosity: flow moves the x-direction, y is the normal direction. The shear force acts perpendicular to the longitudinal axis.

$$\tau_{zx} = -\mu_x \frac{dU_x}{dz}$$

$$\mu_x = \frac{F}{A} \frac{L_y}{U_x} = \frac{ML_xT^{-2}}{L_xL_y} \frac{L_zT}{L_x} = ML_x^{-1}L_y^{-1}L_zT^{-1}$$

Thermal conductivity:

$$k \equiv \frac{Q}{L_yL_zT} \frac{L_x}{\theta} = \frac{ML_yL_zT^{-2}}{L_yL_zT} \frac{L_x}{\theta} = ML_xT^{-3}\theta^{-1}$$

If there are two or three directions, the corresponding variables in those directions should be considered, although, its magnitude is the same.

Palacios [16] in 1964 called the new method Discriminated Dimensional Analysis (DDA) and provided a large number of examples. More recently, Madrid and Alhama [17–26] have promoted the use of this method. The word discriminated is to emphasize that makes a difference in the dimension of length according with its direction.

Example: Determination of the heat transfer coefficient under forced convection.

This example resembles conditions for the cooling rate of furnace walls, spray cooling during continuous casting, etc.

The convective heat transfer coefficient (h) is defined based on Newton's law of cooling:

$$h = \frac{\dot{q}}{(T - T_s)}$$

Conventional dimensional analysis:

The variables that affect the heat transfer coefficient (h) under forced convection are [27]: Fluid properties; density (ρ), viscosity (μ), thermal conductivity (k), heat capacity (C_p) and fluid's average velocity (U); and considering the solid as a flat surface, solid's characteristic length (L). In forced convection, h is independent of the temperature distribution.

$h = f (L, \rho, \mu, k, C_p, U)$.

Variable	Symbol	Units	Dimensions
Convective heat transfer coef.	h	W/m² K	$[MT^{-3}\theta^{-1}]$
Characteristic length	L	m	$[L]$

(continued)

(continued)

Variable	Symbol	Units	Dimensions
Fluid's density	ρ	kg/m^3	$[ML^{-3}]$
Fluid's dynamic viscosity	μ	kg/(m × s)	$[ML^{-1}T^{-1}]$
Fluid's thermal conductivity	k	W/m K	$[MLT^{-3}\theta^{-1}]$
Fluid's specific heat capacity	C_p	J/kg K	$[L^2T^{-2}\theta^{-1}]$
Fluid's average velocity	U	m/s	$[LT^{-1}]$

The problem involves 7 variables and 4 fundamental dimensions. The number of dimensionless groups is three (n − k = 7 − 4 = 3). The variables can be defined in terms of only four dimensions (MLT θ) since Q = ML^{2}T^{-2}. Madrid and Alhama [18] included $\dot{q}$ as another variable and also Q as another dimension.

Using the method of repeating variables, we choose L, U, ρ and k as the repeating variables.

$$\pi_1 = h(L)^{a_1} (U)^{b_1} (\rho)^{c_1} (k)^{d_1}$$
$$\pi_2 = \mu(L)^{a_2} (U)^{b_2} (\rho)^{c_2} (k)^{d_2}$$
$$\pi_3 = C_p(L)^{a_3} (U)^{b_3} (\rho)^{c_3} (k)^{d_3}$$
$$\pi_1 = f(\pi_2)(\pi_3)$$

First π-group:

$$\pi_1 = h(L)^{a_1} (U)^{b_1} (\rho)^{c_1} (k)^{d_1}$$

$$\pi_1 = \left(MT^{-3}\theta^{-1}\right)(L)^{a_1} \left(LT^{-1}\right)^{b_1} \left(ML^{-3}\right)^{c_1} \left(MLT^{-3}\theta^{-1}\right)^{d_1} = M^0T^0L^0$$

Balance:
   M: $1 + c_1 + d_1 = 0$
   L: $a_1 + b_1 - 3c_1 + d_1 = 0$
   T: $-3 - b_1 - 3d_1 = 0$
   $\theta: -1 - d_1 = 0$

Solution:
   $d_1 = -1$
   $b_1 = 0.$
   $c_1 = 0.$
   $a_1 = 1.$

Substitution of these values on the π_1-group yields:

$$\pi_1 = \frac{hL}{k}$$

Second π-group:

$$\pi_2 = \mu\,(L)^{a_2}\,(U)^{b_2}\,(\rho)^{c_2}\,(k)^{d_2}$$

$$\pi_2 = \left(ML^{-1}T^{-1}\right)(L)^{a_2}\left(LT^{-1}\right)^{b_2}\left(ML^{-3}\right)^{c_2}\left(MLT^{-3}\theta^{-1}\right)^{d_2} = M^0T^0L^0$$

Balance:
   M: $1 + c_2 + d_2 = 0$
   L: $-1 + a_2 + b_2 - 3c_2 + d_2 = 0$
   T: $-1 - b_2 - 3d_2 = 0$
   θ: $-d_2 = 0$

Solution:
   $d_2 = 0.$
   $c_2 = -1$
   $b_2 = -1$
   $a_2 = -1$
   Substituting values:

$$\pi_2 = \frac{\mu}{\rho U L}$$

Third π-group:

$$\pi_3 = C_p(L)^{a_3}\,(U)^{b_3}\,(\rho)^{c_3}\,(k)^{d_3}$$

$$\pi_3 = \left(L^2T^{-2}\theta^{-1}\right)(L)^{a_3}\left(LT^{-1}\right)^{b_3}\left(ML^{-3}\right)^{c_3}\left(MLT^{-3}\theta^{-1}\right)^{d_3} = M^0T^0L^0$$

Balance:
   M: $c_3 + d_3 = 0$
   L: $2 + a_3 + b_3 - 3c_3 + d_3 = 0$
   T: $-2 - b_3 - 3d_3 = 0$
   θ: $-1 - d_3 = 0$

Solution:
   $d_3 = -1$
   $c_3 = 1$
   $b_3 = 1$
   $a_3 = 1$

Substituting values:

$$\pi_3 = \frac{C_p\rho U L}{k}$$

Multiplying by μ/μ

$$\pi_3 = \frac{C_p \rho U L}{k} = \frac{\mu C_p}{k}\frac{\rho U L}{\mu}$$

Substituting into the general expression:

$$\frac{hL}{k} = f\left(\frac{\mu}{\rho U L}\right)\left(\frac{\mu C_p}{k}\frac{\rho U L}{\mu}\right)$$

This is equivalent to:

$$\frac{hL}{k} = f\left(\frac{\rho U L}{\mu}\right)\left(\frac{\mu C_p}{k}\right)$$

π_1 is the Nusselt number, π_2 is the Reynolds number and π_3 is the Prandtl number.

$$Nu = f(Re)^{x_1}(Pr)^{x_2}$$

Discriminated dimensional analysis:

The variables are the same. The change is in the dimensions. Instead of 4 dimensions, it increases to 6 dimensions ($ML_xL_yL_zT\theta$).

Variable	Symbol	Units	Dimensions
Convective heat transfer coef.	h	W/m² K	$[MT^{-3}\theta^{-1}]$
Characteristic length	L	m	$[L_x]$
Fluid's density	ρ	kg/m³	$[ML_x^{-1}L_y^{-1}L_z^{-1}]$
Fluid's dynamic viscosity	μ	kg/(m × s)	$[ML_x^{-1}L_y^{-1}L_zT^{-1}]$
Fluid's thermal conductivity	k	W/m K	$[ML_xT^{-3}\theta^{-1}]$
Fluid's specific heat capacity	C_p	J/kg K	$[L_xL_yT^{-2}\theta^{-1}]$
Fluid's average velocity	U	m/s	$[L_xT^{-1}]$

The problem involves 7 variables and 6 dimensions. The number of dimensionless groups is one ($n - k = 7 - 6 = 1$).

π-group:

$$\pi_1 = h(L)^{a_1}(U)^{b_1}(\rho)^{c_1}(k)^{d_1}\left(C_p\right)^{e_1}(\mu)^{f_1}$$

$$\pi_1 = \left(MT^{-3}\theta^{-1}\right)(L_x)^{a_1}\left(L_xT^{-1}\right)^{b_1}\left(ML_x^{-1}L_y^{-1}L_z^{-1}\right)^{c_1}\left(ML_xT^{-3}\theta^{-1}\right)^{d_1}\left(L_xL_yT^{-2}\theta^{-1}\right)^{e_1}$$

$$\left(ML_x^{-1}L_y^{-1}L_zT^{-1}\right)^{f_1} = M^0T^0L^0$$

Balance:

M: $1 + c_1 + d_1 + f_1 = 0$

L_x: $a_1 + b_1 - c_1 + d_1 + e_1 - f_1 = 0$

L_y: $- c_1 + e_1 - f_1 = 0$

L_z: $- c_1 + f_1 = 0'$

T: $- 3 - b_1 - 3d_1 - 2e_1 - f_1 = 0$

θ: $- 1 - d_1 - e_1 = 0$

Solution by Gauss elimination:

$$a1 = 1, b1 = 0, c1 = 0, d1 = -1, e1 = 0, f1 = 0$$

$$\pi_1 = h(L)^1 (U)^0 (\rho)^0 (k)^{-1} (C_p)^0 (\mu)^0$$

$$\pi_1 = \frac{hL}{k}$$

This is the Nusselt number:

$$\frac{hL}{k} = Nu \therefore h = Nu\frac{k}{L}$$

So, DDA shows that the number of dimensionless numbers can be decreased even further.

Madrid and Alhama [18], for this problem, reported two different dimensionless numbers. It should be noticed that they used wrong dimensions in two variables (k and μ).

$$\pi_1 = \frac{k/C_p}{\mu/\rho} \approx \frac{\alpha}{\nu} \approx Pr$$

$$\pi_1 = \frac{h/k}{\sqrt{\mu L/\rho U}}.$$

References

1. H.P. Langtangen, G.K. Pedersen, *Scaling of Differential Equations* (Springer Open, Switzerland, 2016). ISBN 9783319327259
2. R.I.L. Guthrie, *Engineering in Process Metallurgy* (Oxford University Press, New York USA, 1993).
3. J. Kuneš, *Similarity and Modeling in Science and Engineering* (Cambridge International Science Publishing, 2012). ISBN 9781907343
4. J. Hua, J. Lou, Numerical simulation of bubble rising in viscous liquid. J. Comput. Phys. **222**, 769–795 (2007)

5. D.R. Poirier, G.H. Geiger, *Transport Phenomena in Materials Processing* (Switzerland, TMS Springer, 2016).
6. T.M. Dalton, M.R.D. Davies, Dimensional analysis in heat transfer. J. Heat Transf. **121**, 471–473 (1999)
7. Q.-M. Tan, *Dimensional Analysis With Case Studies in Mechanics* (Springer, Berlin Germany, 2011).
8. L. DeVille, Nondimensionalization, scaling, and units
9. O. Ehrich, Y.K. Chuang, K. Schwerdtfeger, Melting of sponge iron spheres in their own melt. Arch. Fur Das Eisenhuttenwes. **50**, 329–334 (1979)
10. C. Seaton, A. Rodriguez, M. Gonzalez, M. Manrique, The rate of dissolution of pre-reduced iron in molten steel. Trans. ISIJ **23**, 14–20 (2011)
11. T. Szirtes, Methods of reducing the number of dimensionless variables, *Applied Dimensional Analysis and Modeling* (Butterworth Heinemann, Elsevier, USA, 2007), pp. 413–462
12. L. Rayleigh, The principle of similitude. Nature 66–68 (1915)
13. P.W. Bridgman, Examples illustrative of dimensional analysis, in *Dimensional Analysis*. (Yale University Press, New Haven USA, 1931), p. 56
14. F.W. David, H. Nolle, *Experimental Modelling in Engineering* (Butterworths, London UK, 1982).
15. W. Williams, On the relation of the dimensions of physical quantities to directions in space. Phylosophical Mag. Ser. **5**(34), 234–271 (1892)
16. J.F. Palacios, *Dimensional Analysis* (MacMillan and Co., London, 1964).
17. C.N. Madrid, F. Alhama, Discriminated dimensional analysis of the energy equation: application to laminar forced convection along a flat plate. Int. J. Therm. Sci. **44**, 333–341 (2005)
18. C.N. Madrid, F. Alhama, Discrimination: a fundamental and necessary extension of classical dimensional analysis theory. Int. Commun. Heat Mass Transf. **33**, 287–294 (2006)
19. F. Alhama, C.N. Madrid, Discriminated dimensional analysis versus classical dimensional analysis and applications to heat transfer and fluid dynamics. Chin. J. Chem. Eng. **15**, 626–631 (2007)
20. C.N. Madrid, F. Alhama, Study of the laminar natural convection problem along an isothermal vertical plate based on discriminated dimensional analysis. Chem. Eng. Commun. **195**, 1524–1537 (2008)
21. I.A. Manteca, A.S. Meca, F. Alhama, Mathematical characterization of scenarios of fluid flow and solute transport in porous media by discriminated nondimensionalization. Int. J. Eng. Sci. **50**, 1–9 (2012)
22. I. Alhama, M. Cánovas, F. Alhama, On the nondimensionalization process in complex problems: application to natural convection in anisotropic porous media. Math. Probl. Eng. (2014)
23. M. Cánovas, I. Alhama, F. Alhama, Mathematical characterization of bénard-type geothermal scenarios using discriminated non-dimensionalization of the governing equations. Int. J. Nonlinear Sci. Numer. Simul. **16**, 23–34 (2015)
24. M. Cánovas, I. Alhama, E. Trigueros, F. Alhama, A review of classical dimensionless numbers for the Yusa problem based on discriminated non-dimensionalization of the governing equations. Hydrol. Process. **30**, 4101–4112 (2016)
25. M. Conesa, J.F. Sánchez Pérez, I. Alhama, F. Alhama, On the nondimensionalization of coupled, nonlinear ordinary differential equations. Nonlinear Dyn. **84**, 91–105 (2016)
26. M. Seco-Nicolás, M. Alarcón, F. Alhama, Thermal behavior of fluid within pipes based on discriminated dimensional analysis. An improved approach to universal curves. Appl. Therm. Eng. 131, 54–69 (2018).
27. M. Thirumaleshwar, *Fundamentals of Heat and Mass Transfer* (Pearson, India, 2006)

Epilogue

The author expects that this book can provide the basic information required to apply dimensional analysis in the study of metallurgical engineering problems. It has been shown the benefits using this tool. Complex problems can be simplified not only in terms of the amount of experimental work but also in the assessment of the independent variables that have a larger effect on the dependent variable as well as the main forces involved.

Dimensional analysis is another tool when applying the scientific method. It has advantages over the conventional way to analyze the relationship among the variables that affect any natural phenomena, however it has been shown that has limitations. Those limitations arise, in principle, from the lack of a true understanding of the nature of the problem and consequently knowledge on the true variables that should be considered. It has been shown that in some metallurgical problems there is not yet universal consensus on the main variables, which leads to different results. In spite of this, dimensional analysis has proven to unify a complex problem involving many dimensional variables into a single relationship involving dimensionless numbers.

© The Editor(s) (if applicable) and The Author(s), under exclusive license   379
to Springer Nature Singapore Pte Ltd. 2021
A. N. Conejo, *Fundamentals of Dimensional Analysis*,
https://doi.org/10.1007/978-981-16-1602-0

CPSIA information can be obtained
at www.ICGtesting.com
Printed in the USA
LVHW061337180821
695582LV00002B/30

9 789811 616013